Dedication

In grateful memory of all those many men and women who have dedicated their careers, and in some cases given their lives, to managing the public trust embodied in the forests of Alberta.

These people have brought us through an astounding evolution from the pioneer hardships of life in the bush 100 years ago to the creativity and innovation of the information age. All this is due to their dedication to the ideal of forests that will provide environmental, economic and social benefits for generations to come.

Those who love our forests and wildland environment today owe a debt of gratitude to these persistent, creative and colourful men and women. The authors hope this book will be part of that tribute, and also an opportunity for generations of the 21st century to discover what it takes to manage 405,758 sq km of public forest – an area almost three times the size of New Brunswick and Nova Scotia combined.

Ranger Jack Thompson, 1920s

Lola Cameron, Secretary to Director, 1950s

Readers are encouraged to provide any feedback or clarification, or additional stories and photographs for possible future publications. Please contact the Forest History Association of Alberta.

The Alberta Forest Service, 1930-2005: Protection and Management of Alberta's Forests
ISBN No. 0-7785-4519-9
Pub No. I/133
©2006 by Alberta Sustainable Resource Development, Government of Alberta, Canada

The photographs used in this book come from a variety of sources and are now in almost all cases in the possession of the authors. In the photo credits used in the book, the specific collection is listed when appropriate, and the letters AFHPC are used to indicate Alberta Forest History Photographic Collection. For more information on the photographs please contact the authors.

Measurements are generally presented in the form, imperial or metric, appropriate to the time period under discussion. A conversion table is provided at the end of this book.

Main cover photo: Willis Millar, Inspector of Forest Reserves, 1911

The ALBERTA FOREST SERVICE
Protection and management of Alberta's forests 1930 - 2005

Authors: Peter J. Murphy, Bob Stevenson, Dennis Quintilio and Steve Ferdinand
Management Team: Bruce Mayer, Anne McInerney, Deanna McCullough, Patrick Loewen
Production Management: David Holehouse, MediaMatch West Communications Inc.
Design: Studio X Design & Illustration; *Proofreading:* Bob Newstead
Champion: Cliff Henderson

Four well-known and highly-regarded veterans of Alberta's forest management and fire protection community present a collation of records, anecdotes, experiences and archival photographs that tell the story of forest management and protection in Alberta from the earliest days of aboriginal use and settlement to 2005. This detailed and informative book will be of value to those interested in forest issues, those who contributed their stories and to all those who share a passion for the great public forests of Alberta. Special thanks go out to all the people who participated in the 'can you tell me who is in the picture' exercise; their time and input is appreciated. They helped identify photographs and people that capture the history of the Alberta Forest Service.

About the authors:

Peter Murphy is a Professor Emeritus in Forestry at the University of Alberta, where he taught and conducted research in forest policy and forest fire management from 1973 to 1995. During this time he also served as Chair of Forest Science and Associate Dean for Forestry. After graduating from the University of New Brunswick in 1953, he worked for the B.C. Forest Service, moving to Alberta in 1954 with the Alberta Forest Service (AFS). In 1956 he became head of the AFS Training Branch that led to establishing the Forestry Training School (now Hinton Training Centre) in 1960. He completed an MScF at the University of Montana in 1963 and PhD at the University of British Columbia in 1985. His particular interests are in forest policy, forest fire behaviour, fire history and history of forestry. He co-authored *Learning from the Forest*, a book detailing 40 years of forest management at Hinton for Weldwood of Canada with Bob Bott and forestry colleague Bob Udell. Peter and Bob Stevenson have compiled a collection of more than 7,400 historic forestry photographs available on CD. Peter's professional contributions include Chair of the Technical Committee (TC) on Sustainable Forest Management for Canadian Standards Association 1998-present and member of TC from 1994; President, Canadian Institute of Forestry 1993-94; President of the Forest History Society Inc. (Durham, NC) 1993-95; President, Alberta Registered Professional Foresters Association 1985-86, and member of the Forest Management Science Council, Alberta, 1996 to 1999.

Bob Stevenson is a retired forester who spent 20 years with the Canadian Forest Service (CFS) from 1960 to 1980, including educational leave for an MSc at the University of Idaho in the early 1960s. His CFS work involved forest protection studies and numerous forest extension programs throughout Alberta and the NWT with industry, government, educational institutions and the public. In 1980 he transferred to the Alberta Fish and Wildlife Division as Director of Public Information and Extension Services and later Head of Commercial Wildlife, retiring in 1992. During this time he was instrumental in preparing and managing numerous high-profile publications, many of which are common today. Much of this effort involved the transfer of information and technology to a variety of user groups including the forest industry. Bob has continued to support the Canadian Institute of Forestry throughout his career and is also a member of the Forest History Association of Alberta. Bob has worked with Peter Murphy to complete a CD containing upwards of 7,400 historic forestry images, most of which pre-date 1930. Bob also serves as custodian of historic forestry memorabilia for the Alberta government.

Steve Ferdinand is a retired forester who obtained his BScF from the University of British Columbia in 1960. He then joined North Western Pulp & Power Ltd. (later Weldwood of Canada and now West Fraser Mills) in Hinton, Alberta, where he worked for 13 years in forest inventory, harvest planning and silviculture. Steve joined the AFS in March, 1974, and worked in various positions for the next 28 years, including liaison with AFS regional staff and in silviculture/reforestation, woods operations and forest recreation. Prior to his retirement in 2002, Steve spent two years in the Integrated Resource Management Division of Alberta Environment.

Dennis Quintilio worked as a Fire Behaviour Specialist in Alberta for 24 years prior to assuming a management position with the government in 1990. From 1967-1974 he was stationed at the Northern Forestry Centre as study leader and worked on early design and implementation of the Canadian Forest Fire Danger Rating System. He was appointed Project Leader in 1975 and continued to refine fire behaviour prediction elements of the system through study of large-scale experimental burns. From 1980 to 1990, he taught at the Forest Technology School which offered a two-year diploma in Forestry, and coordinated all in-service fire management training in Alberta. In addition to his teaching responsibilities, Dennis was also a practising Fire Behaviour Officer and served on the AFS Fire Investigation Team. Dennis became Director of the Forest Technology School in the fall of 1990. In 1995, he assumed the position of Executive Director, Forest Management Division, Alberta Environmental Protection, and in 1999 was appointed Executive Director of the Integrated Resource Management Division (IRM) responsible for implementation of IRM in Alberta. Dennis retired in June of 2001 after 34 years of forestry practice in Alberta. He has a BScF and an MSc degree from the University of Montana, is a member of the College of Alberta Professional Foresters, and has a list of 25 publications to his credit.

Ranger Dexter Champion on horse Brownie with dog Train, on patrol, head of Pincher Creek, 1942
Jay Champion

Morning Sun at Adams Creek Lookout, February, 2005
Artist Robert Guest, Grande Cache

Table of Contents

		Page
One	Early Days	2
Two	Birth of the Alberta Forest Service	28
Three	A Way of Life	42
Four	Momentous Change	80
Five	Lookouts and Communications	120
Six	1966 - 1984	162
Seven	The Use of Aircraft	216
Eight	1985 - 1992	246
Nine	1993 - 2005	282
Ten	Fire in Alberta	346
Eleven	Alberta Forest Service Leaders Reflect	400
	Epilogue	414
	References and Suggested Further Reading	416
Chart 1	Directors of Forestry and Division Heads, 1930 - 2005	424
Chart 2	Executive and Forest Superintendents, 1930 - 2005	432
Graph 1	Area Burned, 1930 - 2004	440
Graph 2	Volume Harvested, 1931 - 2002	441
Graph 3	Area Harvested, 1937 - 2003	442
Graph 4	Seedlings Planted, 1961 - 2004	443
	Ranger Stations and Cabins	444
	AFS Museum	480
	History of Posters	484

CHAPTER 1

Early Days

Forest Origins

The glacial ice that covered much of Alberta began to melt between 10,000 and 12,000 years ago, sculpting the uplands and lowlands, the fertile meadows and rocky outcrops that sustain the native plants and animals of today.

As the ice disappeared, plants such as sedges and willows returned, followed by trees. The predominant tree species, then and today, were coniferous spruce, pine, fir and tamarack, in company with deciduous poplar, aspen and white birch. As the plants returned so did the animals, and human beings were not far behind. Along with trees and people came wildfires. It was under all these dynamic influences of soils, climate, fire and people that the forests of Alberta were established. Today they represent some of the most diverse landscapes in the world.

The first people of the forest were aboriginal hunters and gatherers, living on and with the land. There is evidence that they also managed their environment through the use of fire to clear certain forest areas for ease of travel and to encourage the plants and animals on which they depended for life.

The first Europeans, arriving after the mid-1700s, were mostly interested in furs for the European market. Missionaries and settlers followed the traders, and all were frequently challenged by extensive wildfires that threatened their homes, livestock, timber and water supplies. Concerns about this threat eventually led to the formation of an organized forest service in Alberta.

Forest Use

Most of Alberta's forests are located on Crown (public) lands, meaning governments have the major responsibility for forest management. This designation evolved from early colonial regulations requiring that timber be preserved and kept as a strategic reserve for use by Britain's military shipbuilding industry. In 1826 this Crown reservation of timber was modified to allow public sale of timber that was deemed "not fit and proper" for Britain's Royal Navy. These new regulations for the sale of timber contained four clauses, which set lasting precedents. The clauses included continued Crown ownership of forested land, leasing of harvesting rights, selling timber by auction or tender, and allocation of cutting permits that were renewable if certain conditions were met.

The government of the United Provinces of Upper and Lower Canada enacted *An Act for the Sale and Betterment of Timber upon Public Lands* in 1849. This legislation incorporated the principles of 1826, and formed a model for later forest laws within the provinces and on Dominion lands. Most significantly, it continued the arrangement under which timber-harvesting rights were leased while the forestland remained in public ownership. This fundamental concept remains in force on Alberta's provincially-owned Crown land to the present day (2005).

The land now called Alberta was

Aboriginal family north of Hinton, 1913
Dominion Forestry Branch, AFHPC

partly contained within Rupert's Land, the area granted in 1670 to the Governor and Company of Adventurers of England Trading into Hudson Bay, or what became known as the Hudson's Bay Company. After Confederation in 1867, Canada's first Prime Minister, Sir John A. Macdonald, negotiated the purchase of Rupert's Land to help realize his vision of a Canada stretching from sea to sea. By 1870 he had succeeded in this. Britain granted its other northern lands on the continent and Canada became owner of the North West Territories. Macdonald created a Department of the Interior to manage this huge area, parts of which would become the Provinces of Manitoba, Saskatchewan and Alberta. At that time the founding provinces in central Canada had been granted control of their forested lands and resources. However, Macdonald had been concerned about timber supplies and forest fires. He thought it would be a "very good thing," as he explained to his friend Sandfield Macdonald,[1] that since the Dominion government had no direct interest in the subject, and since forests were a provincial responsibility in central Canada, that Ontario and Quebec should set up a joint commission to examine: 1) the best means of cutting the timber after some regulated plan, as in Norway and on the Baltic, 2) replanting so as to keep up the supply as in Germany and Norway, and 3) the best means of protecting the woods from fires. His concerns were later reflected in policies for prairie forests as well.

In the region that would become Alberta, the strongest initial demand for wood had been for firewood. Construction logs and fence posts were next, with coniferous white spruce and lodgepole pine being the species most often used for these purposes. Demand for boards was limited, and they were mostly sawn by hand in sawpits or frames. A good two-man team could saw up to 25 planks a day. Gradually a few small, mechanized sawmills began to supply lumber for local needs. Oblate missionaries at the Lac La Biche Mission built Alberta's first powered sawmill, a converted gristmill driven by water. The waterwheel was rebuilt to a diameter of 15 feet and the mill began sawing in 1871, producing over 250 planks per day when the water was flowing and the homemade belts stayed on.

Pit-sawing lumber, Fort Smith, NWT, 1900
Provincial Archives of Alberta

The Dominion Lands Act of 1872 enabled the federal government to sell timber cutting rights through Timber Berths and a variety of permits. It also provided authority to exclude timberlands from sale and settlement. This authority was later used to establish the first Forest Reserves. The Act also required timber operators to prevent the ignition and spread of fires.

Wildfire Problems

Despite the good intentions of forestry staff and enforcement of regulations by the North West Mounted Police, fire problems continued to grow. Even in 1883, when Prime Minister Macdonald appointed J.H. Morgan as a one-man commission to "examine into and make a Preliminary report on the subject of the protection of the forests of the Dominion," fire was a major concern. In his sweeping report of 1885, Morgan's remarks included the following:

"Enough has been shown to make it evident that it is the duty of our Government to adopt measures, immediately, to arrest further destruction of our remaining forests (except under some very improved system of supervision), and to replant, where practicable, the high lands which were formerly covered with forest trees, and also to devise or adopt some

Early Days

plan or system of forest plantations for the great stepped [*prairie*] region of the North-West.

"In any system that may be adopted by Canada, special care should be given to see that provision is made for the fullest enforcement of the laws. The Government of the Dominion should, without loss of time, appoint a Forest

Ghost River fire, c. 1915, Bow River Forest
Dominion Forestry Branch, AFHPC

Commission, to co-operate with a similar Commission from every Province in the Dominion, to deal with this all-important question of the protection of our old forests and the production of new forests."[2]

These general aims underpinned Canadian forest management policies for more than a century afterwards. Morgan also commented on four other items which remained key issues for the next 70 years: not knowing the extent and nature of forest resources, the undesirable consequences of uncontrolled cutting and fires, the need to organize a system of forest management, and the importance of forestry schools to train qualified staff.

Prairie fires continued to be a persistent and growing problem. The Department of the Interior first discussed the need for firebreaks, seasonal fire guardians and organized volunteer fire brigades in its Annual Report in 1886. That same year, the Council of the North West Territories passed an ordinance establishing fire districts and appointing fire guardians. The North West Mounted Police were spread very thinly throughout the country, making it difficult for them to effectively enforce the fire ordinances. They also reported that local Justices of the Peace were usually reluctant to prosecute fire violators. For whatever reasons, settlers seemed to have a fatalistic approach to prairie fires, confounding efforts by the department to encourage ploughing of firebreaks.

This growing problem set the stage for formation of a forestry agency within the federal government. Advocates promoted the need to protect forests on Dominion lands. Other goals included setting up Forest Reserves to protect water supply, managing the forests to ensure a supply of wood for settlers, and planting trees on the prairies to provide timber and localized climate benefits.

The Dominion Forestry Branch

On July 24, 1899, the Wilfrid Laurier government in Ottawa passed an Order in Council to create the post of Chief Inspector of Timber and Forestry in the

Early fire control posters   Dominion Forestry Branch, AFHPC

Department of the Interior. This was the start of what would become Canada's largest organized forest service for the next 31 years - the Dominion Forestry Branch (DFB). It was responsible for forests in Canada's interior western region that later became Alberta, Saskatchewan, Manitoba, Yukon and the Northwest Territories.

On August 15, 1899 Elihu Stewart was appointed Canada's first Dominion Forester in Ottawa. During his first year as Chief Inspector he made two trips west through Manitoba, present-day Saskatchewan and Alberta and parts of British Columbia, five months in all to see more of the western forests first-hand.

York boat, Saulteaux Landing, Lesser Slave Forest Reserve, 1911
Dominion Forestry Branch, AFHPC

After his journey, Stewart stated that there would be two great divisions to the work of his branch, both of which deserved careful attention: the protection and management of the present forested areas, and the encouragement of tree planting on the prairies. At the same time, the sale of timber was important for revenue. Timber sales and revenue were handled by the federal Timber and Grazing Branch, while the DFB was responsible for inspections, forest protection and forestland management. The appointment of the first two rangers in what was to become the Province of Alberta took place in 1899. Ranger D.G. McPhail worked under the supervision of C.L. Gouin, who was in charge of the Calgary Timber Agency. The other ranger, John A.C. Cameron, was responsible to Thomas Anderson, head of the agency at Edmonton.[3]

Increasing populations and resultant political activities in what is now Alberta led to a request by the NWT Council for provincial status. The council advocated one new province, to be called Assiniboia, with control of its natural resources. However, in 1905 Parliament instead established the provinces of Saskatchewan and Alberta, while retaining federal control of natural resources.

With natural resources in federal control, an interesting 25-year period of duality of governance began in which forestry, wildlife and public lands activities were handled by the federal government. As a result, the DFB continued to develop its operations in Alberta. For the most part, this worked well, but some disputes between the two levels of government arose when responsibilities collided. For example, DFB staff complained at one point that brush disposal on provincial roads created fire hazards and provincial cooperation to improve the situation was lacking. Dominion staff also commented that settlers in forested areas berated them for not fighting fires in those areas, while in fact the province – whose fire wardens began wearing a special uniform in the spring of 1911 - had responsibility for settlement areas and would not appoint DFB staff as fire guardians. The situation changed in 1921 when Alberta amended the *Forest and Prairie Fire Protection Act* to give DFB staff and fire rangers *ex officio* authority to enforce provincial legislation.

Elihu Stewart, first Dominion Forester
Dominion Forestry Branch, AFHPC

Early Days 5

Forest Reserves

The Forest Reserve system was the most prominent DFB program prior to 1930. Even before the DFB was created in 1899, the federal government had started to set aside five areas in Alberta as possible Forest Reserves. These reserves, identified by federal departmental order, were Cooking Lake, Foothills, Kootenay Forest Park, Louise Lake Park, and Sand Park.

In those early years the forested areas of the North West Territories, including the future province of Alberta, were vast and government resources were very limited. Stewart decided to focus his efforts on a system of Forest Reserves covering the most important areas. His early surveys laid the groundwork for the first *Forest Reserves Act* of 1906. Most of the areas that were previously excluded from settlement by departmental order were now confirmed in legislation as Forest Reserves, and many new ones were also declared. The total area reserved in Alberta in 1906 was 6.2 million acres, including Cypress Hills and Cooking Lake, and the entire southern East Slopes. This was a good start but there were many other candidate areas to examine, so forest surveys were extended. These covered a wide band including the northern foothills and the boreal forests from Lac La Biche west through Lesser Slave Lake and Whitecourt to the British Columbia boundary, and north through Grande Prairie and Peace River. These forest surveys were conducted through 1915 and led by notable foresters such as J.A. Doucet, P.Z. Caverhill, G.H.

Construction near completion on new gateway into the Bow River Forest, c. 1915
Dominion Forestry Branch, AFHPC

DOMINION FOREST RESERVES IN ALBERTA – 1915

Forest Reserve	Forest	Area Approx. mi^2	Headquarters	Total Area mi^2 / million acres
Rocky Mountains Forest Reserve			Calgary	17,529/11.218
	Crowsnest Forest	1,544	Blairmore	
	Bow River Forest	3,089	Calgary	
	Clearwater Forest	4,247	Rocky Mtn House	
	Brazeau Forest	4,633	Coalspur	
	Athabasca Forest	3,861	Hinton	
Cooking Lake Forest Reserve			Edmonton	27/0.017
Cypress Hills Forest Reserve			Calgary	156/0.100
Lesser Slave Forest Reserve			Slave Lake	5,023/3.215
Total Area of Forest Reserves				22,735/14.550

Areas above interpolated from Annual Reports of the Director of Forestry, Department of the Interior, Ottawa, for the fiscal years ending 31 March 1913 and 1914.

The Athabasca Forest, in the foothills region around Hinton/Edson, was sometimes, but not consistently, spelled Athabaska.

Alberta Forest Service

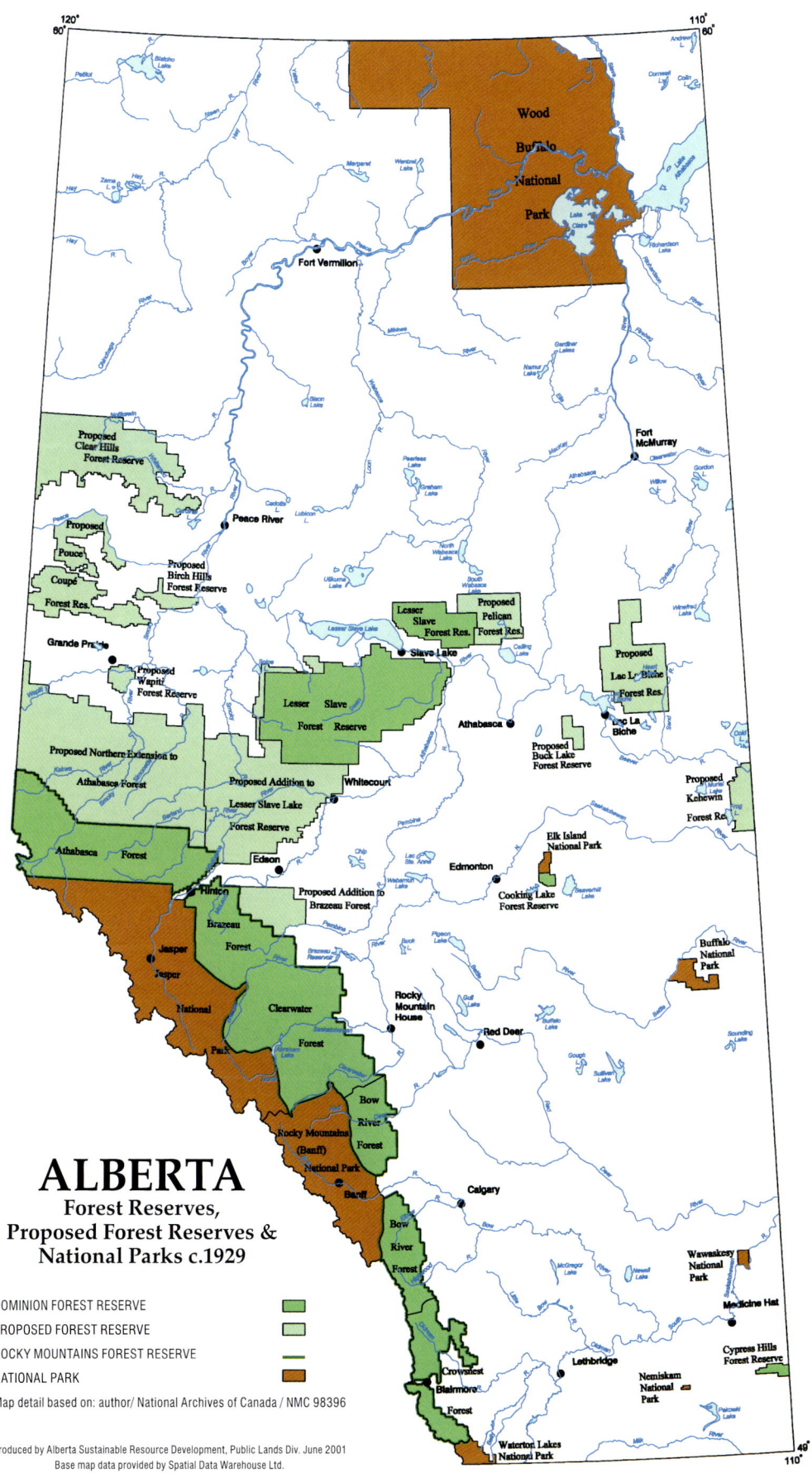

ALBERTA
Forest Reserves, Proposed Forest Reserves & National Parks c.1929

- DOMINION FOREST RESERVE
- PROPOSED FOREST RESERVE
- ROCKY MOUNTAINS FOREST RESERVE
- NATIONAL PARK

Map detail based on: author/ National Archives of Canada / NMC 98396

Produced by Alberta Sustainable Resource Development, Public Lands Div. June 2001
Base map data provided by Spatial Data Warehouse Ltd.

Early Days

Edgecombe and R.H. Campbell.

Elihu Stewart explained the visionary rationale behind this land allocation to Forest Reserves in his 1901 report: "The Dominion government should know in advance of settlement the character of the unsettled districts, so as to direct immigration aright; so that mineral land might be set aside as such, agricultural land developed to the agriculturist and land unsuited for agriculture but on which timber is growing reserved permanently for timber."[4]

The *Forest Reserves and Parks Act* of 1911 added significant lands to existing forest reserves and also defined expansion of the national parks.

The most prominent of these Forest Reserve areas were on the eastern slopes of the Rocky Mountains. In 1911 an additional 2.97 million acres were added to the Rocky Mountains Forest Reserve (RMFR) bringing it to 9.52 million acres. This was in addition to Rocky Mountains Park (Banff), Jasper Forest Park and Kootenay Forest Park (later renamed Waterton Lakes National Park), previously reserved from 1899 to 1906. The RMFR was deemed the most important in the western provinces "as it is on a timbered area lying alongside of a prairie country hundreds of miles in extent which is almost devoid of trees. Also, it forms the watershed for the river systems which water the great plains to the east where the water supply is practically the only limit to anticipated settlement and development."[5]

Forestry float in Calgary Stampede parade, 1920s
Dominion Forestry Branch, AFHPC

Ranger McDonald on makeshift lookout, near Brazeau Forest HQ, c. 1912
Dominion Forestry Branch, AFHPC

In 1913 the RMFR was expanded yet again, and the Lesser Slave Forest Reserve was created in 1914. These were among the last of the Forest Reserves to be established. None of the other northern areas surveyed and recommended was granted this status, although their areas were significant.

The total area of Forest Reserves in Alberta peaked at 14.55 million acres. An additional parcel of land near Wainwright was set aside to house the "Pablo" herd of buffalo purchased in Montana. In 1908, this land was called Buffalo National Park, but it was not given the formal designation of national park until 1913 under the *Forest Reserves and Parks Act*. The area ceased to be a national park and became a military training

Ranger Lyndon and his wife, in the Porcupine Hills, c. 1920. Note DFB badge on vest. Women were important partners at the ranger stations
Sir Alexander Galt Museum & Archives, Lethbridge

Alberta Forest Service

ground in 1940. By this time the park had served its purpose well, with bison well established in several other national parks. The site today is used by the Canadian Armed Forces as Land Forces Western Area Training Centre.

R.H. Campbell became Director of Forestry when Elihu Stewart left the position in 1906. He observed that the western Forest Reserves were together larger than the province of Nova Scotia. The average size of reserves in 1914 was 460,700 acres. In one district in Alberta, he said, the inspector could not cover his whole territory even once a year if he travelled the entire time. Therefore, he decreed, each Forest Reserve was to be divided into ranger districts varying in size according to the needs of the country, from 20,000 to 500,000 acres. Campbell also noted in his report for 1910-11 that the objective of the U.S. national forest organization was to have one ranger for every 100 square miles. After Alberta's Forest Reserves were expanded in 1913 there were 59 rangers for 14.55 million acres, or one ranger for every 400 square miles. Staff had to do more with less, right from the start. The rest of the forested area of Alberta was protected by a Supervisor of Fire Ranging in Calgary who hired seasonal fire rangers. In 1930 there were about 50 of them to patrol more than 98.8 million acres in the Fire Ranging Districts, an average of two million acres each.

Within the Forest Reserves a small number of personnel worked diligently to do the job required

H.R. MacMillan (L) and Willis Millar in Idaho, 1908
Dominion Forestry Branch, AFHPC

Fire prevention sign along the Athabasca River, Fort McMurray, 1922
Dominion Forestry Branch, AFHPC

Pelletier Mill near Coleman, Crowsnest Forest. Pole road in foreground bore carriages on metal wheels to carry logs, 1911
Dominion Forestry Branch, AFHPC

of them. One focus was prevention of wildfires and monitoring of logging operations; another was monitoring of slash burning in the forest and brush burning in settlement areas. Along the forest boundaries rangers encouraged prairie residents to establish and maintain fuel breaks and fireguards around buildings and haystacks. They pursued a fire prevention campaign that focused on education.

Willis N. Millar moved to Alberta from the U.S. Forest Service in 1911 to work as Inspector of Forest Reserves with an office in Calgary. He was a Yale forestry graduate in the same class as H.R. MacMillan, who also worked for DFB from 1906 to 1912. During Millar's time in Alberta, he travelled extensively to organize and extend the Forest Reserves. One of his major contributions was a comprehensive survey and report on wildlife in the Rocky Mountains, published as Bulletin 51, DFB, in 1915. His vision and recommendations for wildlife protection were far-reaching and served as the basis for protected areas along the eastern slopes. Millar left in 1914 to teach forestry at the University of Toronto.

Forest Rangers

Once Forest Reserves were established, ranger districts were set up. Each district had headquarters located within the Forest Reserve. A house and barn, generally made of logs, provided a base from which to patrol each district. Each ranger was expected to protect the district from fire and illegal logging, open up new trails and roads, and build and maintain telephone lines that connected his district with others. All logging operations were monitored to ensure regulations were being observed and the "cut" was within limits.

A major effort involved the development of a network of trails and cabins along patrol routes for fire access and to connect each forest district. The trails' importance was highlighted in 1913 when Inspector Willis Millar wrote specifications for three categories: primary, secondary and auxiliary.[6] The idea was to ensure that the primary trails were "good" ones and that not too much effort was put into the others. Specifications for the primary trails included 10-15 per cent maximum grades, 7-10 foot clearings, overhead clearance of 10 feet and a "tread" of 16-18 inches cleared of all stumps, roots and rock. Many of these trails began as aboriginal travel routes. Some trails followed rivers and were used by fur brigades while others were built to connect one valley route to another. Some of these early trails later became foothill and mountain roads such as the Forestry Trunk Road and the connecting recreational and resource roads. Similarly, many popular camping spots today are also the same locations used by aboriginal people and the first explorers.

Early foresters and rangers used prominent ridges, high hills and mountain-top locations for fire lookouts. Stopover cabins were well-constructed log buildings. These were initially built along major trails and generally at intervals of a day's travel by horse, or about 15 miles. Some of the original cabins are still serviceable, a mute testament to their durability. A few, such as the Gregg River Cabin, south of Hinton, constructed around 1917, have been recognized as registered historical sites.

Pack string nearing summit with lookout construction items, Moose Mountain Trail, Bow River Forest, 1928
Dominion Forestry Branch, AFHPC

In 1911, R.H. Campbell, Director of Forestry, presented a list of what the qualifications of a forest ranger should be. He began by stating that: "the success of forest administration rests to a very large degree on the intelligence, the faithfulness and the practicability of the forest ranging staff. The work of a forest ranger is arduous and requires a man of energy and strong physique. The qualifications for appointment as a forest ranger should be as follows:

"He should be between the ages of twenty-five and forty.

"He should be sober, industrious and physically fit.

"He should be able to read and write and have sufficient knowledge of arithmetic to transact the ordinary business of the reserve, such as calculating the dues on permits.

"He should be able to handle horses and to ride.

Inspector Abraham Knechtel inspects settler's improvements on Cypress Hills Forest Reserve, 1909
Dominion Forestry Branch, AFHPC

"He should be experienced in work in the woods, should be accustomed to handling an axe and should be able to estimate and scale timber."
"He should be able to handle a gang of firefighters or men working on roads or trails."[7]

To this list, forester and inspector of Forest Reserves Abraham Knechtel added the ability to "locate and estimate timber, do a logging job, run a sawmill, build log houses and have education enough to report intelligently to the department."

H.R. MacMillan of the DFB described similar principles, pointing out that forest supervisors were responsible for areas encompassing over a million acres that involved over $10 million-worth of government property. Within these areas, he noted: "the average forest ranger had charge over 200,000 acres estimated to be worth at least 10 dollars an acre for timber." Therefore, the means for choosing good people had to be instituted. MacMillan also recommended in-service upgrading through a training school for rangers, but it would be almost 40 years before this came about in Alberta.

The early Dominion forest rangers were hired from among local Alberta settlers, ranchers, trappers and loggers. They tended to be independent of mind and spirit, and included such stalwarts as "Posthole" Smith, Boer War veteran and rancher in the Porcupine Hills. Others were Fred Nash at Turner Valley, whose horses were always a forestry feature at the Calgary Stampede; Bill Shankland, another Boer War veteran from Nordegg and Bragg Creek who always rode tall in the saddle and who led Alberta's first Forest Ranger training school in 1946; Jack Glen, a First World War veteran who built many of the trails and cabins west of Entrance and captured his experiences in his memoirs, and Albert Foley, the ranger from Swan River in the Lesser Slave Forest Reserve, the first in a four-generation line of Alberta rangers.

Sturm's Sawmill, Cypress Hills Forest Reserve, 1910
Dominion Forestry Branch, AFHPC

Loading Norwegian reindeer on barge at Athabasca Landing for 500-mile trip to Fort Smith, Northwest Territories, September, 1911
Glenbow Archives NA-2788-26

Increasing Use of the Forest

The concept of multiple-use of forestlands was clearly recognized by the DFB. There was a belief that "protected" or well-managed forests would provide many benefits to many people. Watershed protection, wood for settlers, grazing and recreation were among the values considered. Abraham Knechtel described the purpose of the Forest Reserves in 1910:

"The Dominion Forest Reserves are intended to preserve and produce a perpetual supply of timber for the people of the prairie, the homesteaders' needs being considered of the first importance. They are not intended to furnish wood for the lumber trade. Hence the policy of the Department is favourable to small mills rather than to large ones which need large tracts of forest and manufacture lumber beyond the needs of the settlers. To furnish wood is primarily the purpose of Parliament in the creation of the reserves. To be sure, our legislators are not unmindful of other blessings of the forest. They are well aware that forests feed springs, prevent floods, hinder erosion, shelter from storms, give health and recreation, protect game and fish, and give the country aesthetic features. However, the Dominion Forest Reserve policy has for its motto, 'Seek ye first the production of wood and its right use and all these other things will be added unto it.'[8]

There were some specific initiatives in respect to other uses. Recreation in Forest Reserves and national parks was tempered by fire-prevention considerations. The approach was to congregate people in safe camping areas where woody fuels were removed and trees pruned to prevent campfires from escaping and spreading. This resulted in the first designated campgrounds. Posters urged campers and forest travellers to prevent forest fires.

Grazing was also promoted as a fire-prevention measure within the Forest Reserves. Cattle reduced the accumulation of dry grass and created trails that could serve as fuel breaks. Leases encouraged ranchers to become partners in protecting the forest.

In northern areas, contact with aboriginal communities centred on trappers and hunters who might prevent fires and serve as guardians of the forest. Wildlife projects such as the experiment to establish reindeer near Fort Smith, NWT from 1911 to 1915 involved some DFB staff. Later, in the mid-1920s when plains bison from Buffalo Park at Wainwright were moved to the Fort Smith area, DFB staff members were involved in transferring the bison to barges at Waterways on the Athabasca River.

Most of the northern fire rangers were situated to protect the most important routes of travel: the Athabasca River, the Lesser Slave River, the Peace River and the Great Slave River. Some of the country was accessible only by water, and some of the water was so fast that canoe travel was hazardous. River travel was an important aspect of a ranger's duties, but was also a major hazard. To get a reliable patrol on these more inaccessible waterways, the government supplied sternwheeler steamboats. One patrolled the Athabasca River between Mirror Landing and Grand Rapids (SS Rey). The boat was 42 feet in length with a ten inch draught and a powerful engine to force the boat upstream against the fastest current. Another steamer, which was obtained in 1912, patrolled the Slave River between Fort Smith and Great Slave Lake. In each boat the skipper was a fire ranger. His crew consisted of an engineer, a fireman-stoker and whatever other help was occasionally required. The boat patrols were on the alert for lightning strikes and campfires left by travellers who used the rivers as highways to the north.[9]

Ongoing Wildfire Problems

Those surveying for Forest Reserves between 1910 and 1915 reported how dismayed they were with the extent of past wildfire burns in their areas. In the southern east slopes, forester G.H. Edgecombe commented that: "repeated fires have devastated the eastern slopes and the forest type consequently was altered and areas eroded."[10] He estimated that perhaps 60 per cent of the area had been 'fireswept' in the 60 years since 1850. North of the Bow River, forester Peter Z. Caverhill estimated that 80 per cent had burned in the last 50 years. Similarly, J.A. Doucet, a forester surveying from the Smoky River north

Aura Ranger Station (now Ghost Ranger Station), Bow River Forest, northwest of Calgary, 1923. The stone cross was an identifying marker that could be seen by passing aircraft
Dominion Forestry Branch, AFHPC

A ploughed fireguard, Cypress Hills plateau, 1911
Dominion Forestry Branch, AFHPC

to Grande Prairie, reported 65 per cent of the area had burned in the last 50 years.

Many steps were taken in an effort to more effectively tackle the fire problem. Build-up of the administration of the fire control system was often provided in an incremental manner in response to fire problems. This was illustrated after 1908, which was described as a "bad year." In 1909 the number of fire rangers in Alberta was increased from 12 to 34. The year 1910 was another difficult year, resulting in a further increase to 45 fire rangers for 1911. These increases represented the beginning and growth of the DFB as it gradually assumed responsibility and control of the Forest Reserves. This protection role consumed much of the staff's time. Even with this increased vigilance, however, fires continued to cause considerable damage and loss.

There were three particularly severe fire years. In 1910 an estimated 316,295 acres burned in the Bow River Forest. That was also the year of the "Big Burn" in Montana and Idaho. An estimated two per cent of Alberta's Forest Reserves was reported burned in 1914. In 1919 there were two major outbreaks. The first was towards the end of May, with the town of Lac La Biche burning out on May 19. A series of fires stretched east to the area of Prince Albert, Saskatchewan, and may have burned as much as five million acres during that same period. The second event comprised summer fires in the foothills, particularly in the Livingstone, Highwood and Ghost River valleys. Their estimated combined area was 145,795 acres.

Fires along railway lines were also common and persistent. Sparks and cinders from steam engines, along with sparks from brake shoes and hot-boxes were frequent sources of ignition and major problems on busy rail lines. An amendment to the *Railways Act* in 1912 empowered the Railway Commission to require companies to employ fire rangers and patrol the railway lines, and also made the railway companies liable for damage caused by fires started by locomotives. The Grand Trunk Pacific and Canadian Northern Railways were closely monitored during their construction west from Edmonton with the result that few fires occurred. Nine fire rangers, under a senior ranger by the name of J.A. Dunn, devoted their whole time to patrolling the construction areas of the Grand Trunk Pacific and Canadian Northern Railways.[11]

Reports indicated that construction of the Edmonton, Dunvegan and British Columbia Railway presented great problems. To control railway fires the DFB instituted patrols on motorized and hand-propelled "speeder cars" which actually followed each train to ensure

First train into Leslieville, east of Rocky Mountain House, 1912-13
Dominion Forestry Branch, AFHPC

Early Days

prompt detection. Many stories were told about the speeders that became a common method of travel for forestry crews in the roadless mountains and northern forests.

Communications were a problem. The telegraph was used for messages between headquarters and the nearest railway office, but within the forests there was nothing. Without communication, there was no need to build sophisticated lookouts. Construction was limited to the occasional crawl tower, a very rudimentary lookout consisting of a ladder with a safety hoop around it at intervals. The single-line ground-return telephone became available around 1910, and construction and maintenance of phone lines became a major activity during the next 40 years. Construction of lookouts began in earnest after 1912 when they could be linked via telephone to ranger stations. Considerable effort by rangers was directed to construction of telephone lines along trails. These lines were typically strung between trees or on tripods across muskegs and open hillsides. A lot of time was spent in maintaining them as trees or snags frequently fell across the lines or moose and elk caught the line in their antlers, breaking the connections. Wireless sets were introduced when aircraft entered forestry operations in 1920, but did not become fully operational in the field until the 1930s.

The use of aircraft for fire detection patrols began in 1920 with the cooperation of the National Air Board of Canada, the forerunner of the Royal Canadian Air Force. After the First World War experienced aviators and aircraft were available. DFB staff members were convinced that aircraft had a place in forestry operations after seeing a demonstration in the Okanagan Valley of British Columbia. The first aircraft were engaged in fire detection and trial forest surveys, and this experiment was so successful it was expanded and maintained through to 1930. An early report in Alberta noted that one fire was monitored daily with aircraft by the forest supervisor, who upon his return relayed instructions to the rangers via telephone. Without the aircraft a similar trip would have taken a week with saddle and packhorse.

Vehicle used to crank aircraft engine, 1920s, High River, Bow River Forest
Dominion Forestry Branch, AFHPC

Supervisor McAbee with speeder car between Rocky Mountain House and Nordegg, on the Canadian Northern Railway, Clearwater Forest, 1914
Dominion Forestry Branch, AFHPC

The land-based aircraft were stationed briefly at Morley, but the frequency of high winds and existence of gravelly soil made landing difficult. In 1921 the air station was moved to a new facility at High River. From High River, two patrols were conducted daily. One went to the United States - Canada border while the northern flight went as far as the divide between the Red Deer and Clearwater rivers. Supplemental landing strips were built at Pincher Creek and Eckville at the furthest ends of the patrol trips. Reports of fires were first sent by phone after the aircraft landed, and later by wireless from the aircraft itself. When convenient, rangers went along on reconnaissance flights to determine firefighting progress. As an educational thrust, leaflets warning the public of the danger of forest fires were dropped over towns during fairs and sports days and over popular camping areas.

A fire-permit system for the burning of settler's land-clearing slash was finally introduced in 1928. During this time the DFB continued its public education activities through speaking tours. The 'Save the Forest Week' was introduced in 1925, a precursor to today's National Forest Week held in May each year.

Early Industry

The arrival of the railway in Calgary in 1883 led to a demand for building materials. This helped define a monetary value for the timber on the eastern slopes of the Rocky Mountains. The timber was in the Dominion's newly established reserve later named the Rocky Mountains Forest Reserve. Large steam-powered sawmills, using timber floated down the east-flowing rivers, launched the forest industry in the west-central region. Two prominent sawmills in the south were the Northwestern Coal and Navigation Co. at Lethbridge and the Eau Claire and Bow River Lumber Company in Calgary, both established in 1885.

They were followed by several more large, steam-powered mills in Red Deer and Edmonton, as well as a host of smaller local sawmills.

Dominion timber berths were the major source of logs. Timber berth operators had to abide by a number of regulations, largely focused on record keeping for payment of timber dues and fire prevention. Reforestation requirements were not included. There seemed to be a general expectation that since nature grew the forests in the first place, it would replace them naturally. While it soon became evident that the lack of regeneration was a problem, there was no ready solution.

In any case, reforestation efforts took a back seat to fire protection. As explained in the 1914-15 Annual Report: "Reforestation has not as yet been taken up actively on the reserves in general, as the work of protection has been given first consideration."[12]

Abraham Knechtel, Inspector of Forest Reserves, had discussed silviculture as early as 1910, but it was not until 1920 that reforestation

Burned and windthrown timber, Grande Prairie area, 1913
Dominion Forestry Branch, AFHPC

Nursery seed beds at Cooking Lake Forest Reserve, 1921
Dominion Forestry Branch, AFHPC

trials really got started in Alberta. They began on the Cooking Lake and Cypress Hills Forest Reserves since they were "located on the prairie in poorly timbered country."[13]

After establishing beds for seeding and seedlings in both locations, actual planting trials began. A 21-acre planting of 65,478 trees was done in Cooking Lake in 1923, half each of jack pine and white spruce. The next year 200 acres at Cooking Lake were spot-seeded. Planting was also done in Cypress Hills and seeding in the Crowsnest Forest Reserves. Results were variable, and references were made later to problems with rodents, frost and drought. The small DFB nurseries in Cooking Lake and Cypress Hills Forest Reserves were maintained by the Alberta Forest Service after 1930, but soon abandoned in favour of a new central nursery at the Oliver Hospital site, an area now part of northeast Edmonton. In the meantime the federal prairie tree breeding and production program continued at the DFB forest nursery at Indian Head, Saskatchewan.

Reconnaissance forest surveys, which provided information regarding tree species, condition of timber, and types of topography had been carried out in the forest reserves since 1908. Now Campbell wanted to survey the entire boreal forest and woodlands from the Hudson's Bay to the Rocky Mountains. One of his objectives was to determine which public lands were non-agricultural and therefore should become part of the forest reserve complex.

Initially, Campbell decided on exploratory surveys that did little more than locate the main, merchantable forests. In Alberta, he assigned two men to the job: S.H. (Stan) Clark carried out the work north and east of Lac La Biche, J.A. Doucet examined the country south of Lesser Slave Lake and westward to the Rocky Mountains Forest Reserve.

Stan Clark's report of 1912 outlined his explorations: "The bad state of roads from Athabasca Landing to Lac La Biche made it imperative that I should freight via Lamont. [On May 28] we started toward Heart Lake with a cook, packer and seven horses. It required six days to freight the provisions from Lamont to Heart Lake, a distance of about 180 miles. The first two months were spent examining the country north and east of Heart Lake… The inaccessibility of the country west of Lac La Biche made it advisable to leave the pack train in care of Mr. James Spencer and hire his rowboat. By this means we were able to make a hurried trip down the La Biche River until we were stopped by the rapids."[14]

Transfer of Resources

The long-awaited *Alberta Natural Resources Act* of 1930 was created as an amendment to the *British North America Act*. The former was also known as *the Natural Resources Transfer Agreement (NRTA)*. Some Dominion foresters argued, without success, that at least the Forest Reserves should remain under DFB control. The transfer of natural resources to Alberta was made effective October 1, 1930. The DFB shifted its activities to research and information-gathering programs, handing the day-to-day operational activities and management of the forests to the government of Alberta.

Dominion Forestry Branch staff in 1926. Many transferred to the Alberta Forest Service in 1930
Back Row (L to R): Tom Burrows, Forest Supervisor Athabasca Forest; Charles McDonald, Assistant Supervisor Bow River Forest; R.M. Brown, Forest Supervisor Crowsnest Forest; Harry L. Holman, Forester Calgary. Middle Row: C.K. Le Capelain, Civil Engineer Calgary; Harry A. Parker, Forest Supervisor Cypress Hills Forest Reserve; A.G. Smith, Forest Supervisor Clearwater Forest; Don McKenzie, Forest Supervisor Brazeau Forest; J.P. Alexander, Forest Supervisor Crowsnest Forest; Freeman Kelley, Chief Ranger Cooking Lake Forest Reserve; Symen Nelson, Accountant Calgary Office. Front Row: Col. Robert H. Palmer, Head Edmonton Fire Ranging District (E.F.R.D.), Edmonton; James A. Hutchison, Forest Supervisor Bow River Forest; Charles H. Morse, Inspector of Forestry Calgary District; James Smart, Assistant Inspector of Forestry Calgary; Ted F. Blefgen, Forest Supervisor Lesser Slave Forest Reserve.
Dominion Forestry Branch, AFHPC

HOW TO Pack A Horse

by Archie Pendergraft

THE SINGLE DIAMOND HITCH

1. With panniers and top pack in place, lash rope is thrown over top of pack, lash cinch swing under belly.

2. Point of hook *back*, cinch is hooked in loop formed in lash rope by twist as shown. Twist is lifted to top of pack after hooking cinch, to point "A"

3. Tuck second loop "B" under rope crossing top pack, from *rear*, and enlarge to make large loop for right side of pack.

4. Leaving loop "B" hanging on right side of pack, pull third loop "C" from between points "A" and "D." This loop "C" is for left side of pack.

5. Now pull up on rope at "A" and across at section "E" of loop "B," tightening cinch as much as possible. Section "E" of loop "B" is then taken back, down and under rear of right side pack, and continuing up front of right side pack to center of top pack, where slack is pulled from "B" at point "F" of loop "C." Loop "C" then encircles left side of pack from front to rear.

6. Final shaping of the "Diamond" and tightening of the hitch is accomplished by pulling back hard on the end of the lash rope, which is then tied under left pack, above the cinch hook.

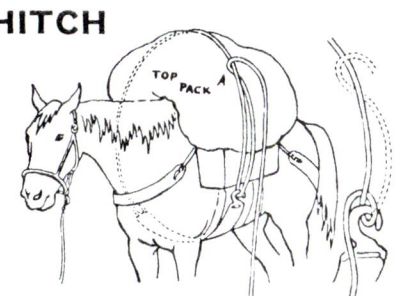
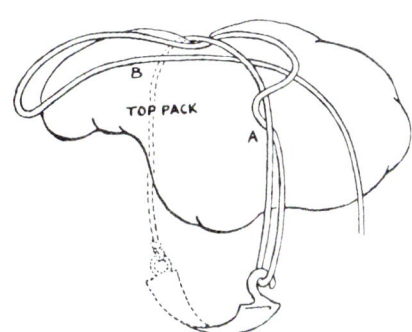
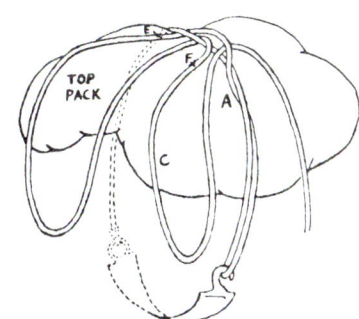

Packing Pokey, 1917, Upper Sentinel area of Highwood River. Rangers Tom Willdigg, Harry Holman and J. Archer
Dominion Forestry Branch, AFHPC

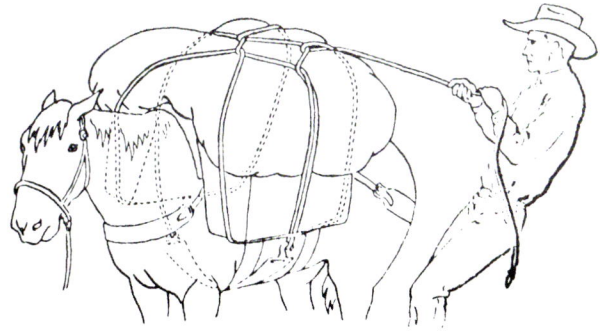

CREDIT: COLORADO OUTDOORS
Alberta Lands - Forests - Parks - Wildlife Vol. 12(2), Summer 1969

Forest Fire Rangers Ted Blefgen (L) and Gordon Ritchie at Lynx Creek Cabin, Crowsnest Forest, 1912. Blefgen was appointed first Alberta Director of Forestry in 1930
Dominion Forestry Branch, AFHPC

Cowboy Camp at Cameron Creek, in Waterton Lakes National Park, 1911
Dominion Forestry Branch, AFHPC

Burned lodgepole pine salvaged for corral construction, Cypress Hills Forest Reserve, 1915. Settlers on the prairies would travel for days with horse team and wagon to get timber from the Forest Reserves
Dominion Forestry Branch, AFHPC

Bow River Forest staff at Jumping Pound Ranger Station for a lesson in using survey equipment, 1923. Forester Harry Holman at front right
Dominion Forestry Branch, AFHPC

Eau Claire and Bow River Lumber Company log drive - sluicing logs through the lower dam on North Ghost River, 1924
Dominion Forestry Branch, AFHPC

Planting crew Peter Ward, Ben Shank and Harry Groves, Cooking Lake Forest Reserve, 1922
Dominion Forestry Branch, AFHPC

Lower flood dam, North Ghost River, Aura District, Bow River Forest. Dam was used by the Eau Claire and Bow River Lumber Company for water transport of logs
Dominion Forestry Branch, AFHPC

Nordegg Ranger Station, Clearwater Forest, Rocky Mountains Forest Reserve, 1928. Nordegg was a major headquarters in the Clearwater Forest. It served a large area north to the Brazeau River and west to what is now Banff National Park
Dominion Forestry Branch, AFHPC

Dominion Forest Service Ranger, early 1900s, Rocky Mountains Forest Reserve
Dominion Forestry Branch, AFHPC

Early Dominion Forest Service employees constructing trails and telephone lines in the Bow River Forest, Rocky Mountains Forest Reserve. Photo shows Elbow Trail crew and camp cook, 1915
Dominion Forestry Branch, AFHPC

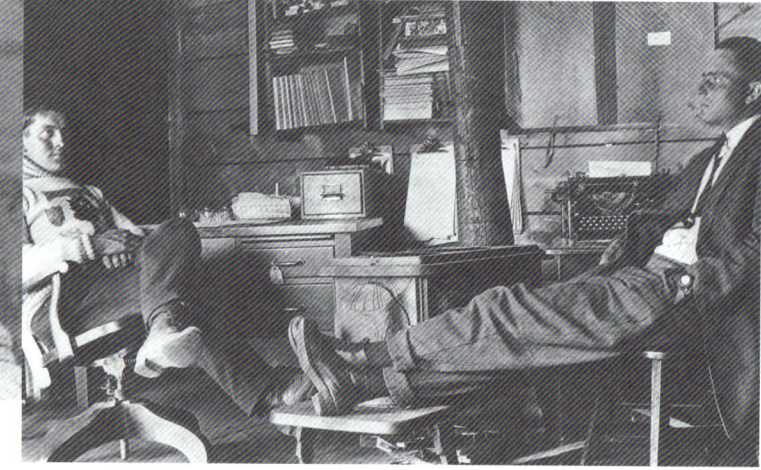

Stan Clark (L), Forest Supervisor of the Athabaska Forest, with Charles Morse, Alberta District Superintendent, Hinton, 1913. Neil Gilliat recalls "Stan Clark owned and operated the General Store in New Entrance when I was there in 1949. He did most of the pioneer work in establishing the Athabaska Forest, Rocky Mountains Forest Reserve."
Dominion Forestry Branch, AFHPC

Forester J.A. Doucet was one of the Dominion Forest Service employees tasked to survey the Forest Reserves between 1910 and 1915. Photo shows Doucet and his survey crew, Athabaska Forest near Entrance, Rocky Mountains Forest Reserve, 1912
Dominion Forestry Branch, AFHPC

Early Dominion Forest Service employees on trail and telephone line construction – having a bite to eat, Bow River Forest, Rocky Mountains Forest Reserve, 1915
Dominion Forestry Branch, AFHPC

Ranger tent camp for telephone line construction, Athabaska Forest, Rocky Mountains Forest Reserve, 1915
Dominion Forestry Branch, AFHPC

Temporary field telephone installation, Bow River Forest, Rocky Mountains Forest Reserve, 1916
Dominion Forestry Branch, AFHPC

Rangers install first pole on the North Trunk Telephone Line. Phone lines were a vital link for rangers between their cabins, headquarters and lookouts. Bow River Forest, Rocky Mountains Forest Reserve, 1922
Dominion Forestry Branch, AFHPC

Poles used to build tripods in the construction of telephone line system. Construction of phone lines through open areas like this lessened maintenance caused by trees falling across the line. Bow River Forest, Rocky Mountains Forest Reserve, 1920s
Dominion Forestry Branch, AFHPC

Early Days

Forest Supervisor McAbee saddling up for the day's ride, Clearwater Forest, Rocky Mountains Forest Reserve, 1912
Dominion Forestry Branch, AFHPC

'Slinging the 2nd pack' (loading pack horse) for day's ride, Brazeau Forest, Rocky Mountains Forest Reserve, 1913
Dominion Forestry Branch, AFHPC

Foresters and Rangers at the Porcupine Hills District, Crowsnest Forest, Rocky Mountains Forest Reserve, 1921
Front Row (L to R): E.B. (Eb) Walker, Assistant Ranger; Lloyd van Camp, Forester Pincher Creek; W. Antle, Assistant Ranger; J. A. (Jock) Frankish, Forest Ranger. Back Row: H. G. (Harry) Nash, Forest Ranger Livingstone (Gap); R.J. Prigge, Assistant Ranger; J. H. (Harry) Boulton, Forest Ranger; J. H. McLeod, Forest Ranger Crowsnest Pass; J.P. (Jack) Alexander, Supervisor Pincher Creek; H. B. (Posthole) Smith, Forest Ranger Porcupine Hills; G. A. Ritchie, Forest Ranger; T. D. (Tom) Best, Assistant Ranger; F.T. (Fred) Monk, Assistant Ranger; W. A. Lyndon, Forest Ranger
Mrs. T. Vickerman

E.H. Finlayson and crew member wake up to early snowfall at Monaghan Creek, Athabasca Forest, Rocky Mountains Forest Reserve, 1916. Finlayson was the Chief Inspector for the Alberta District of the Dominion Forestry Branch from 1915 to 1920
Dominion Forestry Branch, AFHPC

Chief Forest Ranger Margach on horseback, Crowsnest Forest, Rocky Mountains Forest Reserve, c. 1908
Dominion Forestry Branch, AFHPC

Dominion Forestry Branch Fire Rangers, Rocky Mountains Forest Reserve, early 1920s
Dominion Forestry Branch, AFHPC

Dominion Forestry Branch Supervisors meeting, Canmore, 1920s
Jack Janssen, middle front row, was head of AFS Forest Protection until the early 1950s
Dominion Forestry Branch, AFHPC

Ranger Fred Nash, Sheep Ranger Station, leading pack string at the Stampede Parade, Calgary, 1920s
Dominion Forestry Branch, AFHPC

Early Days 23

Peter McLaren Lumber Company bush crew, Camp 3, Crowsnest Forest, Rocky Mountains Forest Reserve, 1910
Roy Campbell

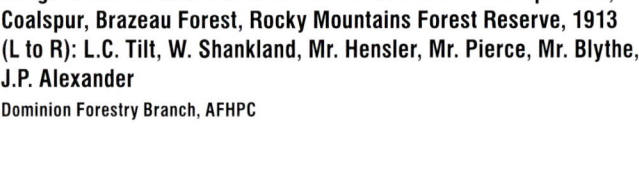

Rangers and Foresters outside the Brazeau Forest Headquarters, Coalspur, Brazeau Forest, Rocky Mountains Forest Reserve, 1913 (L to R): L.C. Tilt, W. Shankland, Mr. Hensler, Mr. Pierce, Mr. Blythe, J.P. Alexander
Dominion Forestry Branch, AFHPC

Pack crew camp, west of Rocky Mountain House, Rocky Mountains Forest Reserve, c. 1920. Packing equipment and gear into the Nordegg mine from Innisfail. Crew stopped for the night half way between Rocky Mountain House and Nordegg. Wilford Gray on the right (grandfather of Howard Gray) and his crew also hauled equipment and gear for the Dominion Forest Service for trail construction west of Rocky Mountain House
Howard Gray

Peter McLaren Lumber Company, Blairmore, Crowsnest Forest, Rocky Mountains Forest Reserve, 1912. The background slopes show evidence of fires from either the early 1890s or 1904, or a combination of both
Roy Campbell

Dominion Forest survey crew at Camp 1 along the Athabasca River, T71, R26, W4M, Athabasca Forest, Rocky Mountains Forest Reserve, 1911
Dominion Forestry Branch, AFHPC

Peter McLaren Lumber Company log flume, Crowsnest Forest, Rocky Mountains Forest Reserve, c. 1912
Roy Campbell

Peter McLaren Lumber Company Camp 3, office and bunkhouse. Camp was at 5701 feet above sea level, Crowsnest Forest, Rocky Mountains Forest Reserve, 1912
Roy Campbell

Early Days

Rangers with pack string traversing a 10% grade using a switchback, Brazeau Forest, Rocky Mountains Forest Reserve, 1915
Dominion Forestry Branch, AFHPC

Prevent Forest Fires sign along the Elbow Road, Bow River Forest, Rocky Mountains Forest Reserve, August, 1925
Dominion Forestry Branch, AFHPC

Making a road grade on the Coal Camp Hill, Bow River Forest, Rocky Mountains Forest Reserve, July, 1929
Dominion Forestry Branch, AFHPC

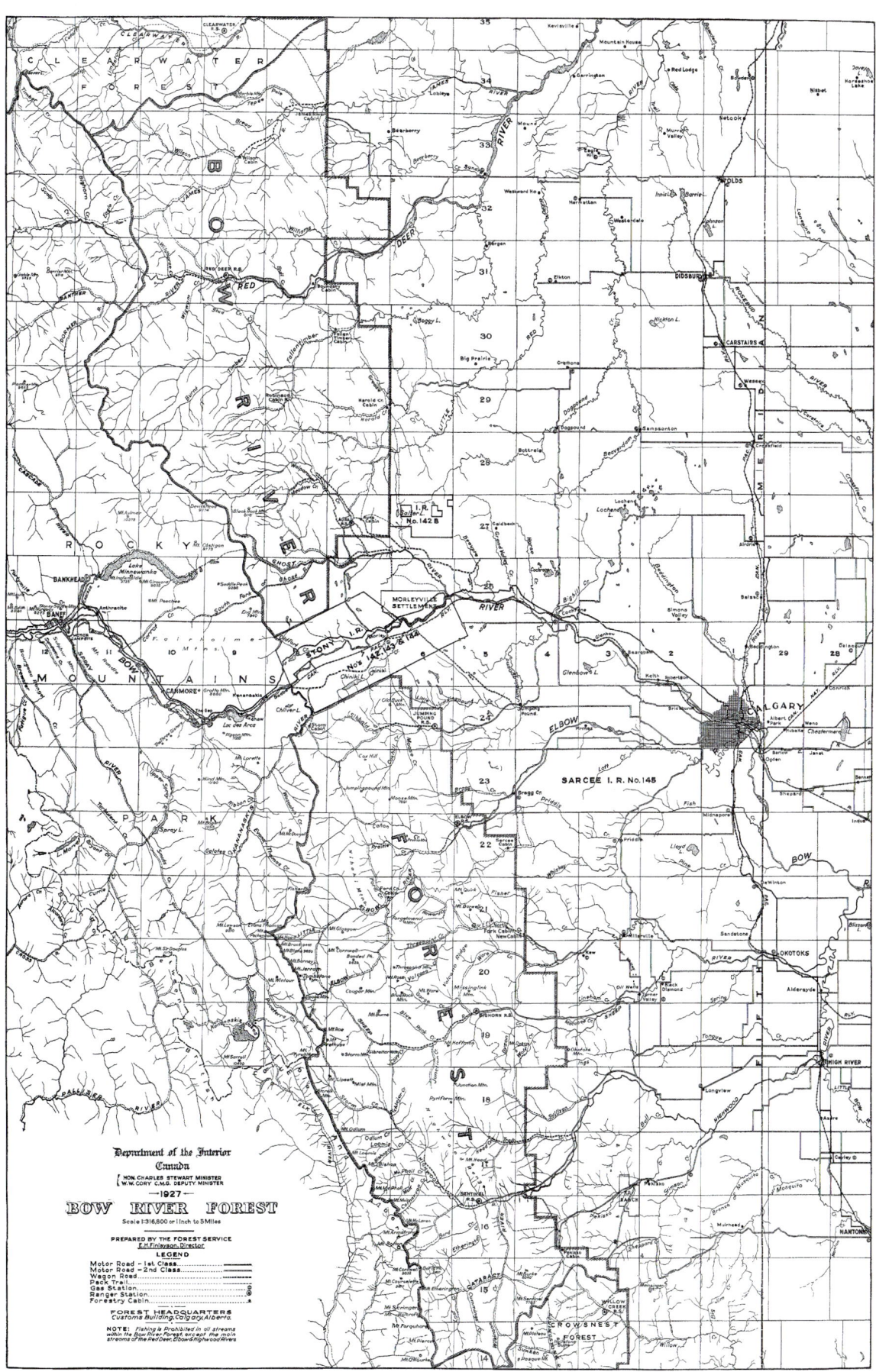

Map of the Bow River Forest, Rocky Mountains Forest Reserve, 1927
Alberta Government, AFHPC

CHAPTER 2
Birth of the Alberta Forest Service

20-21 GEORGE V.

CHAP. 3.

An Act respecting the transfer of the Natural Resources of Alberta.

[Assented to 30th May, 1930.]

HIS Majesty, by and with the advice and consent of Senate and House of Commons of Canada, ena follows:—

1. This Act may be cited as *The Alberta Natur sources Act.*

2. The agreement set out in the schedule he hereby approved, subject to the proviso that, in addi the rights accruing hereunder to the province of A the said province shall be entitled to such further r any, with respect to the subject matter of the said ment as are required to be vested in the said pro order that it may enjoy rights equal to those which conferred upon or reserved to the province of Saska under any agreement upon a like subject matter h approved and confirmed in the same manner as agreement.

A constitutional amendment was made allowing for Federal (left) and Provincial (lower) legislation transferring responsibility for natural resources to Alberta in 1930. The Provincial Act is commonly known as the *Natural Resources Transfer Agreement*

1930

CHAPTER 21.

An Act respecting the Transfer of the Natural Resources of Alberta.

(Assented to April 3, 1930.)

HIS MAJESTY, by and with the advice and consent of the Legislative Assembly of the Province of Alberta, enacts as follows:

1. This Act may be cited as *"The Alberta Natural Resources Act."* — Short title

2. The agreement set out in the schedule hereto is hereby approved, subject to the proviso that, in addition to the rights accruing hereunder to the Province of Alberta, the said Province shall be entitled to such further rights, if any, with respect to the subject matter of the said agreement, as are required to be vested in the said Province in order that it may enjoy rights equal to those which may be conferred upon or reserved to the Province of Saskatchewan under any agreement upon a like subject matter hereafter approved and confirmed in the same manner as the said agreement. — Approval of agreement

3. *The Transfer of Public Lands Act*, being chapter 69 of the Statutes of Alberta, 1926, is hereby repealed. — Repeal

4. This Act shall come into force upon a date to be fixed by Proclamation of the Lieutenant Governor in Council. — Coming into force of Act

Signing of the *Natural Resources Act*, transferring control of Alberta's natural resources over to the province, Privy Council Chamber, Ottawa, December 14, 1929

Seated (L to R): Dr. James H. King, Charles Stewart, Prime Minister Mackenzie King, Alberta Premier John Brownlee, George Hadley, John F. Lymburn. Standing: Col. Oliver Mowat, Robert Forke, James Malcolm, J.C. Elliot, J.L. Ralston

Provincial Archives of Alberta, A10924

Birth of the Alberta Forest Service

The province of Alberta took ownership and responsibility of public lands, forests and other natural resources from the federal government on October 1, 1930. The change, which came 25 years after the move to provincehood within the Canadian confederation, meant Alberta now had control of its public lands, forests, wildlife (except fish and migratory birds) and minerals.

The transfer included responsibility for all of the publicly-owned forestland outside the national parks. This land included 12.44 million acres in Forest Reserves previously managed by the Dominion Forestry Branch (DFB), and 91.43 million acres designated as the Edmonton Fire Ranging District, as cited in the 1930 - 31 Annual

Truck hauls sleighs loaded with ties and logs for Erith Tie, Brazeau Forest, 1935
Alberta Government, AFHPC

Report of the new Department of Lands and Mines. The total area amounted to 63.5 per cent of the province's total land base of 163.8 million acres.

A new Alberta Department of Lands and Mines was created to manage this landbase and its natural resources, under the direction of Minister Richard G. Reid and Deputy Minister John Harvie. Harvie came to the position with experience in lands administration gained in service to the federal government. He characterized the transition by saying: "The administration changed, but the policies stayed essentially the same."[1]

During that first year, however, staff members were required to draft six new Acts and 48 Orders in Council to transfer natural resource management to provincial administration. It was a major task! As a result, few innovations were incorporated. Also, times were tough. A worldwide economic depression was settling in, money was scarce and unemployment was common. Canada as a whole was experiencing "hard times."

The Department of Lands and Mines contained the newly formed Alberta Forest Service (AFS), headed by T.F. (Ted) Blefgen. Blefgen had worked his way through the DFB in Alberta, from his initial appointment as a ranger in the Crowsnest Forest in 1912 to his eventual service as forest superintendent of the Lesser Slave Forest Reserve. J.A. (Hutch) Hutchison, who was Assistant Director of Forestry from 1934-47, had also come up through the federal field service.

Alberta's population in 1930 was just 730,000, and it was not a wealthy province. The economy was based largely on agriculture and coal. There was a scattered forest industry, mostly producing

Ranger Danny Fraser was one of those who transferred from the Dominion Forestry Branch to the Alberta Forest Service. He was stationed in Whitecourt from 1928-1932. He was well-known as a fiddle player and was much in demand at local dances
Alberta Government, AFHPC

lumber, railroad ties and mine props for use within the province. Timber was sold through sealed tender, as set up under the previous federal administration.

Although the province's fiscal resources

ALBERTA FOREST SERVICE
1932

Minister - Richard Reid
Deputy Minister - John Harvie
Director - Ted Blefgen
Asst. Director - James Hutchison

Administration Office
Chief Clerk - W. Ronahan
Timber Clerk - A. Peart
Clerk - D. Florence
Stenographers - Miss M.E. Reid, Miss K. Filyk, Miss M. Fowler

Chief Timber Inspector
F.W. Neilson
Timber Investigator
E.S. Huestis
Timber Inspector (Relief & Yards)
C. Ranche

Forest Service

Brazeau - Athabaska Forest

Supervisor
F.G. Edgar
Ranger
L.J. Main
Clerk
J.A. Bailey
Seasonal Rangers
S.V. Scoble
J.W. Walker
W.H. McCardell
A. Crawford
C. Hughes
A.H. Hammer
R.W. Holgate
W. Smith
T.F. Coggins
T.R. Hammer
A. Reimer
Lookoutmen
D. Griffiths
H.W. Lendrum
R. Thompson
Temporary Rangers
W. Adamson
H.H. Cochrane
E.A. Harrison
R. Lendrum
T.C. Burrows
J. Glen

Clearwater Forest

Superintendent
J.P. Alexander
Ranger
W. Scott
Clerk
A.H. Taylor
Seasonal Rangers
J.A. Reynar
C. Sawyer
T. Weaton
H. Crabtree
C.E. Earl
F.G. Bradshaw
E.L. Whidden
W.E. Fisher
Lookoutmen
G.B. Elliott
N.W. Justinen
Telephone Operator
F.H. Jackson
Temporary Rangers
W. Bell
J.R. Engebretson

Crowsnest - Bow River Forest

Superintendent
A.G. Smith
Rangers
P. Campbell
J.H. Boulton
Clerk
J.E. Redden
Stenographer
E. Roper
Seasonal Rangers
J. Kovach
T. Clark Jr.
J. Cardinal
G. Copithorne
J. Reid
S.R. Measor
C.G. McKenzie
W.H. Griffiths
J.S. Gowland
R.J. Steeves
E.F. Creed
Patrolman
L. Measor
Telephone Operator
S.O. Swan
Temporary Rangers
T. Howard
A.H. Bryant
J.A. Atkinson
J.A. Frankish
T. Harvey
F. Nash
W. Antle
R.J. Prigge
T.D. Best
F.T. Monk
J.E. Bell
J.H. McLeod
L.L. Waikle
Seasonal Lookoutmen
J. Lardinois
E. Gamache
R.C. Lloyd
G. Davis
J.D. Champion
A.B. Shantz
G. Pearce
A. Fraser

Cypress Hills Forest

Ranger
G.R. Ambrose

Northern Alberta Forest District

Edmonton

Timber Inspector
R.S. Wyllie
Seasonal Rangers
G.M. Beattie
F. Smith
J. Bowman
A.H. White
H. Burden

Athabaska

Timber Inspector
A. Smith
Seasonal Rangers
M.J. Doucette
W. Richardson
C. Carter
A.E. Parker
C.H. Jones

Peace River

Timber Inspector
D.H. Minchin
Seasonal Rangers
A. Stevenson
B. Broughton
R. Gicquel
F. Freeborn
W.R. Hawkes
B.H. White

Slave Lake

Timber Inspector
C.H. MacDonald
Seasonal Rangers
O. Schroder
D. Trindle
J.L. Janssen
S. Johnston
R. Hubley
H. Haight
H. Wileman
B. Watkins
A. Foley
C. McKinley
Lookoutmen
A. Clark
W. Thompson
S.E. Hatcher
A.D. Craddock
Telephone Operator
C.D. MacDonald

Bonnyville

Timber Inspector
D.A. McKay
Seasonal Rangers
W.E. Brown
J.W. Allen

Edson

Timber Inspector
J.R.H. Hall
Seasonal Rangers
R. Chamberlin
F.H. Shearn
T. Gowdie
C. McDiarmaid
J. Fandrich

Fort McMurray

Timber Inspector
H.D. McDonald
Seasonal Rangers
F. Parker
J. Foley
H.K. Graham
L.R. West
E. Wylie
R. Fraser
J.N. Fournier
E. Hogue

Grande Prairie

Timber Inspector
D. Buck
Seasonal Rangers
D. Harrington
T. Walters
D. McMillar
A. Sherman
D.G. O'Brien
R.L. Everrett
V.W. Mitchell

Alberta Forest Service organization as at September 1, 1932. The AFS was part of the Department of Lands and Mines under Minister R.G. Reid and Deputy Minister John Harvie.

were very limited, the Alberta government resolved to maintain the management standards set by the DFB. As Ted Blefgen stated in his first Annual Report:

"On October 1st, 1930, the Forest Service of the Department of Lands and Mines, Government of the Province of Alberta, became responsible for the major forestry activities within the province of Alberta. In line with this change, practically all the officials of the Service previously responsible for this work, under the Department of the Interior, were taken over by the Alberta Forest Service, the exceptions being principally those technical officers employed by the forest service of the Department of the Interior engaged in research work – research and investigative work being on the programme of the Dominion service. Retention of these men by the Dominion service depleted the administrative staff to some extent, in view of the fact that these men had all, from time to time, taken an active part in administrative work."[2]

There is ample evidence that those who did transfer to the AFS from the DFB remained wholeheartedly committed to protecting Alberta's forest and rangelands. Alberta was their home.

Alberta's Forest Reserves, originally created by the federal government, were considered to be the most important forests at the time – both for watershed and timber values. Of the four Forest Reserves, the largest was the Rocky Mountains Forest Reserve comprising the Crowsnest, Bow River, Clearwater, Brazeau and Athabasca Forests. The other three were the Lesser Slave, Cooking Lake and Cypress Hills Forest Reserves. Forest protection and timber management were priorities in the Forest Reserves. Of necessity, former Dominion policies were retained, though this changed over the next 20 years as awareness of the extent and importance of Alberta's forest resource increased and the economy began to grow.

The *Alberta Natural Resources Act* of 1930 provided an opportunity to adjust boundaries between federal and provincial lands. For example, the community of Entrance was then the last railway station east of the boundary of Jasper National Park, which also included Brule and some of the land east of Brule Lake. As Blefgen explained in 1931:

"Some time before the transfer of resources a survey of the Banff

Fire Ranger René and Mrs. Gicquel, Upper Landing, Vermilion chutes, Peace River, 1936
Alberta Government, AFHPC

Whitecourt Lookout, 1938. This 10-metre tower was built with local timber
Alberta Government, AFHPC

and Jasper National Parks was undertaken with a view to eliminating from the National Parks, those areas 1) which were, or would likely become, areas required for industrial development, 2) unsuitable for park purposes, or 3) for the purposes of establishing a more definite and more satisfactory boundary than had previously existed."[3]

Blefgen noted that in total 1364.46 square miles

were added to the Forest Reserves, and that 821.5 square miles were added to the Parks, particularly from the Clearwater Forest to Banff. His comments referred to a "considerable area added to the Athabasca Forest in the vicinity of Brule, and to the Brazeau Forest northwest of Luscar."

In his first Annual Report in March, 1931, Ted Blefgen described how the AFS was immediately involved in setting up unemployment relief work camps to tackle forestry projects. It was a sad reflection of the economic conditions of the day, but the camps did provide the labour required to extend work on forestry roads, trails and buildings. As Eric Huestis later recalled, the AFS was required to lay off about a third of its staff in 1932 as a budget-cutting measure. Deputy Minister John Harvie reflected that: "… it was a terrible depression, you know, and we … just had to cut in every way, shape and form."[4]

Hard Times

Provincial budget cuts were not long in coming, along with staff layoffs and changed working conditions. While the federal Annual Report for 1928 listed 155 people in provincial forestry positions in Alberta, the 1931 list showed 123, a reduction of 32 people. Most of the ranger positions were subsequently made seasonal – meaning most staff were laid off during the winter for financial reasons. Eric Huestis, Director of Forestry from 1948 to 1963, carried with him a dog-eared list with the names of the eight permanent rangers the AFS counted upon for the Forest Reserves in 1939. These were G.R. Ambrose at Cypress Hills, Jack Bell in Calgary, J.H. Boulton in Coleman, R. Lendrum at Coalspur, J.H. McLeod at Willow Creek, Fred Nash at Turner Valley, Eric L. Whidden at Rocky Mountain House, and L.L. Waikle at Entrance.

The 1930s generally were a most difficult time for the fledgling AFS. The years were characterized by the Great Depression, drought and the onset of the Second World War.

The AFS was reorganized in 1932 when the budget cuts were made. Until that time, timber inspectors in the Lands Division were responsible for timber berths, doing the job of the former dues-collecting Timber and Grazing Branch. Now, timber inspectors were made part of the AFS,

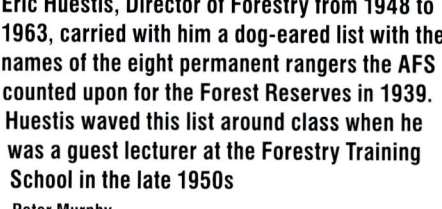
Eric Huestis, Director of Forestry from 1948 to 1963, carried with him a dog-eared list with the names of the eight permanent rangers the AFS counted upon for the Forest Reserves in 1939. Huestis waved this list around class when he was a guest lecturer at the Forestry Training School in the late 1950s
Peter Murphy

Rock Lake cabin, late 1930s - more than a ton of gear to be moved up the mountain trail by horse. Ranger Jack Glen with pipe to the right of cabin door
Alberta Government, AFHPC

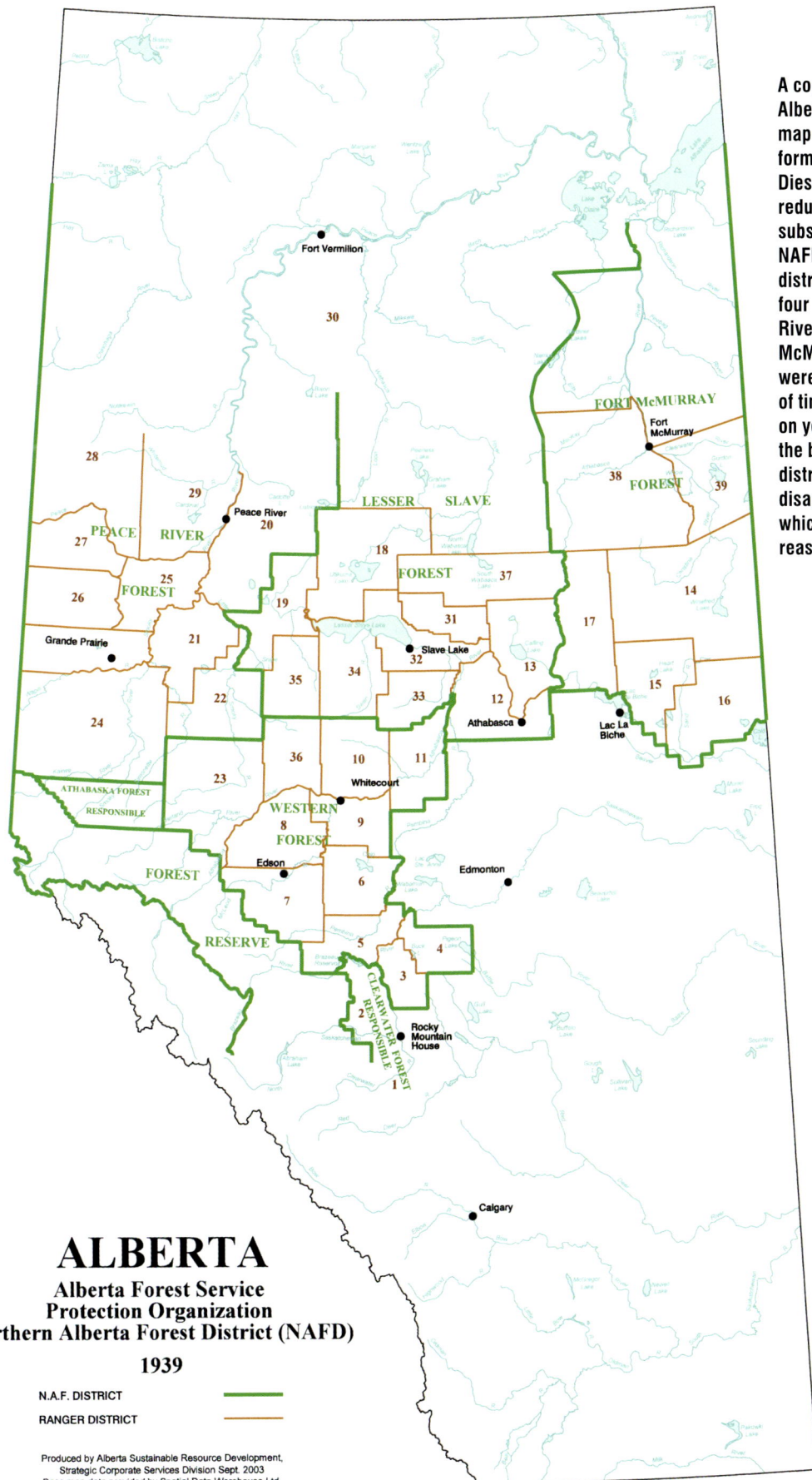

A copy of the 1939 Northern Alberta Forest District (NAFD) map provided by the widow of former northern Ranger Bob Diesel. After the serious staff reductions of 1932-33, and a few subsequent reorganizations, the NAFD included about 39 ranger districts consolidated into four Forests - Western, Peace River, Lesser Slave and Fort McMurray. Most of the rangers were seasonal, while a handful of timber inspectors stayed on year-round. Of interest are the boundaries of the northern districts that just run up and disappear into the bush, which was beyond any hope of reasonable access effectively integrating timber with protection and forest management. The Edmonton Fire Ranging District was renamed the Northern Alberta Forest District (NAFD). It was divided into five areas called Forest Divisions with timber inspectors (later to be called forest superintendents) in charge of protection and inspection of timber operations.

The Forest Divisions that later emerged initially comprised Rocky Mountain House, Whitecourt, Lac La Biche, Grande Prairie and Peace River. Forests within the Rocky Mountains Forest Reserve were re-combined into three, with a superintendent in charge of each – the Bow-Crow, Clearwater and Brazeau-Athabasca (headquartered in Edson). Slave Lake also continued its Forest Reserve status, although with reduced staff. Two smaller Forest Reserves were disbanded: the Cooking Lake Forest Reserve east of Edmonton, parts of which augmented Elk Island National Park, with the balance becoming the Cooking Lake-Blackfoot Recreation Area, and the Cypress Hills Forest Reserve which later (in 1951) became a provincial park.

By 1934 the former DFB had reassessed its role. Having given up virtually all of its land management responsibilities in the western provinces in 1930, it now focused on forest research. In August, 1934, the Canadian Forest Service, the DFB's successor organization, negotiated a lease from the AFS to establish the Kananaskis Forest Experiment Station within the Crowsnest-Bow River Forest. Its objectives included research in silviculture and other phases of forestry including protection, measurement and watershed management. The Science Service

Youth Forestry Training Program class, Highwood River, 1937-38. The class included Pat Donnelly, Larry Gauthier, Jack MacGregor with Bill Shankland as the Ranger in charge
Alberta Government, AFHPC

National Forestry Program enrolee learns the forester's craft, Bow River Forest, 1939. Jack Turner recording measurements
Alberta Government, AFHPC

of the federal Department of Agriculture conducted surveys of forest insects and diseases.

With continued unemployment, a national Youth Forestry Training Program (YFTP) was initiated and funded through the DFB to provide work and training to unemployed

Oliver Tree Nursery, Edmonton, 1933. This was the primary nursery for the AFS until 1975
Alberta Government, AFHPC

youth. In Alberta the program was run by the AFS. The first camp was set up on the Highwood River in 1937 with 60 trainees under the direction of district ranger Bill Shankland. At least two of those trainees later joined the AFS – George Deans and L.N. (Doonie) Donovan. Donovan served on Carrot Creek tower in 1939 and 1940, launching a career with the AFS that included service in the radio division prior to his retirement. W.J. (Jack)

MacGregor, L.P. (Larry) Gauthier and Gordon Fowlie were three who attended the course in 1938. Fowlie started as a lookout observer and later became a radio technician. MacGregor was Superintendent in the Slave Lake Forest before transferring to Edmonton and retiring in 1974 as the Supervisor of Construction and Buildings. Gauthier was Superintendent in Peace River Forest before transferring to Calgary, retiring in 1974 as Superintendent of the Bow Crow Forest.

In 1939 the youth program was rolled into a National Forestry Program (NFP), still with federal funding, at which time 130 young people were employed by the AFS. In addition, there were many similar camps in the national parks. Numerous participants pursued careers in forestry. For example Stan Hughes, who later became Director of Forest Protection, started his career in the Jasper camp. The camps were closed in 1940 with the advent of the Second World War, when most of the trainees joined the Armed Services.

Reforestation

In the fall of 1931, the AFS selected a site next to the Oliver Hospital near Edmonton for a forest nursery. The Oliver Tree Nursery was initially stocked with nursery materials moved from plots established earlier at the old Cooking Lake Forest Reserve. The provincial government administered the nursery in connection with the nearby psychiatric facility. The thought was, as Eric Huestis later explained, that patients could help with nursery work as a form of therapy. Although the original objective was to provide planting stock for windbreaks and shelterbelts on farms, the nursery also supported the establishment of a provincial reforestation program. However, it was not for another 30 years, until the early 1960s, that budgets allowed reforestation to receive serious attention. Changes in forest management regulations that took effect in 1966, with the introduction of the quota system, included conditions that cutovers on permanent forestland must be restocked to provincial standards.

"Turk" Bailey (L) and Stan Hughes, National Forestry Program field office, Jasper National Park, 1939. Bailey later went on to become Deputy Minister of Lands and Forests, Ontario, while Hughes served with the Royal Canadian Air Force in many combat zones, later entering a rewarding career in Alberta. He retired in 1975 as Director of Forest Protection
Alberta Government, AFHPC

Tony Earnshaw with portable radio, Crowsnest Forest, Rocky Mountains Forest Reserve, 1949
Tony Earnshaw

Wildfire

Despite the cutbacks in the 1930s, new fire towers and lookouts were needed and several were constructed, including the first AFS tower on Buck Mountain near Breton in 1931 and then Whitecourt Mountain and Mayberne in 1934. The wooden towers were 40 feet to 60 feet high, and were built by rangers with locally-cut poles and logs. The DFB had constructed a 60 foot pine 'crawl tree' with ladder rungs on Whitecourt Mountain during the federal days for use as an observation point on their patrol route. It is believed that Carrot Creek was the first steel tower built by the AFS with used steel from an old Cooking Lake Forest Reserve tower.

The early towers were linked by phone line, but in 1935 radios were installed in towers situated between the Brazeau and Athabasca Rivers. The arrival of radio occurred in stages. In the Department's 1935 Annual Report, Blefgen speaks of growing interest in the use of radio. The 1938-39 Annual Report shows that in the NAFD (the large forest area outside the Forest Reserves) five towers had already been equipped with radios, while the rest still used telephones.

A 1948-49 report shows that the east slopes lookouts were still using the single-line ground-return telephone system, in which phone lines were strung between standing trees. This continued until the 1950s when the AFS installed an FM radio system. In 1942, radio frequencies used at 45 northern Alberta sites had to be changed to avoid interference from the U.S. Army Signal Corps, which was working on construction of the Alaska Highway. A new radio headquarters was established on the western edge of Edmonton on 146 Street and 103 Avenue, with Tony Earnshaw as Radio Superintendent. By 1941, prototype two-way radios had been installed in the cars of timber inspectors and other staff.

Despite the advances in radio communication, radio operators were required to have a basic proficiency in telegraphy into the mid-1940s. Radio operator and towerman Sam Fomuk said that he "sent countless messages over the years," including daily weather reports. He used telegraphy on towers until 1959, and AFS operators used radio-telegraphy for messages between Forest Headquarters and Edmonton into the early 1960s, when the teletype was introduced. The last of the telegraphers included Bill Norton in Calgary, Joe Wuetherick and Helen Ledingham in Edmonton, Sam Fomuk and Larry

Forest Service radio car outside AFS radio headquarters, 10322 - 146 Street, Edmonton, early 1940s. Note antenna on car's front fender
Alberta Government, AFHPC

White spruce was a favoured material for food containers
Bob Stevenson

36 Alberta Forest Service

Lookout man sights through Osborne firefinder
Alberta Government, AFHPC

Huberdeau at Footner Lake.[5]

The drought years of the 1930s set the stage for more extensive fires, some of the most serious occurring in the Crowsnest and Bow River forests. It was suspected that in some northern areas, many fires were started by people as a way of creating employment – even though the going rate of pay was only 15 cents per hour.

Second World War

With the declaration of war and the "call to arms" in September, 1939, many staff left for active duty in the armed forces. By 1940, 39 men and three women from the AFS were serving in the armed forces. Among the first were J.P. Alexander, A. Craig, J.A. Hutchison, F.V. Keats, R. Krause, B. Longson, W.H. McCardell and W.J. MacGregor. The war effort also stimulated demand for forest products, including mine timbers required for accelerated coal production. Large-diameter paper birch trees were logged in the Lesser Slave Lake area for construction of the Allies' Mosquito bomber aircraft, thanks to reports commending the species for its "required strength with the minimum weight."[6]

There was also a strong demand for white spruce as crate material for shipping food since "no odour passed on to meats and butter."[7] During the war, some of the incendiary balloons launched by Japan reached Alberta, although none was reported to have started a fire. Forestry staff members were warned about the balloons, but this information was kept from the general public as a military secret.

Registered traplines were introduced in 1939 to regulate the burgeoning fur industry, and in response to concerns about the decline in beaver populations as a result of disease and over-trapping. AFS rangers were given the responsibility for game and trapline administration. They also assumed the task of tagging beaver pelts, which were restricted by quota to one animal per beaver house on the trapline. The process of issuing a tag, known as "sealing beaver," was a conservation measure to control the harvest, curtail poaching among trappers and prevent the "bootlegging" of beaver pelts. The beaver tags consisted of a lead seal pressed over the ends of a cord running through the eyeholes of the pelt. Rangers began a closer involvement with trappers as registered trapping areas were gradually granted north of the Brazeau and North Saskatchewan rivers. The rangers also helped trappers map the locations of the beaver houses.

At the end of the war in 1945 a period of reflection and recovery began. Blefgen presented the customary plea for additional support for the AFS in his Annual Report for 1945-46, adding the comment: "during the depression years we were definitely informed that no money could be made available, and during the war years the necessary labour could not be secured." The published lists

Ranger John Currat (R) visits with outfitter/ trapper at Moberly Ranger Station, Athabasca Forest, 1940s
Alberta Government, AFHPC

of AFS staff offer evidence of this frustrating situation. It was not until 1947, 17 years after resources were transferred to the province, that staffing levels again reached those of 1929. By 1952, staff levels had increased to about 2.5 times those of the low year of 1933. The long-hoped-for period of recovery and development had begun.

Post-war Recovery

The recommendations of the Alberta Post War Reconstruction Committee bolstered the gradual recovery of AFS resources. The Deputy Minister reported in 1946:

Rangers travel the Athabasca River, Whitecourt Division, 1947
Alberta Government, AFHPC

"The report of the sub-committee of the Alberta Post War Reconstruction Committee appreciates the vital need for preserving our forest areas and watersheds. Now would appear to be the opportune time to implement some of the main findings and recommendations of the sub-committee particularly in regard to:
• making a physical inventory of the forest resources of the Province;
• expansion of fire prevention services;
• instituting a long range programme of reforestation of cut over and burn over lands and afforestation of marginal and sub-marginal lands;
• inauguration of a training programme for men already in the forestry service and those wishing to join it which would give courses in timber cruising, insect and disease control, reforestation, wildlife, forest protection, etc.;
• establishment of additional tree nurseries in different parts of the Province to enable a study to be made of the species most suitable for planting in the area to be supplied both from the commercial point of view, as well as providing forest cover and trees for farm planting."[8]

Fire Detection

This was also a period of innovation in the areas of science and technology. One example was development of effective firefinders for use in towers. Firefinders are devices that enable lookouts to pinpoint the location of a fire by triangulation, using a scope or eyepiece attached to a circular base marked with the points of the compass. The first firefinders were homemade, using cardboard or wood, with bearings etched on the base.

American towerman William "Bush" Osborne developed and refined the Osborne Firefinder in the U.S. between 1913 and 1934. The model eventually adopted for common use in Alberta was the Osborne 1934A. Several types of range-finding equipment were tested and rejected. The most common reason was the large areas and long distances dealt with at Alberta towers. Military technology such as the artillery range finder was tested and found to be too cumbersome. The Davis Range Finder was tested in the Whitecourt area but could not give

Jack MacGregor, Forest Superintendent, Slave Lake, 1960
Bruce MacGregor

consistent accuracy over long distances.

Several other types of firefinders were looked at (i.e. the Michigan Model 1968) and two were utilized. These were the Canadian Penitentiary (so named because it was made in federal prisons) and the Osborne. The Canadian Penitentiary was originally constructed of hardwood with a brass alidade (a topographic surveying and mapping instrument used for determining directions). Further refinements were made to this and the wood replaced with metal. Several of these AFS Modified Canadian Firefinders are still in use, with the last 15 being purchased in 1968.

The Royal Canadian Air Force, which had used vertical aerial photographs for wartime mapping, began trials in Alberta to investigate the peacetime application of aerial mapping and forest inventory. At the same time war veterans were returning to fill out the growing AFS staff ranks. They included some of the well-recognized early rangers such as Ernie Ferguson, Phil Nichols, Harry Edgecombe, Bert Coast, Ray Smuland, Larry Gauthier, Dick Radke and Jack MacGregor. Many veterans were also interested in opportunities for homesteading on or near forested lands, which prompted discussions about a possible exclusion of prime forested lands from settlement. There was also discussion about a possible joint federal-provincial board to increase fire control capability on the former Rocky Mountains Forest Reserve and to intensify protection of the eastern slopes' watershed values.

Rangers Ben Shantz and Ken Wheat on inspection of Bigstone Fox Creek pipeline 1965. Forestry truck with snowmobile to left of right-of-way
Alberta Government, AFHPC

A Major Oil Strike - Leduc #1

Perhaps the most significant single event contributing to the economic and social development of Alberta took place on February 13, 1947, when Leduc Oil Well No. 1 blew in. This event stimulated a great increase in exploration and development for oil and natural gas, created a demand for good maps and, most importantly, started to generate some significant revenues for the provincial government. Increased funds enabled the government to better support provincial programs, including those of the AFS. A more negative impact was the extensive disturbance that later occurred within the forest from exploration seismic lines, roadbuilding and general oilfield construction.

These events, pressures and conflicts in land use set the stage for the next period, significantly changing the nature of the work and responsibilities of the AFS.

Oil rig in Fort McMurray region, Lac La Biche Forest Division, 1960
Jack Roy

Students Frank Muldoon (L) and Larry Gauthier attending the Youth Forestry Training Program, Bow River Forest, Rocky Mountains Forest Reserve, 1937-1938. Larry Gauthier went on to a successful career with the Alberta Forest Service, employed as Forest Superintendent of the Peace River Forest in the mid-1960s and then retiring as Forest Superintendent in the Bow Crow Forest in 1974
Alberta Government, AFHPC

Youth Forestry Training Program students building road in the Highwood District, west of the Sentinel Ranger Station, Bow River Forest, Rocky Mountains Forest Reserve, 1937. L.P. (Larry) Gauthier is the student holding the water dipper
Alberta Government, AFHPC

Lookout and Radio Technician training at Tony Earnshaw's cabin at Pigeon Lake, early 1940s
(L to R): Not Identified, Not Identified, Don Bruce, Charlie Curran, Bill Norton, Not Identified, Gordon Fowlie
Freida Earnshaw

Digging stake truck out of snow, National Forestry Program, Highwood River, Rocky Mountains Forest Reserve, 1939
Alberta Government, AFHPC

Ranger Dexter Champion, with carpenter's assistant Jay Champion, building camp shelter at the Kananaskis Lakes, Bow River Forest, Rocky Mountains Forest Reserve, 1938
Jay Champion

Kitchen and bunkhouse buildings used for the Youth Forestry Training Program, Highwood River, Bow River Forest, Rocky Mountains Forest Reserve, 1938
Alberta Government, AFHPC

Students attending the National Forestry Program, Highwood River, Bow River Forest, Rocky Mountains Forest Reserve, 1939. The Alberta Forest Service ran the Youth Forestry Training Program (YFTP) in 1937 and 1938. The program later merged with the National Forestry Program (NFP) in 1939. These programs were initiated and funded through the Dominion Forestry Branch to provide work and training to unemployed youth. The first camp was set up in 1937 under the direction of Highwood district ranger Bill Shankland. Many students of the YFTP and NFP later joined the ranks of the Alberta Forest Service. This photograph was taken at the Kananaskis Forest Experiment Station
Alberta Government, AFHPC

Testing and evaluation of a Johnson H.O.K. fire pump, mid-1930s
(L to R): John Harvie, Deputy Minister of Lands and Mines; Jim Hutchison, Assistant Director of Forestry; R.G. Reid, Minister of Lands and Mines; Ted Blefgen, Director of Forestry
Alberta Government, AFHPC

Ranger Dexter Champion on horse Brownie and with dog Train, on patrol, head of Pincher Creek, Castlemount District, Crowsnest Forest, 1942. Mount Victoria in background. Jay Champion, son of Dexter, said "The saddle that is pictured was given to my Dad on his 21st birthday by my grandfather and was the only saddle that Dad ever used during his time with AFS."
Jay Champion

Alberta Forest Service Radio Technicians office party, Edmonton, early 1940s. The radio technicians under Tony Earnshaw were responsible to ensure radio communications existed between lookouts, headquarters and some vehicles
(L to R): Tony Earnshaw, Head Radio Branch; Don Bruce, Buck Mountain Towerman; Miss Jeans; Gladys Earnshaw; Helen Ledingham (Wenerstom), Continuous Wave (CW) Operator (Morse Code Operator); Jim Ledingham; Pam Earnshaw (cigarette); Don Riggan (between Jim and Ron), Radio Operator Edson; Eva Latiff (front centre) Stenographer Edmonton Radio Station; Joe Sotcky (front centre); Ron Linsdell (back row) Radio Technician Edmonton; Peggy Linsdell (partially hidden); Howard Traxter, Radio Technician and Operator Calgary; Louis (Doonie) Donovan, Radio Operator; Audrey Davis, Stenographer; Pat Donnelly, Radio Operator
Ron Linsdell

Ranger Dexter Champion with packstring hauling telephone wire from Canmore to Spray Lakes and south for a new telephone line to Kananaskis Lakes, Bow River Forest, Rocky Mountains Forest Reserve, 1938
Jay Champion

CHAPTER 3

A Way of Life

Alberta Forest Service (AFS) staff shared the landscape with a great variety of creatures, wild and domesticated, helpful and dangerous. Stories of relationships and encounters between human and beast occupy a colourful part of the recollections of any ranger or tower person.

AFS staff members were called upon to deal with a serious wildlife problem when a major rabies epidemic broke out during the winter of 1952-53. Northern rangers were placed on the front lines of activity as the government launched a full-scale program of predator control. A control line was set up across the entire province, from east to west, in an attempt to prevent rabid animals from migrating to the south. Rangers worked with trappers to encourage trapping and poisoning along this line. The program effectively held the epidemic to the northern areas, and ran for several years until the problem subsided in the natural course of events.

Former northern Ranger Bob Diesel was one who worked with trappers to reduce the number of animals susceptible to rabies. He recalled that trappers specified their territory on a map, and then ran traplines for 20 or 25 miles using poisoned bait. They were required to check the line once or twice a week, maintain the supply of bait, and register animals caught in this manner.[1]

Former Ranger Jack Grant told a story about a trip he made with trapper Emil Ducharme.[2] He was on patrol in 1953, checking a rabies line and at the same time scouting out a site for the Clear Hills tower. "That was the way it was done in those days [*identifying tower sites*] - we had no airplanes or anything like that - you just climbed the highest spruce tree you could find and mapped out the visible area," Grant said. "I had one of the rabies trappers (Ducharme) with me and for travelling we had snowshoes and two pack dogs.

"The trapper was walking ahead of me and he had his rifle over his back on a sling on one

Heading out with the dog team - always reliable in cold weather
Dominion Forestry Branch, AFHPC

Ranger Fred Nash and his dog team, Highwood Ranger Station, Bow River Forest, 1926
Dominion Forestry Branch, AFHPC

Rangers Dexter Champion (L) and Ernie Ferguson with specimens from rabies control program, 1956
Alberta Government, AFHPC

arm and on the other arm he had his snowshoes, because there was a good enough path that it was easier to walk bare-shod rather than on our snowshoes. My pack dog was walking right up against the trapper's heels. His pack dog was walking just ahead of me, and I was bringing up the rear, and kind of watching around to see where I could take off to look at possible tower points. While we were walking down this line, I saw the trapper's dog pick up one of these rabies baits in his mouth. It was made up of fat with a cyanide pellet inside. I had to do something, but my hands were full with all the maps, and I had my snowshoes over my back, so I gave him a swift kick between the hind legs - it was a male dog - and I gave him a dandy. He yelped and jumped ahead and he hit my dog, who in turn hit the trapper and ran between his legs, and of course the dog couldn't get between the trapper's legs because he had his packs on both sides. The trapper's feet went up in the air and he landed on his back behind the first dog! Everybody was amazed because they didn't know what had precipitated all this. But the dog spat out the bait!"

Bob Diesel recounts an encounter with an aggressive bear – thankfully a rare occurrence. "Only once in all the years I was with the Forest Service did I ever have to use my hand gun, and that was against a black bear. It was on the Athabasca River where the Lac La Biche River joins the Athabasca. We went down river on a timber cruise from the Town of Athabasca to a cabin that we stayed in – it was about a day's trip. We pulled the canoe up on the bank that night and the assistant ranger and I went into the cabin and made our camp and spent the night. Just at daylight the next morning this doggone black bear is coming through the window into the cabin.

"We made enough commotion trying to find the gun that he backed out, but by then we were up for the day so we started making our breakfast and my assistant went down to the river for a pail of water. He came back and told us the bear had ripped the canoe. Sure enough he'd ripped a hole maybe a foot and a half long in the side of the canoe because we had turned it over and the bear wanted to see what was in there. So we had to call a halt for the day and I spent the time patching the canoe. My assistant was down by the river bank and he starts hollering at me to beat heck. I ran to the edge of the river and saw

Fish and Wildlife biologist Bob Webb, with government Courier aircraft, refuelling in north-central Alberta during rabies control survey, 1961
Alberta Government, AFHPC

him standing on one side of a tipped-over tree and this black bear was on the other side. The bear was trying to get over this tree after my assistant and the assistant is swatting at it and just hollering to beat heck so I hollered as well. The bear looked up, saw me and came straight up the bank right at me. He was just coming right for me. By the time he was four or five steps from me I was down on one knee and shooting and after about the fifth shot he rolled. They are awfully hard to kill. But there was no place to run. There were no trees to climb. You either do it or you don't do it. I saw

Ranger Herron ready to skin black bear, Brazeau Forest, Rocky Mountains Forest Reserve, 1913
Dominion Forestry Branch, AFHPC

lots of bears through the years, but usually you could talk to them and just reason with them and they would get out of your way or you could get out of theirs. This was the one and only time I ever had to shoot a bear – and I'm awful glad that Mr. Huestis had got me a permit to carry firearms."

Chuck Rattliff,₃ an AFS forester stationed in Grande Prairie, does not believe those who say bears only become aggressive when threatened. "If you fool with them long enough you will get one that is cross. So you want to respect every last one of them because you don't know which one is on the prowl.

"I had no fear of grizzlies. I remember chasing five of them in Swan Hills one day with a little camera, trying to get pictures of them. I was running after a sow and four cubs. She must have had two families with her. They were yearlings.

"On another occasion I took the front end out of a 4x4 truck and broke the hub on it while pulling a person out of the Otauwa River south of Slave Lake, so I had to go into town. A couple of mechanics and myself came out to put a new front housing into this 4x4 truck. It was about supper time. I said, 'I've got some flies here. I'll go down to the river and catch us a few grayling and I will cook them,' because it was going to be midnight before we got home.

"So I walked down about a quarter of a mile or so to a really good hole that I knew about and got me a great big alder branch. Thank goodness it was a big one. The Lord must have been with me that day. I only had a few feet of line to tie this fly on and I was hocking the grayling out of the pool. I heard rustling in the bush but it was fall and the moose were on the rut. I didn't pay much attention until all of a sudden I hear this thing. So I looked up and there is a grizzly standing right on top of the bank behind me about 30 feet away, looking straight at me. I didn't really get too concerned about this. He was just looking at me. But anyway this went on and on and on but it got my nerves up.

"Finally I said something to him and he just exploded. He just flew out of there. I could see him just coming down through the air

Meat house and means of safeguarding provisions from bears, near Waterton Lakes National Park, 1939
Alberta Government, AFHPC

Ranger patrol visits northern trapper, 1940
Alberta Government, AFHPC

with his mouth open at me and I figured the jig was up. I bailed into the river. There was a riffle above the hole and I was on the riffle with water up to my knees and when the grizzly came down he somehow slipped but he hit the hole. There was a little ledge there. He went right under the water and when he came up I thumped him right between the ears so hard with the pole and he took off out of there like he had been shot at.

"He went up one bank and I went up the other. I came clean across the river. I ran so hard on the way home. I was scared. It was an awful fright. I got back to the truck and I couldn't tell them what happened. I was so exhausted from that running, that fright, that I laid down and I can still remember laying right there on my back trying to get my wind back so I could tell these guys. The game warden from Westlock came along. I finally got up and crawled in the back of the truck and I pulled out my rifle. The warden asked what I was going to do. I said I was going to shoot a grizzly bear, and he said I couldn't do that, and I replied yes I could. I was going to go back and do that son of a gun in. He was not going to get away with that. When I got back there the trail was still wet - but he was gone."

A story in the department's Land Forest Wildlife magazine$_4$ brings together the topics of radio communications and wildlife. Speaking about radio systems used by the AFS in 1939, the story says: "A forest ranger, who shall remain unknown, had difficulty remembering the exact time when he was supposed to make scheduled radio contact with his head office. In order to remind himself, he purchased an alarm clock which he set to ring at the appointed hour. Whenever he left his cabin he tied the clock to his belt.

"A grizzly bear had been reported in this man's district and the ranger was anxious to have the bear's hide to add comfort and decoration to his cabin. Thus it was that he set out early one morning equipped with a new .22 rifle and with his faithful 'Big Ben' tied to his belt.

"The bear was sighted and our intrepid hunter crept toward it from down-wind until he was within 50 yards of his prey. Since he had only a small calibre rifle and it was a large bear, these close quarters were indeed necessary. As a further precaution he climbed a tree before bearing down with his rifle. Tensely he waited for the bear to become still and carefully he centred his sight on the bear's ear. He had just commenced to squeeze the trigger when – you guessed it – the alarm clock went off with a clang. Our hero nearly fell out of his tree and the bear promptly took off for other places, encouraged, no doubt, by the insistent clanging of the faithful clock.

"To this day the ranger has never ceased to blame the loss of his prize rug on the radio communications system of the department. The bear, of course, was unavailable for comment."

Jack Grant talks about the ranger's task of sealing beaver (see chapter 2). "Every spring when I was at Keg River I had to make a trip to Carcajou to seal beaver. At that time it was just a

Inspector Willis Millar (L) and Ranger C. Bremner – breakfast on the Clearwater River, 1913
Dominion Forestry Branch, AFHPC

Coyotes taken in the winter of 1950-51 by Frank Jones, Castlemount Ranger Station, Crowsnest Forest (note skis used for backcountry patrols)
Alberta Government, AFHPC

A Way of Life

Inspector Millar's pack outfit, Clearwater Trail, Clearwater Forest, 1912
Dominion Forestry Branch, AFHPC

wagon trail through the bush and that was how all the supplies came to Keg River. Carcajou on the Peace River was the old stopping point for the river boats. I would make my trip down there once a year. The trappers would collect there, after I'd sent word out a couple or three weeks ahead of time, with all the beaver pelts for the season.

"I would go down to the end of the wagon road on the east side of the river and then I would send up a flag and they'd know somebody was waiting to cross the river - they all knew the date I was coming. The trappers would come across in a boat and pick me up and take me back over. So then the main event, the sealing, would take place and at that time you'd pick up all the news and all the activity - illegal activity on the traplines, and everything like that. I didn't stay for any of the frivolities or whatever happened afterwards. I just did my job and left, because I'd have to stay down there overnight amongst the mosquitoes and flies and whatever, and it wasn't a very pleasant place to be."

The horse was the constant companion of many a ranger, but the relationship wasn't always sweetness and light. Ranger Bill Balmer recounts the rocky relationship between him and his horse back in the early days:[5]

"That horse probably taught me more things about horses than all the rest of them put together. He was a green broke horse when I got him and it was an education for both of us, except that he was smarter than me. We were going down the trail one day and I just got out of the south gate at Kananaskis. There was dense black spruce and a narrow trail. We went around the first corner and we met a black bear. When the dust settled, the bear was going one way, I was sitting on my butt on the ground, and the horse was going the other way. Well, he only went as far back as the gate and I caught him. I got back on him, there was no problem but you could see that so-and-so, his ears would start flipping back and forth. You could see the gears going in that brain of his. When I'd relax, he'd dog his head and fire me off.

"He dumped me more times and the thing is, you're miles out in the bush, you don't want to walk, you don't want to kick the heck out of him. With him

Nosebags were used to keep flies out of the horses' noses, Cypress Hills, 1911
Dominion Forestry Branch, AFHPC

Ranger brings in hay for winter, north of Entrance, Athabasca Forest, c. 1934. Putting up hay was one of the ranger's duties
Charles Clark

it was a game. I'd get back on again and as long as I rode the stirrups and rode alert, no problems. But I'd start to relax a little bit and he'd know it. Bingo! Sometimes, it would be two or three times a day that he'd unload me. Sometimes he'd go a couple of weeks and never try nothing. You never knew with him. Finally one day - I had the strings behind the cantle - I looped them up. I tried grabbing the saddle horn, nothing helped. I'd never rode broncs and like this wasn't my thing. But I got hold of that string and I kid you not, by the time he got through with me, my thighs were black and blue. He hammered but I stayed on board. I had him for many more years and he never dogged his head again."

Forest rangers in the foothills and mountain country of Alberta relied heavily on their home base, often a cabin in the woods with some adjacent hay meadow to keep the horses fed. It was common practice to turn out the saddle horses and pack animals after a hard day's ride, letting the animals roam free so they could rest and graze and be ready for travel the next day. Normally a bell was attached to one or more of the horses so the ranger could find them easily the next day.

Forestry pack outfit grazes west of Cut Off Creek, Clearwater Forest
Dominion Forestry Branch, AFHPC

Forest rangers with pack horses, Oyster Creek, Crowsnest Forest, Rocky Mountains Forest Reserve, 1909
Dominion Forestry Branch, AFHPC

The horses had been turned out this way on August 7, 1952, at the Meadows Patrol Cabin, 30 miles due west of Rocky Mountain House. Ranger Ronnie Lyle[6] was out looking for the animals south of Rough Creek (north of Ram Mountain). After hiking in a cool drizzle for some time he sighted the horses. He was ready to follow the usual practice, which was to lure one of the animals with oats, secure a halter and bridle on it, then ride the horse to round up the others.

As Lyle was making his way through dense willow and young pines, he startled a large black bear at close range. The bear immediately began to growl and become aggressive, putting the chase on Lyle who promptly made for the closest tree - a slender lodgepole pine. Lyle was unarmed and had no choice but to climb quickly to avoid the bear. The bear bolted straight up the pine with Lyle scrambling ahead of it through the branches. The attacker was extremely quick and soon swiped Lyle's boot heel. He in turn kicked back at the bear, causing it to crash to the ground and giving him time to climb even higher. Not to be deterred, the bear came back up the tree, hooking its claws into Lyle's boot and tossing the ranger 30 feet to the ground, whereupon Lyle passed out.

When he came to, some time later, he found himself bundled under the low branches of a spruce tree. Apparently the bear had dragged him there and left him for further attention at a later time. Lyle was badly bruised and suffering pain from injuries to his neck, shoulder and back, but he heard the horses' bells and staggered out to snag one of them. He was able to catch Dusty, one of the shyer horses. He hung his belt through the strap that held the bell on Dusty's neck, and used this to steady himself as he limped alongside the horse back to the Meadows Patrol Cabin. It was no easy task to convince Dusty to leave the supper table and picked up in a forestry truck for the ferry crossing of the North Saskatchewan River at Saunders. The ferry wasn't working, however, so their only means of crossing the swift river was a flimsy and precarious old cable car. Inside the cable car's basket, McDonald fought to brace Dr. Greenaway between his knees while he cranked on a lever that had to be jerked back and forth to propel the shaky carriage forward. It was a slow process, with the basket and occupants swaying wildly above the roaring waters of the North Saskatchewan. With great determination and a lot of pluck the two reached the far shore, where Ranger Wiedeman was waiting with a small Fordson tractor.

Their apparent good success now took a downward turn, because recent heavy rains had inundated the primitive trail to Meadows Cabin. After grinding through the mud, and even with endless pushing by the men, the heavily loaded little Fordson

Rangers in their Model A Ford truck pulling a stalled car out of the Pembina River. One of the hazards of fording was getting the spark plugs wet - removing the fan belt was a typical precaution. Brazeau Forest, 1930
Dominion Forestry Branch, AFHPC

other horses, but Lyle's desire to live was greater than the challenge of overcoming the horse's reluctance as well as his own pain.

Fortunately Lyle's assistant ranger Ed Wiedeman saw Lyle and Dusty enter the pasture at the Meadows Cabin and telephoned Ben Shantz at the Shunda Ranger Station, 15 miles north of the Meadows. Shantz in turn called Forestry Headquarters in Rocky Mountain House, where Ranger Don McDonald immediately began to pull together the necessary people and equipment to reach the cabin and evacuate Lyle to hospital.

At 6 p.m. Dr. Greenaway was roused from his

Ranger Ron Lyle on patrol, Prairie Creek District, Clearwater Forest, 1966
Alberta Government, AFHPC

became mired in a large mudhole two miles out from the cabin. The doctor and McDonald had no option but to tackle the rain-soaked trail on foot, with the ranger reaching the cabin first. Immediately McDonald dispatched Don Fregren, a young visitor staying at the cabin, to go back and help the Doctor. Fregren took the trusty forestry horse Dusty to carry Dr. Greenaway the last two miles to the Meadows Cabin.

Dusty and Fregren brought Dr. Greenaway through the wet and cold night to the ailing Ranger Lyle at 2 a.m. The doctor's examination revealed Lyle had sustained crushed upper vertebrae and could only be moved by aircraft. Immediately, the call went out via the old phone line and was later repeated by radio. The small group of men waited out the night for a response and some much-needed aid from Rocky Mountain House forestry headquarters.

Soon after dawn a Beaver aircraft left Edmonton and flew over the cabin to evaluate the adjacent hay meadow for a medivac landing. The Beaver is renowned for its short take-off and landing capabilities, but the Edmonton pilot was concerned about the extremely small landing area and the large trees at each end. It was determined a smaller and lighter aircraft, and perhaps a more daring pilot, would be required.

Forestry officials recalled the bush flying skills of a pilot from the farming community of Olds. This pilot was Mel Clipperley, who operated a local car dealership and flew his Aeronca Super Chief on numerous trips throughout the western foothills and mountains. The gravity of the situation prompted Clipperley to offer his services without hesitation. His guide would be another veteran ranger, Harry Edgecombe, from the Clearwater Ranger Station. By this time Dr. Greenaway, Wiedeman, McDonald and Fregren were removing hay for a makeshift landing strip and cutting away a number of tall trees that might endanger the aircraft.

The area is in the high foothills, directly against the front range of the Rocky Mountains. The weather is very changeable and wind currents can be extremely strong and completely unpredictable. With all the experience and skill at his command, Clipperley brought his aircraft down in the meadow, arriving unharmed with Edgecombe despite some of the freshly-cut hay becoming tangled in the landing gear. It was apparent the hay could be a serious impediment to a take-off, so it was decided all of the cut hay should be removed from the meadow.

The Meadows Patrol Cabin, Clearwater Forest, where Harry Edgecombe started his career and caught his first wild horses
Alberta Government, AFHPC

Harry Edgecombe, shown when he was head of fire control training in Hinton (1969-1979)
Alberta Government, AFHPC

The only horse available for the hay rake was the venerable Dusty. Despite some initial unwillingness, Dusty was harnessed and put into service with Ranger McDonald holding his head while young Fregren drove with the reins. A clear strip was freed of hay in short order. At the same time, Dr. Greenaway and Rangers Edgecombe and Wiedeman laboured with clumsy crosscut saws to remove more large trees at the windward end of the meadow.

All was now ready. At Clipperley's insistence there was a "test run" of the runway before Ranger Lyle could be placed in the aircraft. Clipperley gauged the calm conditions and the quiet evening air and decided it was now or never. The men held their breath and said a prayer as the tiny aircraft sped down the grass meadow and cleared the trees, circling the meadow and coming back for a textbook landing. Now it was time for the real test. Dr. Greenaway directed the careful placing of Ranger Lyle in the small Aeronca Super Chief. Straps were used to secure the patient and all non-essential gear was removed to make the aircraft as light as possible. A check ensured the fuel tank carried only enough gas to get to Rocky Mountain House. The aircraft was ready. Clipperley revved the engine up to full throttle while the others held the machine back to allow a "rapid take-off" under maximum power. When they let go, the aircraft streaked off down the hay meadow, barely cleared the tall trees, and then climbed steadily into the gray sky under the falling darkness. You could hear the sigh of relief as those who stayed behind returned to the cabin to wait by the telephone for news of Ranger Lyle's situation in Rocky Mountain House.

In the small town of Rocky Mountain House, meantime, residents were called to bring their cars and trucks to the airport so their headlights would illuminate the landing strip. Clipperley landed safely and Ranger Lyle was taken to hospital. He underwent a two-month rehabilitation before getting back to his beloved foothills and mountains.

Shortly after his rescue, and while still in hospital, Ranger Lyle asked Wiedeman to revisit the scene of the bear attack, saying he wanted to get his service cap, bridle, halter, and oat pannier back. At the site of the attack, Wiedeman located the remains of the forestry cap and its badge, the horse bit and portions of the oat pannier. Within a short distance he found evidence of a bear's cache containing remnants of a mule deer carcass. It seemed that while pursuing his horses, Ranger Lyle had inadvertently roused the bear from its food cache, sparking the attack. The bear was further frustrated when Lyle kicked it out of the tree and down to the ground. It was enough to result in Ranger Lyle being shoved under a spruce tree for later disposition.

Ranger Bill Shankland at Mountain Park Station, Brazeau Forest, 1940
Alberta Government, AFHPC

Ranger Angus Crawford in 1962, Edson Forest
Alberta Government, AFHPC

After Ranger Lyle retired from the AFS, he continued working on a seasonal basis with the B.C. Forest Service. Ed Wiedeman relocated to Prince George and worked for the forest industry. Don Fregren continued his interest in the AFS, graduating with a degree in forestry and retiring in 1993 after numerous senior positions in the AFS. Harry Edgecombe served in a variety of AFS positions, most of which involved fire protection roles, ending his career at Hinton where he was in charge of the Provincial Fire Control Training Program. Pilot Mel Clipperley continued his car and machinery business until retiring in the Olds area.

Jack and Doris Gosney tell another rescue story involving Doris's father Angus Crawford. Angus was the ranger at Mountain Park in 1948.[7]

Angus and his assistant ranger Jim Bradshaw were packing up their horses at the Grave Flats cabin, ready to return home after the last fall trip of the year. Angus was going to ride Buddy

Malloy's "town horse" back to Mountain Park, a distance of 20 miles. The horse had different ideas, however, and went into a violent bucking fit, finally throwing Angus and breaking his pelvis in the process. He was seriously hurt and it was only with a lot of effort that Jim got him back to the cabin. Jim phoned the ranger's wife Winnie in Mountain Park to report the injury. Doris recalls that her two brothers, Ken and Don, hooked up the dog team and sled and departed for Grave Flats immediately with a supply of pain killers from the local doctor.

The trip was difficult because of the rough terrain and patchy snow cover, but they made it to the cabin after nightfall and administered the medicine to Angus. In the meantime Doris's husband Jack came off shift at the local coal mine. After hearing about the plight of Angus, Jack started the hike to Grave Flats to help the boys on the return trip, carrying a bottle of Planters Punch rum in his jacket pocket. He saw how much pain Angus was in when he met the party at the Red Cap cabin and asked him if he would like a drink. Angus agreed that a shot of rum would help and said he would keep the bottle with him in the dogsled. The dogs and the boys were tired and the roughest country was still ahead, but Jack added a strong hand on the hills. They made it past Mackenzie Creek and at Mile One met townspeople who had got a 4x4 truck up the mountain to carry Angus the rest of the way. After a short discussion, however, Angus decided not to attempt the transfer to the truck and the dog team continued on to Mountain Park, arriving at midnight. Jack's relationship with his father-in-law certainly went up a notch after the rescue and provision of the "medicinal" rum.

(L to R): Jack Gosney, Len Allen and Carl Leary, Forestry Training School in Hinton, November, 1960
Alberta Government, AFHPC

Phil Nichols, raised as a cowboy in southern Alberta and steeled as a commando during the Second World War, was forest ranger at Salt Prairie for his entire career with the AFS from 1945 to 1972. He preferred getting around on his horses, but during the winter he travelled by dog team. He told this story about a fisheries inspection he was doing one winter at the west end of Lesser Slave Lake as part of his departmental responsibilities. [8]

"I was checking the commercial fishermen. There was a wind blowing and with that snow drifting, you couldn't see anything but the sun was shining up above. I was going along standing on the tail end of the toboggan and all of a sudden the dogs stopped. I yelled at them to go and they just laid there. We always had a rope from the front of the toboggan, a big old rope. We called it a tail rope and sometimes if you were stuck you'd tie it around a tree so the dogs couldn't run away on you. I had that in my hand and I started running up alongside the dogs and there was an air hole in the ice. I went through, and when that cold water comes in the front of your parka you know you're in trouble!

"As I went down I yelled 'mush!' and the old lead dog was pulling sideways and I was going down the hole and hanging on to that rope. The dogs took right off – though the lead dog just swung sideways instead of going ahead where he wanted to go. They took off and I went down in the lake and then came back up hanging onto that rope – and when I came out I was skidding along on my belly on the ice.

"So then I just let the old dogs go, figuring to get to the north shore and back to White Creek where Joe McDermott had a camp. Shorty swung around and he just headed for Joe's camp. They started a-legging 'er and I was running behind and my pants broke right across the back of the knees because of the ice on them. It must have been 30 or 40 below zero. I got to Joe and Mary Rose's camp and I asked, 'You got a fire on - I fell in the lake and I'm kinda wet.' My mitt was frozen on that rope so they just took the axe and chopped it off.

"I stood by the heater till I thawed out, taking off one thing after the other. The old parkas with damned zippers on them - you couldn't open them all the way so you got it a little bit loose and pulled it over your head. Mary Rose warmed up my sleeping bag and I dived into that. I never even had the sniffles the next day. But boy, that's a sickening feeling when that cold water comes pouring in the front of your parka."

Despite the perils and adventures of the Forest Ranger's life, it was not uncommon for the job to be carried on by younger members within a family. In the Foley family, four successive generations have served Alberta's forests with pride and distinction. Albert worked as a ranger in the Swan River Ranger District of the Lesser Slave Forest Reserve. In the initial years he was only paid for the summer months but was allowed to use the Forest Service's log house for the winter. He died in an accident when returning to Kinuso by train from an Edmonton Ranger meeting in 1943.

His son Pat worked as a ranger in the Slave Lake area from 1945 to 1967. Pat was a well-known figure to residents of the area, known for his large, burly build and his abilities in the bush. Pat worked during the transition period that saw the transfer of the forests from federal to provincial hands. He experienced the introduction of new equipment and technology, and was the first to map a fire from an airplane.

Pat's son Lou spent 32 years working for the AFS throughout the province. He made a name for himself in wildfire suppression, and after retirement in 1996 he took on a number of senior consulting contracts with government and industry.

Lou's son Hudson completed the Forest Technology program at NAIT in 1994. He has worked as a logging supervisor and forest planner with Zeidler Forest Products (now Alberta Plywood Ltd.) in Slave Lake, and most recently with ATCO Electric as Hazard Reduction Coordinator for the company's service areas.

Women in Forestry

Women have always been an important part of the organization, initally serving as unpaid and usually unrecognized partners of rangers. Anne Dixon has highlighted this "silent partner" role in her book of the same title, profiling the roles of wives of western national park wardens.

The wife of a ranger provided a

Ranger Phil Nichols, High Prairie, 1998
Peter Murphy

Ranger Pat Foley, one of a long line of Forest Rangers
Lou Foley

critical link for the ranger while he was on patrol. She could transfer vital information to regional headquarters about forest fire situations and other items pertinent to public safety. She knew how to operate the telephone system and the radios and knew who to contact both in the Department and the nearby communities. These functions were generally performed without any fanfare and became an integral part of being a ranger's wife.

Quite often in the old days the ranger's wife conducted correspondence courses and home-schooling lessons for children of the family. For some, access to schools was severely limited by distance compounded by unpredictable weather and poor roads. Each winter, once heavy snows restricted access, many ranger districts were out of reach. There are many examples of students using the correspondence programs through to grade nine then going to live with friends or family in nearby towns for high school. In some instances, school buses would meet the ranger's children at the forest reserve boundary gate. Again, the role of driving the children to meet the bus usually fell to the ranger's wife.

Typical contributions of forestry wives are illustrated by personal experiences. For example, in 1924, Ethelwyn Octavia Alford, wife of Nordegg Ranger "Bertie" Alford, in her 50s and suffering with arthritis rode to the head of the North Saskatchewan River to build the cabin at Camp Parker. They left on the fifth of August and returned on the third of October, experiencing the onset of winter snowstorms in the high country before they finished the job. Bill Shankland was also a ranger at Nordegg in the 1920s to 1936 when they moved to Bragg Creek. His daughter Jessie described how Nordegg was at the end of steel on the railway from Rocky Mountain House that served the coalmines. Forestry officials visiting the station had to wait four days between trains so their house was always full of visitors staying over and she and her mother produced meals in the kitchen. She recalled that at Bragg Creek, then the end of the road, it was even busier with campers and tourists who found refuge at the ranger station when their cars broke down, got stuck or bad weather settled in. Jessie married Ranger George Deans and raised her own family at a succession of ranger stations.

Louella Krause lived in a tent with her husband Rein during their first winter together while they built their cabin in 1935-36. She recalled how she would get radio messages about fires from Rein on patrol and would dispatch fire crews and get supplies to him wherever they were needed. Wanda Edgecombe, like so many of the ranger's wives on the remote districts, taught their children at home with correspondence courses from the Alberta Department of Education, giving them a great start in their schooling. At this time in the early 1950s weather stations were being set up at ranger stations with the expectation that wives would take the twice- or three-times-daily readings to pass on by radio, a task they did faithfully and well.

In 1937, the Calgary Herald published an article that described the living and working conditions of Ranger Dexter Champion and his wife Louise. In the spring of 1936 Ranger Dexter Champion, accompanied by his wife Louise and son, John 'Jay' Dexter, journeyed up the Elbow, past Elbow Lake to Kananaskis Lakes where Champion was posted. They covered 16 miles on horseback in one day, something of a record for any nine-month-old youngster. On reaching their destination, it was found necessary to build a small corral covered with mosquito netting to protect the young man from any tendency to wander off and also to protect him from the mosquitoes. [9]

Jay Champion recalls that when his mother looked back on the early years as the wife of an AFS employee and added the pluses and minuses she was quite happy with that part of her life. Even though she grew up in Calgary as part of a large family, she was the type of person who was always comfortable with her own company, so living somewhere remote where there were no neighbours did not turn out to be a big problem for her. She also adapted to the lack of conveniences without too much difficulty, such as lack of plumbing, electricity, refrigeration, shopping, etc. "She did have a bit of exposure to 'country life' prior to her marriage, as she was a schoolteacher who taught in one-room country schools, so she had learned to ride a horse, and became accustomed to outdoor biffies. However, coping with ornery packhorses, campfires, wet tents etc. was certainly something new for her. Of course it was accepted that she pitch in and help with those types of things. I don't believe she enjoyed those irritating events but did accept them as things that had to be done. She used to talk about some incidents that stood out in her mind. The difficult incidents revolved around "critters" both two legged and four legged. Dad of course had patrols to make, and also spent a lot of time maintaining the various trails that had been cut through the bush mainly for possible firefighting access. When he was doing trail work, if it was any distance

Dexter Champion transferred to the Kananaskis Lakes Ranger Station, Bow River Forest, after serving on Moose Mountain Lookout. Champion is shown accompanied by his wife Louise and son, Jay. On reaching their destination, it was found necessary to build a small corral covered with mosquito netting to protect the young man from any tendency to wander off and also to protect him from the mosquitoes
Jay Champion photo; Calgary Herald article 1937

from home, he would take a packhorse or two and stay out there for a week or sometimes more. While at Kananaskis, even going to Canmore for supplies was a three or four day trip. Once while he was away, a black bear was hanging around the cabin. One night it finally decided that there was something to eat in the cabin and climbed up on the roof and was attempting to get in. Of course my mother was terrified as she was not only there alone but had me there as a baby. She was convinced the bear would get in, so she put me in the oven of the stove (it was cold) as there was no other place to hide me. She then prepared to do battle with the bear. Fortunately the bear gave up and climbed off the roof. When daylight came the bear was still around so my mother got the single shot .22 and opened the window a bit and shot the bear dead with one shot. Of course fresh meat was a luxury so she tied a rope to the bear's leg, caught a horse and pulled the bear up to hang from a tree limb. Unfortunately, she did not know the animal should at least have been bled, so it was spoiled by the time dad returned. Later when we were at Castlemount, about 1942, a grizzly bear followed some range cattle that had come by the Ranger Station. It spent some time sniffing around the house and other buildings before wandering off, which caused some fright and worry. Of course dad was away then too.

"Some of the public who came by and wanted some service from the Ranger were also sometimes difficult. Some of them seemed to think that mother had the same duties and authority as dad did. Some of them were less than pleasant when they learned otherwise. She had one unfortunate experience with a fellow camped nearby who learned that she was there alone and that dad would be away for some time. Whether he was just that kind of person or he was on the booze, mother didn't know. He made several trips to the cabin trying to convince her that they should indulge in some fun and games. After being continually rebuffed the fellow became mean and threatening. At that time we had a big Airedale dog that had a very nasty temper if he didn't know someone, so mother tied the dog near the door so the man couldn't get near without dealing with the dog. End of that problem.

"Another continual worry was the possibility of something happening to one of the children, as of course medical attention was a long way away under the best of circumstances. About 1941, my three-year-old sister broke her arm in the middle of winter when we were completely snowed in. So a Reader's Digest magazine was securely taped around her arm as a splint. The arm healed perfectly. The elements were rather nerve wracking as well. Mother talked about electrical storms when a close lightning strike would cause the stove lids to jump and 'fire' to run along the wire fence."

The Calgary Herald finished its 1937 article with a summary of life in the Kananaskis Lakes district. "When work is some distance from home, the ranger takes his equipment and camps by his work until completion. On one such occasion, Mrs. Champion was alone five days with her son, but her fine garden and household duties kept her busy and happy. She is keenly interested in her husband's work and her love of the wild things, the mountains and lakes compensate for the lack of luxuries of city life. Both the ranger and his wife are extremely happy and contented in this quiet life of protecting the forest and its denizens."

Janet South, wife of Ranger Ken South, recalls that in the early days of the AFS the Ranger in charge of a District was the only person employed by the Forest Service.[10] It wasn't until the 1940s that Assistant Rangers started being hired for some of the districts. In the interim the Ranger's wife often filled the role of part time radio operator, office clerk, receptionist, dispatcher, weather observer and a host of other duties.

"The Ranger was not just the Fire Ranger or the Timber Inspector, he was also the Game Guardian. With the office attached to the house and with only a sliding door between the office and the kitchen, that had a one-inch space under the door, this posed a problem of sorts. In the spring the trappers would bring their beaver pelts to the forestry office to have them "stamped" or "sealed" prior to selling them locally or shipping the pelts to an outside fur market. Not only would the odour of fresh beaver pelts drift into the kitchen and mix with the smell of a meal being prepared, but the dog, in our case a German Shepherd, would have his nose at the space under the door checking out both the trapper and his winter catch of fur."

South said that in the northern districts the Cree language was still the first language and broken English often made it difficult to communicate clearly. In an attempt to fit into the community, families did their best to learn the language. It helped them and the children to use Cree words for food while the family was around the table at meal times. Words like totosapoo (milk), totosapowepime (butter) and nipe (water) were quickly learnt by the children, as well as how to count in Cree.

"Most days the Ranger would leave on a patrol first thing in the morning and be away all day, leaving his wife to handle any public visitors who came to the Ranger Station. If the office was closed because the Ranger was out, the locals never failed to knock on the door of the house. The most common requirements were to get their beaver pelts sealed, get a receipt for the renewal of their trapline for the coming year or to get a fire permit. The forestry radio would always remain turned

Ranger Dexter Champion and his wife Louise and son Jay at the Kananaskis Lakes Ranger Station, Bow River Forest, Rocky Mountains Forest Reserve, 1937. The ranger in charge of this station has an altogether different type of work from that of the Moose Mountain station. Bridges are to be repaired and game preserved, as no firearms or hunters are allowed in the district. Rivers are to be patrolled for fishing. The last two summers [1935 and 1936], both the Upper and Lower Lakes have been stocked with cut-throat and Dolly Varden trout
Jay Champion photo; Calgary Herald article 1937

on, the volume turned up and the office door into the residence left open so as to monitor it while looking after the children, doing house work and other household duties. The forest headquarters office and staff were quick to take advantage of the wives by expecting them to answer the regular radio "skeds" (regular radio check-ins – usually four times daily – including weekends), and send and receive radio messages. The radios were the only communications link the Ranger had should a family emergency occur or should he be called back to the district headquarters to action a new fire or similar unexpected task."

South said that the Ranger was responsible for a number of lookouts or towers within Ranger District boundaries. In the Keg River District there were four towers. At month end there would be shopping to do for these four tower people, often the Assistant Ranger and ourselves included. As it was 63 miles, one way, to the grocery store, on a gravel road that could often be muddy, it made sense to shop just once a month. Shopping for six families made for a long day. To forget anything meant a wait of another month for an item that may make a difference to that individual. This extra duty had its benefits. Having to rely on one another, the towerman and the Ranger's family would often become good friends. A Sunday drive could often mean a visit to the tower, an opportunity for the children to climb the tower and look around, then a treat of cookies and juice.

"Before the Forest Service began building new housing for all stations, Ranger families lived in some unique homes. The logs of the old log house at Keg River had seen better days and if one was to look beneath the kitchen window, while sitting at the kitchen table, it was possible to look outside, through the parting between the windowsill and the log below the sill. Wintering in such a house brought additional problems. This particular house had lost a lot of chinking between the logs allowing snow to blow in and pile up on the floor. Needless to say this did not make for a warm home. The only heat source was a wood burning kitchen stove and an oil heater in every room except the bedroom. Going away for the day without ensuring the oil stoves were filled beforehand meant returning to find the pail of water or quart of milk frozen solid. There was no need to have a fridge.

"The old well behind the house was a real antique. It had been dug by hand and was cribbed with a windlass on top. To get water for the home one had to go to the well and lower the pail on a rope from the windless. When the pail was at the surface of the water a quick flip of the rope caused the pail to tip on its side and fall into the water allowing the pail to fill. The windless was then turned by hand, the water poured from the well pail to another pail for carrying the water into the house. An advantage in the summertime was that the water was always cold; a nice drink on a hot day."

For the wives in the north, with children and many duties and many inconveniences, life was lonely. They lived many miles away from family and roads were not what they are today, so it was a long time between visits. Friends were important for the adults and the children. Our children played with their siblings and looked forward to any visits by friends.

In 1958, Marilyn Peter joined her husband Art on his first job at Fox Creek, living 11 months in the forestry cache – a frame building 12 by 16 feet with no water – because the assistant ranger's house had not yet been built.[11] She also recalled dispatching fire crews, equipment and supplies by prearrangement with Art who could reach her on the radio. At this time, in the 1960s, the office was part of the Ranger's house. Marilyn and Audrey O'Shea recalled how they were always on the radio and answering the phone when their husbands, Art and Kelly, were out. In addition the ladies also responded to visitors at the door such as lumbermen, farmers, settlers, trappers, hunters and fishermen. Their lives were disrupted less often when offices were separated from the residence.

Members of the Alberta Forest Service (AFS) lived and worked in every type of environment, from fly-infested muskeg to craggy mountaintop. They saw momentous changes in technology, from early aircraft and telephones to all the refinements and comparative comforts of the late 20th century.

Following are some photographs from the early years, giving a flavour of the type of life, and the type of person, encountered in the service of the AFS. All photos, unless otherwise identified, are from the Dominion Forestry Branch or AFS, and in the possession of the Alberta Forest History Photo Collection (AFHPC).

A home in the forest

Office and HQ building at Coalspur, Brazeau Forest, Rocky Mountains Forest Reserve, 1912; frame buildings were built if a railway or local sawmill was close enough to supply materials

Mrs. Millar inside the newly-built Wilson Ranger Station cabin, Clearwater Forest, Rocky Mountains Forest Reserve, 1912. The log walls had not yet been chinked. This area is now under water with construction of the Bighorn Dam in 1970

Wilson Ranger Station outfit ready to be packed, with Mrs. Millar and Betty Millar, wife and daughter of Chief Inspector Forest Reserves Willis Millar

A Way of Life

Ranger and wife at Nordegg Ranger Station, Clearwater Forest, Rocky Mountains Forest Reserve, 1914. Logs were cut locally for the first buildings. Lumber was used later when roads were built and construction materials could be hauled in

Kitchen addition at Sentinel Ranger Station on the Highwood River, Bow River Forest, Rocky Mountains Forest Reserve, 1925

Isaac Creek cabin in Jasper National Park, 1925; Ranger W. Shankland (L) and Inspector C.H. Morse

Lynx Creek cabin, Crowsnest Forest, Rocky Mountains Forest Reserve, 1911

Pekisko cabin in the Bow River Forest, Rocky Mountains Forest Reserve, 1922. This cabin is now located at the Cartwright Ranch west of High River

Athabasca Forest HQ house at Entrance, 1917. This was a busy ranger station, located on the railway and trail head for both the Mountain Trail and Lower Trail that met again at Grande Cache on the Big Smoky River

The Moberly cabin on the Lower Trail, Athabasca Forest, 1940 - now preserved at the Hinton Training Centre

Yarrow cache, Crowsnest Forest, Rocky Mountains Forest Reserve, 1911. Caches were very rudimentary buildings designed to store firefighting materials and food. In a pinch they could be used as overnight shelter, though many had a rather ripe smell to them because of packrats

A Way of Life

Hose Line Trail cache, Athabasca Forest,
Rocky Mountains Forest Reserve, 1911

Boundary cache, Red Deer District, Bow River Forest, Rocky Mountains Forest Reserve, 1914

Scott cache, Kananaskis District, Bow River
Forest, Rocky Mountains Forest Reserve, 1914

Forest survey tent camp, Slave Lake, 1959.
Photo taken from the top of Flattop Tower
Cliff Smith

On the trail

Corduroy bridge, McLean Creek, Bow River Forest, Rocky Mountains Forest Reserve, 1914. Corduroy was made of small logs laid tightly together to help horses cross soft ground

Ranger with native people, Assiniboine Trail, Lesser Slave Forest Reserve, 1911

On the Assiniboine Trail, Lesser Slave Forest Reserve, 1911

Supervisor McAbee (R) at native Stoney camp and meat rack on the Brazeau River, drying meat so it could be used for months on the trail. Brazeau Forest, Rocky Mountains Forest Reserve, 1913

A Way of Life

"Hardy Lodge" - forestry staff visit with hunters in the bush, Clearwater Forest, Rocky Mountains Forest Reserve, 1909

Coal Creek logging camp on the Highwood River, Bow River Forest, Rocky Mountains Forest Reserve, 1911. The camp served Lineham Lumber Company's sawmill at High River

"Sub camp" - shelter under a tarp at Deep Creek, Lesser Slave Forest Reserve, 1911

Camp on Baptiste River - sleeping tent and social quarters. Athabasca Forest, Rocky Mountains Forest Reserve, 1912

Interior, Baptiste River forestry cabin, Clearwater Forest, 1914. The logs on this cabin were hewn flat with a broad axe, making the interior walls smoother

Rangers at Mount Bess camp, enroute to Mount Robson Athabasca Forest, Rocky Mountains Forest Reserve, 1912

Trail builders' campground - horse blankets drying on bushes - Brazeau Forest, Rocky Mountains Forest Reserve, 1915

Supervisor's teepee, Sentinel District on the Highwood River, Bow River Forest, Rocky Mountains Forest Reserve, 1914. The teepee-style tent was used for a long time. It had enough headroom in which to stand and could be heated with a small fire inside. Note saddle kept up off the ground at left, to protect it from rodents

A Way of Life 63

Brushing teeth at beaver pond, Bow River Forest, Rocky Mountains Forest Reserve, 1912

Ranger Freddie Nash and Assistant Ranger L.L. Waikle on summit of Mount Burke, Bow River Forest, 1928. AFS staff named Nash Meadows at the Bighorn Ranger Station in memory of Ranger Nash

Supervisor McAbee and Ranger Muncaster prepare lunch near Junction Creek, Elbow River, Bow River Forest, Rocky Mountains Forest Reserve, 1912

Inspector Willis Millar with Clearwater River trout caught with darning needle and some linen thread. Clearwater Forest, Rocky Mountains Forest Reserve, 1913

Native Stoney guide skins out mountain goat on the Siffleur River for fresh meat, Clearwater Forest, Rocky Mountains Forest Reserve, 1912

Traveling the road to Aura (Ghost) Ranger Station, northwest of Calgary, Bow River Forest, Rocky Mountains Forest Reserve, 1921

Car on narrow trail cleared by hand and horse (Gunnery Grade) above Sentinel (Highwood) River, Bow River Forest, Rocky Mountains Forest Reserve, 1925. Note phone line strung along the river

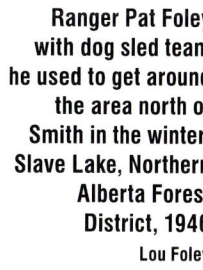

New bridge span over Gorge Creek, Highwood District, Bow River Forest, Rocky Mountains Forest Reserve, 1925

Spring break-up – water on the trail at Ram Lookout, Clearwater Forest, Rocky Mountains Forest Reserve, 1956

Roadbed slumping near Livingstone Gap, Crowsnest Forest, Rocky Mountains Forest Reserve, 1940

Ranger Pat Foley with dog sled team he used to get around the area north of Smith in the winter, Slave Lake, Northern Alberta Forest District, 1946

Lou Foley

A Way of Life 65

Forest survey crew trying to upright a rolled Bombi, 1959. A Bombi was a tracked personnel carrier made by Bombardier
Cliff Smith

Brand new snow cruiser at Waterways, near Fort McMurray, Lac La Biche Division, 1948. Machine was used for winter timber cruising along the Clearwater and Athabasca River valleys
Jack Roy

Repairing track and wheel assembly on tracked vehicle, fall, 1959, Slave Lake Forest. Doug Lyons (L) and Fermen L'Hirondelle. Lyons was the Forester-in-Charge Survey Parties
Cliff Smith

Seismic inspection in the Meekwap Lake area (north of Fox Creek), Whitecourt Forest, spring, 1972. Forest Officer Don Podlubny beside Nodwell tracked vehicle
Don Welsh

Forest Officer Lou Foley driving a J-5 Bombardier tracked vehicle on timber cruising trip on corduroy trail across creek, 1966
Bruce MacGregor

Superintendent Bert Coast uses speeder for patrol on the Northern Alberta Railway near Fort McMurray, early 1960s. Speeders were used to patrol rail lines and follow trains to detect and extinguish fires caused by cinders and sparks from steam locomotives

Forestry crew uses man-powered speeder to move camp on the Coal Branch, Brazeau Forest, Rocky Mountains Forest Reserve, 1912

Patrolling on foot powered speeder, Slave Lake, Northern Alberta Forest District, 1930s

Rangers Greenwood and Rance patrolling the Brazeau Forest, Rocky Mountains Forest Reserve, on DFB speeder, 1913

Firefighters transported on speeders along the Northern Alberta Railway between Lac La Biche and Fort McMurray, Cheecham siding, Lac La Biche Division, 1960

A Way of Life

Ranger John Elliot used St. Bernard dogs to help carry supplies while on patrol, Hinton, 1949
Neil Gilliat

Fort McMurray Air Service plane covered against winter cold, Fort McMurray, 1952. The plane was being used for flying surveys for the rabies control program
Jack Roy

Forest Survey crew crossing river with Bombardier, Slave Lake Forest, 1959. Standing (L to R): Church, Joe McNamara, Stu Cameron, Malcolm Broatch. Seated: Bob Fraser (cook), Fermen L'Hirondelle, Doug Lyons
Cliff Smith

Ranger Dale Huberdeau used this Snow Scoot for transportation in the winter months around the Fort McKay Ranger Station, Lac La Biche Division, mid-1960s
Corinne Huberdeau

Work horses

Rangers on the Sheep Lick, Storm Creek, Bow River Forest, Rocky Mountains Forest Reserve, 1924

Government team, Red Deer Wagon Road, Bow River Forest, Rocky Mountains Forest Reserve, 1914

Smudge behind horses helped to deter flies in camp, Clearwater Forest, Rocky Mountains Forest Reserve, 1912

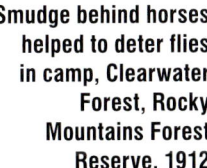

Ranger class - packing contest, Slave Lake headquarters, 1925

A Way of Life 69

Above the timberline, head of the Snake Indian River, Athabasca Forest, Rocky Mountains Forest Reserve (now in Jasper National Park), 1914

Sometimes two horses were packed in tandem with lengths of lumber. Materials were for construction of Blue Hill Lookout, Red Deer District, Bow River Forest, Rocky Mountains Forest Reserve, 1927

Hauling supplies on the Black Rock Trail, Devil's Thumb in distance, Bow River Forest, Rocky Mountains Forest Reserve, 1928

Grueling duty - horses drag lumber to summit for construction of Coliseum Lookout, near Nordegg, Clearwater Forest, 1927. Hauling lumber by this travois method was easier on the horses than the tandem method, but needed a wider trail

On the trail to Cameron Lookout, Bow River Forest, Rocky Mountains Forest Reserve, 1929

Ranger Neil Gilliat packing "Oats" outside the Moberly Ranger Station, Brazeau Forest, Rocky Mountains Forest Reserve, 1951
Neil Gilliat

Moving water barrel and supplies to Cameron Lookout on Burke Mountain, Bow River Forest, Rocky Mountains Forest Reserve, 1929

Supervisor McAbee puts blanket on horses, Brazeau Forest, Rocky Mountains Forest Reserve, 1929

Moke the trusty pack mule, Brazeau Forest, Rocky Mountains Forest Reserve, 1913

A Way of Life

Radio Superintendent Tony Earnshaw uses horse to pack propane bottles to Adams Creek Lookout, Athabasca Forest, Rocky Mountains Forest Reserve, 1940s

Rangers Jack Glen (L) and Ted Hammer at Rock Lake, Athabasca Forest, north of Entrance, c. 1940

Pack outfit and construction supplies near summit of Black Rock Mountain, Bow River Forest, Rocky Mountains Forest Reserve, 1928

Horses used to remove air-crash victim near Nordegg, Clearwater Forest, 1953. The rescue was organized by Bill Shankland, ranger at Nordegg
Jessie Deans

River travel

Forestry crew crosses the Athabasca River by raft at Entrance, Athabasca Forest, Rocky Mountains Forest Reserve, 1911

Indian house and lobstick tree, across from Fort McKay, Fort McMurray District, Northern Alberta Forest District, 1922. A lobstick (associated with northern Cree nations) is a tall conspicuous tree stripped of all but its top branches. It served as a living landmark, monument or mark of honour for a friend

Crossing the Red Deer River, Bow River Forest, Rocky Mountains Forest Reserve, 1913. Fords were chosen at widened points where the water was shallower and footing firm for horses

A Way of Life

Fording Prairie Creek, Clearwater Forest, Rocky Mountains Forest Reserve, 1911

Moving camp on the Lower Athabasca near the Firebag River during the forest inventory in 1954. Bill McPhail (bow) and Jack Robson (stern) later became forest rangers. Jack started nearby at Embarras Portage with Jack Plews
Peter Murphy

R.G. Lewis on portage, Lesser Slave Forest Reserve, 1911. A tump line over the forehead helped carry heavy loads over portages and long distances

Tracking (pulling canoe) up the Athabasca River, Lesser Slave Forest Reserve, 1911

Canoes and boats at Nine Mile Creek, Lesser Slave Forest Reserve, 1911

Crossing the North Saskatchewan River in canvas boat, Clearwater Forest, Rocky Mountains Forest Reserve, 1914

Jumping the Vermilion chutes on the Peace River, 1906

A Way of Life

Abandoned dugout canoe, Lesser Slave Lake, Lesser Slave Forest Reserve, 1911

Temporary ferry crossing the Red Deer River near Sundre, Bow River Forest, Rocky Mountains Forest Reserve, after a flood washed out the bridge in 1915

Fire patrol boat Rey, Athabasca River, 1916. The boat was based at Athabasca Landing and patrolled between Grand Rapids and Mirror Landing

Ranger Jack 'Wink' Plews patrolling in northern Alberta. The outboard motor or kicker made travel faster and easier

York boat, Athabasca River, Lesser Slave Forest Reserve, 1911

AFS airboat used on the Lower Athabasca River in the Fort Chipewyan - Fort McMurray region was known for its intense noise levels. Forest Officer Frank Platt was testing this boat at Cooking Lake, c. 1955

Ranger Gordon Watt from Entrance crosses Big Smoky west of Grande Cache in his home-made boat, Athabasca Forest, Rocky Mountains Forest Reserve, 1948
Gordon Watt

A Way of Life

Moving horses across the Wild Hay River, northwest of Hinton, Athabasca Forest, Rocky Mountains Forest Reserve, 1951
Neil Gilliat

Ranger Ernie Stroebel with M.V. Calling River (nicknamed Inchworm), Athabasca River, Lac La Biche Division, 1961
Ernie Stroebel

Air Foil boat designed by Rein Krause, Whitecourt Forest Superintendent, Athabasca River, Whitecourt Division, 1962

Unloading a truck and trailer on to barge for transport on the Athabasca River, Fort McMurray area, Lac La Biche Division, 1960s

Alberta Forest Service boat fleet on the Snye, near Fort McMurray, Lac La Biche Division, 1963

The 42-foot forestry tug M.V. Athabasca, built in 1967, moved crew and gear on the lower Athabasca River out of Fort McMurray

A Way of Life

CHAPTER 4
Momentous Change

Momentous Change

The years 1948 and 1949 saw momentous changes. Financial support for the Alberta Forest Service (AFS) was at last becoming available after years of constrained visions and plans.

Unfortunately, Ted Blefgen, Director of Forestry, had to retire due to ill health. It was disappointing for him since his retirement coincided with a major – and positive - turning-point in government support for the AFS. However, Blefgen was able to follow and appreciate all the advances and improvements that ensued in the following years.

Blefgen was succeeded as director by Eric S. Huestis, a native Albertan who studied forestry at the University of British Columbia. He started in 1923 with the Dominion Forestry Branch (DFB) and worked on most of the forest reserves before moving to Edmonton as Assistant Director of Forestry in 1940. When the Game Branch was transferred from the Department of Agriculture to the Department of Lands and Mines in 1941, Huestis received the additional responsibilities of Fish and Game Commissioner (without an increase in salary!). His knowledge and experience along with his determination and firm resolve guided the major developments within the AFS for the next 14 years. He finished his career in the position of Deputy Minister of Lands and Forests in 1966.

Eric Huestis
Provincial Archives of Alberta, PA2720-11

Land Use Planning

One of the most far-sighted pieces of legislation was the Order in Council of January 29, 1948, defining the "Green Area" of Alberta. Requests for homesteads had greatly increased after the war, a result of returning veterans as well as immigrants looking for new opportunities. Experience with unplanned homesteading had highlighted problems such as settlers failing on lands unsuitable for agriculture, land-clearing fires escaping into forests, game poaching, and the cost of providing infrastructure such as roads, schools and utilities. Two of the strongest reasons for creating the Green Area were included in the preamble to the Order in Council:

"It is desirable to prevent settlement and indiscriminate squatting on these lands that are incapable of providing sufficient sustenance for a settler and his family. Many of these lands have a nucleus for

Alberta Forest Service Executive meeting November 22, 1946. Meeting held at the Edmonton Royal George Hotel
Back Row: (L to R) Ed Noble, Jack Janssen, Ted Keats, Not Identified, Not Identified, Eric Huestis, Tony Earnshaw, Jack Rogers, Herb Hall, Ted Hammer, Not Identified, Bill Woods. Middle Row: Bill Cronk, Fred Smith, Frank Neilson, Ted Blefgen, Not Identified, John Harvie, Harry Taylor, Jim Hutchison, Vic Mitchell. Front Row: Not Identified, Not Identified, Tony Urquhart, Donald Buck, Not Identified, Walter Ronahan, Not Identified, Scottie Lang, Not Identified
Provincial Archives of Alberta, BL1242

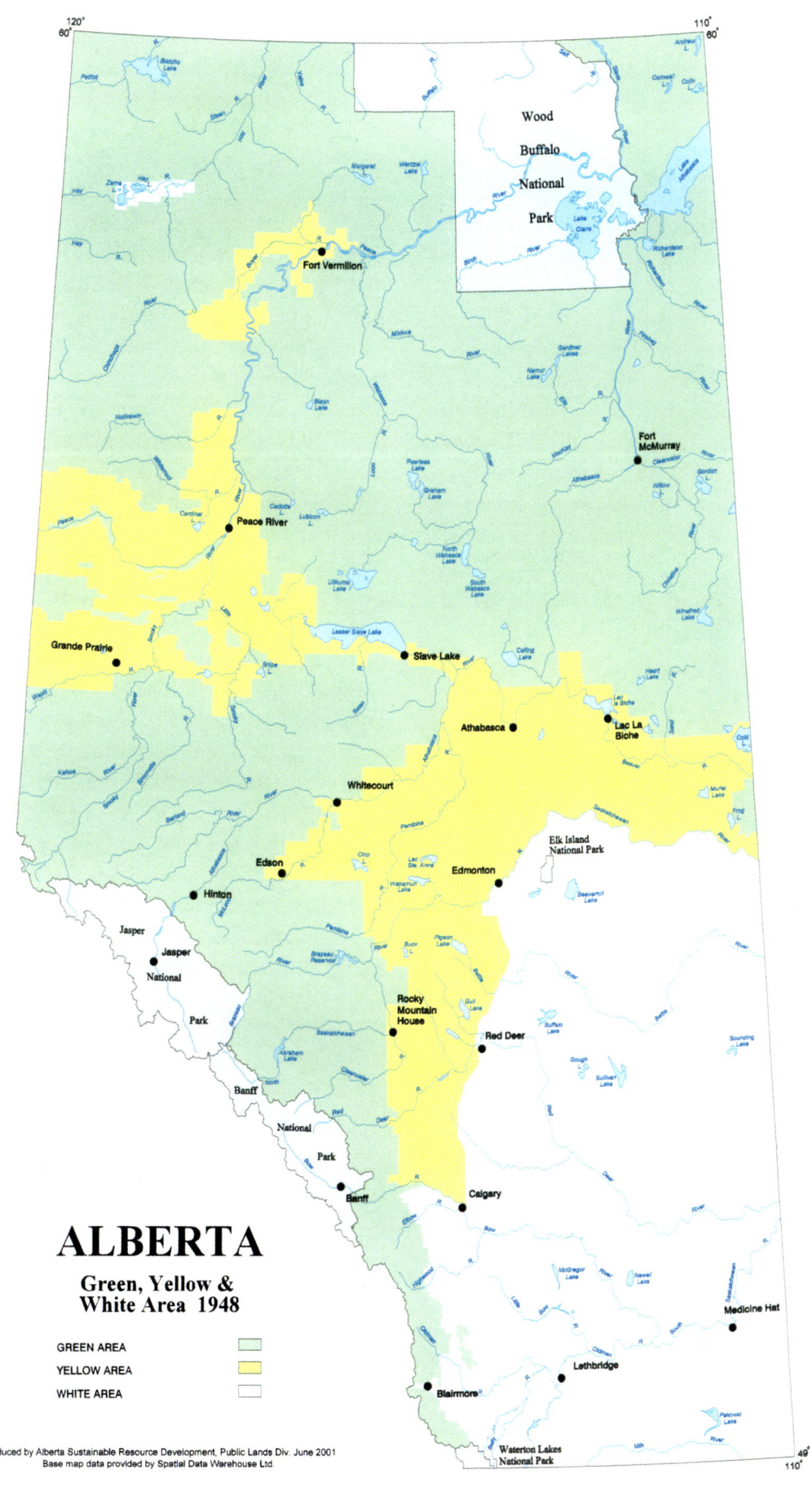

establishment of a valuable forest cover, which if given the essential protection will replenish to a very substantial degree in future years the timber cut to meet the requirements of the Second World War."

The Green Area map showed the reserved forest in green, potential agricultural lands in yellow, and predominantly agriculturally-suited lands in white. The Green Area was intended to reserve forested lands from settlement. The Yellow Area was also reserved from settlement, at least until soil surveys were done and lands suitable for agriculture released in an orderly way. The AFS was thus able to focus its efforts on the Green Area, and to base its forest management planning on those lands most likely to remain forested. The AFS was also responsible for timber and prevention and suppression of fires on forested lands in the Yellow Area.

In 1949 the former Department of Lands and Mines was split, reflecting the greatly increased activity in oil and gas, as well as interest in public lands and forests. Minister Nathan Tanner and his Deputy John Harvie stayed on in charge of both new departments - Mines and Minerals, and Lands and Forests. Eric Huestis remained Director of Forestry, continuing to play a strong leadership role both within the AFS and in his relationship with the Minister and government. A second major development saw final agreements approved for the Eastern Rockies Forest Conservation Board (ERFCB) in 1948.

This was a joint federal-provincial agreement in which both governments recognized that the three southern forests - Crowsnest, Bow River and Clearwater - were important watershed areas and that both governments shared concern about Alberta's financial ability to protect and manage them. The rationale for federal participation was based on the Saskatchewan River system that connected the three prairie provinces. It was estimated that 85 per cent of the flow in the South Saskatchewan River originated on the 15 per cent of the headwaters area lying in the Forest Reserves and National Parks. As a start, the federal government contributed $6 million in capital for roads, ranger stations, fire control facilities and equipment.

Building the Forestry Trunk Road, Mist Creek, Bow River Forest, 1951
Alberta Government, AFHPC

Cat Creek portion of Forestry Trunk Road, Highwood District, Bow River Forest, 1950. Note snags from 1936 fires
Alberta Government, AFHPC

Both governments shared the cost of operation for 14 years. Policy was set by a joint federal-provincial board, on which federal members were initially the majority. The major legacy of the ERFCB was the Forestry Trunk Road running north from Blairmore to Kananaskis and eventually to Nordegg, providing for fire, grazing and timber access and supporting tourism. Initial construction was sound and the road has served well for more than 50 years. The forestry trunk road now extends further north linking Hinton and Grande Prairie. Administrative control of the trunk road was returned to the AFS in 1959 after 12 years of operation under the ERFCB.

Forest Management Planning

Eric Huestis had long advocated an inventory of the forest as a basis of planning for sustained yield forest management, but funds were previously unavailable. AFS staff assessed timber stands as timber sale applications were received, but there was no clear overview of the extent of the northern forest. There was also a need for up-to-date maps, especially for the forested areas.

As Huestis later explained, the lack of maps was also a problem for the exploration activities of the oil and gas industry. He explained the problem to the Minister and was able to convince him, based on the concerns of the oil and gas sector, to proceed with a contract to have aerial photographs taken of the province. These photos were used to produce a set of base maps showing geographic features, roads and other improvements at a uniform scale.

Huestis arranged to insert a clause in the contract allowing for a forest inventory for that part of the province lying south of latitude 54 (east-west line that runs approximately from St. Paul to Grande Cache) excluding the area controlled by the ERFCB. A contract for this broad scale (later called the Phase I) inventory was issued on November 2, 1949, to the Photographic Surveys Corporation (PSC) in Toronto. Although the entire province was photographed, this initial forest inventory was completed for the southern portion of the province only. To make sure the inventory was done properly, Huestis hired Reginald D. Loomis to manage the project. Loomis had previous experience with the DFB in developing photogrammetric techniques (measuring forest cover from photographs), and had also applied these techniques with forest companies in eastern Canada. During his career with the AFS Loomis made a tremendous contribution to forestry, starting with the forest inventory and quickly extending his vision to forest management, silviculture and land use.

Huestis also visited the University of British Columbia in the spring of 1949 to explain his plans for the AFS to the graduating class. He hired nine of the graduates. Bob Steele, John Hogan and Trevor Charles were brought in and loaned to PSC for the forest inventory. Owen Bradwell, Stan Hughes, Jim Clark and Charlie Jackson were hired to be assistant forest superintendents at Blairmore, Calgary, Rocky Mountain House and Edson. Victor Heath and Bill Bloomberg worked on various assignments out of Edmonton, including establishment of the Forestry Training School at Kananaskis. These men constituted the start of a revitalized service working to meet evolving demands and pressures in Alberta.

As the first forest inventory project neared

Rocky Mountains Forest Reserve, staff meeting, Calgary, 1954
Alberta Government, AFHPC

(L to R): Not Identified, Rangers Jack Naylor and Harry Jeremy conduct tree measurements at Kananaskis, 1953
Alberta Government, AFHPC

Mine props salvaged from the 1936 Galatea fire, Bow River Forest, 1941
Alberta Government, AFHPC

1948 - 1965   83

completion, Huestis and Loomis agreed they should extend it to include the rest of the province.

Following passage of the *Alberta Natural Resources Act* in 1930, the *Alberta Forests Act* had been hastily pasted together from the previous federal legislation. However, after 18 years of experience and a great deal of thought, a new Alberta-focused *Forests Act* was passed in 1949. An interesting new clause permitted the government to enter into agreements for areas able to support pulp mills. It had been noted in 1948 that there was much timber which was too small for sawlogs and ties, but which would be suitable for manufacture of wood pulp. The new clause stated that the government might: "enter into an agreement, to be described as

John Hogan, in charge of Forestry Training School, Kananaskis, 1953 - 1955
Alberta Government, AFHPC

a forest management license ... for the management of public lands ... reserved for the sole use of the licensee for the purpose of growing continuously and perpetually successive crops of forest products to be harvested in approximately equal annual or periodic cuts adjusted to the sustained yield capacity of the lands." This was the first time that the term "sustained yield" had been used in Alberta legislation. Sustained yield embraces the principle of harvesting only as much timber as the forest can replace through growth.

Inquiries about pulp mill possibilities had been noted in Annual Reports as early as 1946. An application was submitted by Edmonton Pulp and Paper Mills Ltd in 1949, the same year in which the Act was passed. In 1954, the first agreement was successfully concluded for North Western Pulp and Power Ltd. (NWPP), which subsequently built a pulp mill at Hinton. The new *Forests Act* marked the beginning of serious discussions about constructing major forest products mills in Alberta.

Training Programs

With all forestry practices becoming more technical, Huestis had been promoting training programs to keep field staff up to date on new techniques. It was also becoming difficult to recruit new staff members who already possessed woodsmanship, timber cruising and forest fire control skills. Ranger Bill Shankland conducted the first formal course starting in January, 1947 for returning war veterans. The plan was to select potential new recruits who would spend the winter doing coursework in Calgary, then move to the Kananaskis Forest Experiment Station for fieldwork during the spring. Students were assigned to Ranger Districts for the summer and more coursework was planned for the fall. The first parts went well, but there was a high

Bill Bloomberg (L) and Victor Heath, instructors at Forestry Training School, Kananaskis, 1951
Alberta Government, AFHPC

Jack Macnab (L), instructor, and Peter Murphy, Forester in charge, Training Branch, 1960
Alberta Government, AFHPC

dropout rate during the summer so the fall session was cancelled. Dick Radke was one of the persistent and successful candidates, finishing his career as Forest Superintendent at Whitecourt.

Then, in 1950, a joint AFS and National Parks Service eight-week course that Huestis supported was held in the Banff School of Fine Arts. This class of 20 was drawn equally from both agencies. The course was successful in that it clearly showed the value of in-service training. Huestis asked foresters Victor Heath and Bill Bloomberg to organize an AFS program that would focus on the more specific needs of AFS Forest Rangers.

The first Forestry Training School was held under their direction in the fall of 1951, with a class of 20 rangers. Heath and Bloomberg had put together an intensive 10-week program using the residential facilities at the Kananaskis Forest Experiment Station, the former prisoner-of-war camp. The results were so successful that this basic in-service program was continued into the 1970s when it was replaced by the NAIT (Northern Alberta Institute of Technology, Edmonton) Forestry Program.

The Forestry Training School (FTS) continued each fall at Kananaskis until 1959. John Hogan ran the school from 1953 to 1955 and Peter Murphy took over in January, 1956. The FTS program was extended to over 12 weeks, and a new program for Fish and Wildlife officers was introduced during the summer of 1959. Training courses for lookout staff were developed and run at field headquarters during the spring. Increased training to upgrade staff plus a need to ensure personnel were familiar with new activities in the forest led to construction of a new Forestry Training School at Hinton. It was officially opened by Minister Norman Willmore in October, 1960, with a 20-man basic ranger course as its first offering. Additional in-service training programs quickly followed, including ones for tower staff and a host of new fire control training courses. Training was extended outside the AFS to add First Nation and Métis trainees, and also non-forestry-trained personnel such as farmers who served as fire bosses and patrolmen.

Planning began in 1963 for an Advanced Forestry course and a cooperative two-year Forest Technology program with NAIT. The latter program was to be the first post-secondary

First class of Fire Control Officers, Chief Rangers and Aircraft Dispatchers at Forestry Training School in Hinton, about 1962
Back Row (L to R): John Benson, Ted Bootle, Art Lambeth, Neil Gilliat, Irv Frew, Jack Naylor, Joe Kirkpatrick, Bernie Brower, Ernie Ferguson, Rex Winn, Del Hereford, Bill Kostiuk, Bob Diesel, Ben Shantz, Dick Mackie. Middle Row: August Gatzke, Bernie Simpson, Jim Hereford, Bert Prowse, Lou Babcock, Lou Boulet, Harry Edgecombe. Front Row: Carl Larson, Dick Radke, Pat Donnelly, Jack Grant, John Booker
Alberta Government, AFHPC

Norman Willmore opening the Forestry Training School, Hinton, October, 1960
Alberta Government, AFHPC

Ranger and instructor Sam Sinclair instructs one of his trainee straw bosses. A straw boss is the working supervisor of an eight-person firefighting crew. The straw boss reported to a crew boss, who was in charge of three crews plus cook, cook's helper and timekeeper (total of 28 people). Sinclair headed the first AFS native firefighter training program in the 1960s
Alberta Government, AFHPC

1948 - 1965

Pilot with Forest Superintendent Hank Ryhanen and Des Crossley, NWPP Chief Forester, 1963
Alberta Government, AFHPC

forestry program in Alberta. Huestis and NAIT president Al Saunders discussed the concept in 1963, then Steele and Murphy followed up with detailed planning. The unique arrangement was that students would attend NAIT in Edmonton for the first year to take advantage of the college's technical curriculum and facilities. George Ontkean and Joe Rickert both moved from the AFS to take over that responsibility with NAIT for the first class in the fall of 1964.

During the second year, students came to the Forest Training School (which became the Forest Technology School in 1965) to build on the academic education received at NAIT. This made it possible to hire four new instructors to add strength to the Hinton school: John Wagar, Dick Altmann, Stan Lockard and John Morrison.

The Advanced Forestry course was an intensive one-year program designed to build on the Basic Ranger course to upgrade ranger staff to an academically equivalent level. In the fall of 1965 both the first Advanced Forestry course and students from the first second-year class of the NAIT Forest Technology Program moved to FTS to complete their academic programs. Some of the NAIT students who graduated in the spring of 1966 joined the AFS.

First AFS Junior Forest Wardens camp, Forestry Training School, Hinton, 1960
Back Row (L to R): S.J. Macnab (School Supervisor), Douglas Davidson, Jim Wilson (Hinton Group Leader), Noel Armstrong, John Ross (Slave Lake Group Leader), Ricky Dempsey, Jim Affolter, Wayne Michener, Ricky Christie, Lou Foley, Floyd Collin (Edson Group Leader) Front Row: Gerry Kirkpatrick, Allan Wahlstrom, Gordie Sinclair, Bruce MacGregor, Dennis Maine, Teddy Armstrong, Charles Litke, Dennis Calvert, Terry Caswell. Lou Foley is a third-generation ranger; Bruce MacGregor is a second-generation ranger
Alberta Government, AFHPC

Youth Programs

The first Junior Forest Warden club in Alberta was started by Des Crossley and his staff at North Western Pulp and Power Ltd. at Hinton, and developed under the leadership of former AFS ranger Robin Huth in 1957. Two more clubs in Edson and Wabamun were formed later, and the government assumed leadership for a provincial network in 1960. The program was headquartered at the FTS at Hinton and directed first by Jack Macnab and later by Terry Whiteley. Lou Foley and Bruce MacGregor were two young Junior Forest Wardens in 1960 who later made a career in the AFS.

During the summer of 1965 the Junior Forest Rangers (JFR), a work program for high school students 17 and 18 years old, began. It also was operated by the Training Branch out of the FTS at Hinton. Three camps were run that summer. They were led by three AFS rangers (Emanual Doll, Larry Huberdeau and Horst Rohde) all of whom had opted to take the two-year Forest Technology program through NAIT rather than wait to take the in-service Advanced Forestry course later. They ensured that the JFR program, which continues to the present day, was successful.

Administration

Forest rangers had been provided with uniforms for the first time in 1949 with staff contributing to the cost of the winter jacket. Prior to this the only "uniform" was a badge, so the introduction of the uniform was a welcome event that stimulated a great deal of pride.

With the expansion of ranger staff into the northern districts during the early 1950s, rangers often found themselves to be the sole government agent in their communities. By this time rangers had become involved in the administration of traplines and in enforcing game and fisheries regulations as part of their duties.

Rangers were still responsible for sealing beaver pelts in the late 1940s when beaver numbers began to increase. A serious decline had occurred previously due to a combination of over-trapping and an outbreak of tularemia, an infectious disease. With the population increase, trappers were allowed to take a limited quota, usually based on one animal per beaver house in the trapping area. In addition, rangers in the north became the primary contact for aboriginal people and others when dealing with government and served as the "go-between" with industrial activities. Most rangers had a good rapport with residents in their districts.

(L to R) Ben Shantz, Fire Control Officer, Whitecourt; Dale Huberdeau, Forest Ranger, Fox Creek and Ross Ewing, Office Manager, Whitecourt Headquarters. The men were attending a firefighter training course, 1966
Alberta Government, AFHPC

Advanced Ranger Course Forest Technology School, first graduating class, 1966
Back Row (L to R): Al Walker, Colin Campbell, Karl Altschwager, Harold Enfield, Ray Hill. Middle Row: Howard Morigeau, David Schenk, Dick Girardi, Oliver Glanfield. Front Row: Fred Facco, Harry Jeremy, Hyrum Baker
Alberta Government, AFHPC

Ranger Frank Jones with the new AFS uniform, Castle Ranger Station, Crowsnest Forest, 1949
Alberta Government, AFHPC

Forest Protection

Prompted by concern over inadequate funding for forestry and forest fire control, Minister Tanner in 1950 commissioned an outside study by Wallace Delahey, a forestry consultant from Toronto. Delahey offered suggestions for increased support for and efficiencies within the AFS on which Huestis and his staff were able to draw as the service grew and activities in the forest continued to expand.

Delahey began his report[1] by describing the multiple values and uses of the forest. However, the focus of his report was on forest fire protection. As he stated: "The prime problem which must be solved above all others is that of giving Alberta's forests reasonable protection against the criminal carelessness with fire that has obtained to date and which again reached such disastrous proportions in the summer of 1949."

Delahey made 43 recommendations, 26 of which dealt directly with fire, plus eight closely related to fire and administration. Seven dealt with forest inventory and timber, one with forest insects and diseases and one with forest research. Delahey paid particular tribute to staff members of the AFS: "... it has been encouraging to experience the enthusiasm for the cause of better forest protection and management displayed by almost all members of the Forest Service, and this in spite of the discouraging fact that repeated recommendations and requests, over the years, for additional personnel and equipment, construction of lookout towers, roads, aerial patrols, etc., seldom received any tangible recognition.

"The present staff make up an excellent foundation on which to develop the expanded organization that is required and recommended herewith. Their years of experience and their enthusiasm for better protection and management will act as a real stimulant to bring out the best in the men who will be appointed to the expanded organization.

"Practically no members of the NAFD [Northern Alberta Forest District] staff have had technical training in forestry. In the past this has not been a handicap as their work has been almost entirely on administration. However, the recommendations contained herein provide for progressive steps towards Forest Management and to make for sound progress this will necessitate that some technical foresters be added to the staff."

Additional types of equipment - such as bulldozers, which had been refined during the war years - were now being employed in firefighting. In 1951, the first year in which their use was authorized for fires, it was stipulated that ministerial approval of the "unusual" expense must first be obtained. Bulldozers had proven effective in building fire guards in other areas, and were demonstrably faster than men with hand tools. The government of Alberta, however, initially felt that the cost of renting them for firefighting was much too high when compared with wages that were still in the range of 15-25 cents per hour. The serious 1950 fire season in northern Alberta, and most notably the Chinchaga River fire, prompted a reassessment of firefighting policy and led to two important changes. The first change enabled the hiring of bulldozers to fight fire, with ministerial permission. In 1952 this was changed to allow the forest superintendent to hire the first bulldozer. Ministerial permission was required to hire additional machines.

Footner Lake fire lookout, 1950
Alberta Government, AFHPC

Forest survey crew at Imperial Mills, Lac La Biche Forest, 1950. (L to R) Bill Collins, Jim Keenan, Trevor Charles and Bob Steele. Charles and Steele were two of the foresters hired by Huestis in 1949
Alberta Government, AFHPC

Hauling a bulldozer across the Athabasca River by raft, near Dutchman's Creek, Whitecourt District, July, 1956. The raft was nicknamed S.S. Hammer after Ted Hammer, Chief Timber Inspector and Director of Forest Protection. The dozer was being transported to Fire 36-1-56
Mel Willis

The second change concerned the Protected Area. In 1950, fire suppression in the northern part of the province (above a line running north of the towns of Peace River, Slave Lake and Lac La Biche) could only be undertaken for fires within

10 miles of a town, navigable river or railway. These limits were removed in 1952, although a shortage of resources and access remained a limiting factor.

The Rocky Mountain Section of the Canadian Institute of Forestry, formed in 1949, provided a forum for foresters and rangers to meet and discuss state-of-the-art forestry. One of the major

North Western Pulp and Power Ltd. mill at Hinton. Construction of Alberta's first pulp mill started in 1955 with first production in 1957
Bob Stevenson

Firefighting crew and builders of the S.S. Hammer, named after Ted Hammer who became head of Forest Protection. The raft was built during Fire 36-1-56, in July, 1956
(L to R): Ed Jackman, Charlie Duncan (Sr.), Jack Macnab, Frank Harvey, Clarence Weeks, Fred Lewis, Jack MacGregor, Mel Willis, Nick Nickalatian, Gordon McKin, Rein Krause (Superintendent, Whitecourt Forest), Peter Parranto, Harry Wedow, Joe Beeman
Mel Willis

concerns among foresters was the perceived inadequacy in fire control, so the Section compiled a comprehensive brief and presented it to the Minister in 1954. It was strongly critical of the low levels of funding for ranger staff, equipment, access and applied technology. Huestis welcomed the intervention, but was unable to get the government to respond until after the major fires of 1956.

Ted Hammer was appointed head of Forest Protection to replace Jack Janssen, who retired in 1954. Frank Platt, a returned war veteran and timber inspector at Entwistle, was brought into Edmonton in 1953 to work with Hammer. Together, they began building a case for increased support for fire control and field administration. The catalyst for change was the particularly severe fire season during the spring and early summer of 1956.

Frank Platt initiated many improvements in fire control
Alberta Government, AFHPC

The season included three serious fires that burned within or into the new lease of North Western Pulp and Power Ltd. These fires threatened the viability of the multi-million dollar investment in Alberta's first pulp mill, which was also located in the riding of the Minister of Lands and Forests, Norman Willmore. At the suggestion of Huestis, the mill staff wrote an analytical but critical brief to the govenment, pointing out their concerns and suggesting options to consider. Huestis arranged a meeting between mill staff and the AFS that fall, and urged both parties to work together to make workable and supportable recommendations.

These actions led to increased funding for a marked build-up in fire control capability. Another serious fire year in 1961 further tested the system, and also resulted in increased support. Under the leadership of Ted Hammer and Frank Platt, the AFS fire control organization grew considerably. These were dynamic times for the AFS and provided a good foundation for continuing political support and development.

Primary elements of the expanding fire control program included splitting the large northern ranger districts into smaller ones, building new ranger stations and rationalizing and modernizing existing towers to create a network that covered the entire forested part of the province. Access to fires was improved by the purchase of the first aircraft in 1957, leasing of helicopters, construction of landing strips and extension of roads. All of this meant adding new staff members for the districts, along with specialists to direct construction activities, aircraft management and extension of the radio system.

The mascot Bertie Beaver appeared on the scene in 1958 as a symbol to encourage public awareness of forest fire prevention programs. Bertie Beaver was a gift from Walt Disney in gratitude to Eric Huestis and the AFS for their cooperation and for lending facilities in Kananaskis which Disney used while making two films, one of which was "Nikki, Wild Dog of the North." The Bertie Beaver symbol caught on and became Alberta's unique mascot for forest fire control and forest management efforts.

Forest Management

Alberta's forest management capability and activity increased dramatically with the leadership of Reg Loomis. North Western Pulp and Power Ltd. was the first pulp mill company in Alberta. Staff of the AFS and the company pioneered many of the principles of present-day forest practice. The mill had its beginning in 1951 when Frank Ruben registered the company and signed a conditional Forest Management Licence with the Alberta government. Ruben owned a coalmine near Robb and envisaged combining the energy from coal with timber resources to manufacture pulp. He eventually formed a partnership with the St. Regis Paper Company of New York, together signing an agreement with the Alberta government in 1954 that led to construction of the mill at Hinton in 1955. Des Crossley, previously a research scientist with the

Part of the head office 'team' in the late 1950s, (L to R): Frank Platt, Charlie Jackson, Ted Hammer, Herb Hall, Eric Huestis and Lola Cameron
Bruce MacGregor

Forestry float, Rocky Mountain House parade, Clearwater Forest, 1952
Jack Roy

Walt Disney, creator of the Bertie Beaver mascot, at McCall Airport in Calgary, 1965
Alberta Government, AFHPC

Canadian Forest Service, became Chief Forester that same spring. Initial plans had called for the mill to be built outside Edson, by the McLeod River. Detailed technical studies indicated, however, that the McLeod's water supplies were insufficient for the mill's needs so they moved the site to Hinton.

The first Forest Management Agreement (FMA) was a distinctively Alberta innovation. Originally called a Forest Management Licence or Pulpwood Agreement, the FMA was intended to create partnerships with industry to share protection and management responsibilities. The FMA required the holder to construct a pulp mill, develop a Forest Management Plan and specific annual operating plans, build and maintain their own roads, pay timber dues and fees as part of their commitment to forest management, and maintain a head office in Alberta.

Staff of both NWPP and the province took these responsibilities seriously. Crossley, together with Loomis and Charlie Jackson of the AFS and their respective staff members, worked out details that set a notable precedent in Canadian forestry. Among their innovations was the concept of ground rules that enabled professional judgement to meet agreed-upon objectives instead of setting arbitrary rules. In the spirit of what is now called 'Adaptive Management,' the results of those rules and practices were monitored and changes made as required. Ground rules have since evolved to provide rules and guidelines for logging practices - planning, roading, watershed protection, soil conservation and other objectives.

While this was going on, Loomis, Bob Steele and staff in the AFS were completing the northern forest inventory and were preparing to put the rest of Alberta on the path to sustained yield forest management. Out of the negotiations and inventory work came a great deal of mutual cooperation, exchange of knowledge and understanding among all parties involved in getting modern forest management under way. The first set of forest inventory maps

NAFD patrol cabin at Breton, Northern Alberta Forest District, 1952
Jack Roy

Ranger Jack and Mrs. Roy with Forestry Jeep on Yellowhead grade at Nojack (now Highway 16), Edson Forest, 1954
Jack Roy

Reg Loomis (sitting) Senior Superintendent of Forest Management and his assistant Charlie Jackson study the Department's cruise data, 1964
Alberta Government, AFHPC

was completed, and in 1957, Forest Surveys Branch staff prepared a composite map for the province. By the use of colour codes it graphically illustrated the pervasive influence and extent of forest fires, providing information that had a profound influence on subsequent forest policy.

The 1958-59 Annual Report of the Department of Lands and Forests noted two

7447 T.A.

Department of Lands and Forests
PUBLIC NOTICE

NOTICE OF SALE

LICENCE TIMBER BERTH NO. 4575.

The undersigned will offer for sale by SEALED TENDER at 2:30 o'clock in the afternoon of the thirteenth day of November, 1958 at the Rotunda, Fifth Floor, Natural Resources Building, Edmonton, Alberta the right to cut timber under licence from the following lands.

The South East quarter and Legal Subdivisions 9 and 16 of Section 21, Legal Subdivisions 12 and 13 of Section 22 in Township 73, Range 23, West of the 4th Meridian; containing .506 square mile, more or less.

The Berth is estimated to contain approximately 375,000 F.B.M. merchantable Spruce and 14,000 F.B.M. merchantable Balsam Fir timber suitable for the manufacture of lumber and other forest products, but no warranty is given as to the quantity or quality of the timber on the Berth.

The Berth will be awarded to the person tendering the highest rate of dues on sawn lumber of a species other than poplar over and above the rate as may from time to time be prescribed by the Lieutenant Governor in Council. Bids must be in units of five cents. The dues on all other products, except dry timber, shall be at the rate as may from time to time be prescribed by the Lieutenant Governor in Council plus the commensurate increase in dues on lumber. Dues on dry timber shall be at the rate as may from time to time be prescribed by the Lieutenant Governor in Council.

Tenders enclosed in a sealed container and marked "Tender for Berth No. 4575" will be accepted from an individual, body corporate or registered partnership, (partners to submit proof of registration when tendering) and may be submitted in person or by registered mail, or through an agent whose authorization in Form A of the Schedule to The Forests Act must be filed with the Director prior to the sale. No tender may be withdrawn once it has been submitted.

Any number of tenders may be submitted which must be in Form B of the Schedule to The Forests Act. The first tender must be accompanied by a deposit of $450.00 which shall be enclosed in the sealed container in the form of cash, marked cheque payable to the Provincial Treasurer, or Bearer Bonds of the Province of Alberta or the Government of Canada. If more than one tender is submitted the guarantee deposit accompanying the initial tender shall be deemed to accompany any of the additional tenders.

The deposit of the purchaser shall be retained as a guarantee of compliance with the terms and conditions of the licence and shall be automatically forfeited if the purchaser fails to complete the contract.

- 2 - Licence Timber Berth No. 4575.

The purchaser must sign a contract in Form C of the Schedule to The Forests Act agreeing to carry out the terms and conditions of the sale and at the same time apply for a licence for the current year paying the sum of $103.44, being the cost incurred in cruising, surveying and other incidental charges, together with the rental, licence fee and fire-guarding charges.

The licence shall be for a term not exceeding one year and shall be renewable for two years while there is on the Berth a sufficient quantity of the kind and dimensions of timber specified in the licence. Each licence shall expire on the 31st day of July following its itssue.

The cutting of timber from this Berth will be subject to the following conditions:

No green Spruce timber of a diameter less than 18 inches measured 12 inches from the ground or as is marked or otherwise designated by a Forest Officer shall be cut.
All merchantable Balsam Fir shall be cut and utilized.
All merchantable fire-killed timber shall be cut and utilized.
All merchantable windfallen timber shall be utilized.
Any method of logging causing undue damage to the residual stand, the land surface or creating a fire hazard may be prohibited by a Forest Officer.
All slabs and edgings produced during the close fire season shall be piled on a cleared area and surrounded by a fireguard. If not utilized, they shall be burned within the calendar year in which they accumulate as directed by a Forest Officer. All slabs and edgings produced during the open fire season shall be disposed of in accordance with The Forests Act.
The licencee shall submit a plan of any access road to the Director of Lands and make application for and obtain a Licence of Occupation before using or undertaking construction of such road. All brush and debris resulting from the cutting out of any roadway shall be burned or otherwise destroyed to the satisfaction of a Forest Officer.

All forest products taken from the Berth, except round timbers and pulpwood shall be manufactured at a properly equipped sawmill.

The Minister may, in his discretion, reject any or all tenders for the Berth.

Further particulars may be obtained upon application.

E.S. HUESTIS,
Director of Forestry.

EDMONTON, Alberta,
October 24, 1958.

NOTE: To qualify for appeal from an award under Section 20 of The Forests Act the appellant must have tendered at least 50% of the high tender.

Typical public notice of timber berth sale, 1958. Rights to cut timber were usually sold by sealed tender to the buyer with the highest offer
Alberta Government document

disparate events that symbolized the remarkable changes taking place during this period. Dr. V.A. Wood, Director of Lands, reported that there had been six wild horse roundup permits issued in 1958 that had resulted in the capture of 146 horses, of which 128 were taken in the Edson-Hinton area. This seems to have been the last of the "wild horse" roundups. Rangers in both the DFB and AFS had frequently obtained their saddle and pack horses this way. (See Harry Edgecombe's poem at the end of this chapter. Edgecombe's perennial contention was that a third-class ride beats first-class walking any time.)

The second, more far-reaching, event was the statement in AFS Director Huestis' report that the policy of the AFS was to implement as quickly as possible sustained yield management on forest lands held by the Crown. He referred to the agreement with North Western Pulp & Power at Hinton as an example of an area already being managed for sustained yield. This commitment was made possible with completion of the forest inventory that laid the foundation for science-based forest management in Alberta.

Forest management units averaging 1,930 square miles were established. As sustained yield annual allowable cut calculations became available for these forest management units, it became evident that a rationalization in the allocation of harvesting rights had to be made to prevent over-harvesting of some areas and to encourage utilization of mature stands in others.

Ranger Bill Adams and fire prevention sign in Edson Forest, 1954
Jack Roy

Ranger patrol, Bill Brackman (L) and Bill Adams, Edson Forest, 1952
Jack Roy

Serious discussions with forest industry led to agreement on what was to become the "Quota System," in which forest companies would be allocated a quota of the allowable cut based on past production. This system is described in detail in Chapter 6.

In the late 1950s initial forest management planning had been extended to every forest management unit. The management philosophy within these units was explained by Huestis in his Annual Report (Department of Lands and Forests) of 1959-60:

"There has been a continued effort to establish the best possible methods in the control of timber cutting operations. The policy of the Department is to manage forests on a sustained yield basis, cutting only the yearly increment and making sure that forest areas cut over shall be reforested for the next crop. Management plans control the production in all forest areas accessible for timber harvesting operations. In addition, operators are required to prepare, submit for approval and carry out yearly cutting plans for their licensed timber berths. Present policy is to cut overmature and decadent stands of timber before cutting mature stands that are still increasing their yield through good growth."

References to the impact of oilfield development on the forest began to appear during the 1950s. The Department's Annual Report of 1959-60 stated: "For some years now the oil industry has been pushing farther into our forested area. Exploration work has created many thousands of miles of road with a resultant loss of timber. After an oil field is located, many more miles of roads, pipelines, well sites and battery sites are cleared with further loss of timber. Every effort is made to salvage as much of this timber as is economically possible."

These activities, along with grazing and recreation and concerns about watershed integrity, led in 1960 to creation of a new Land Use Section in the Forest Management Branch. With increased industrial activity in the forest and growing public concerns, this Section soon became a branch in its own right. Its mandate was to "ensure that uses of land in the forested areas of the province are controlled to the best possible interest of conserving the forest as well as soil and water. Particular attention will be given to the activities of the development of petroleum and natural gas in the forest zone in order to reduce the possibilities of unnecessary damage to these other resources, i.e. forest, soil, wildlife and water."[2]

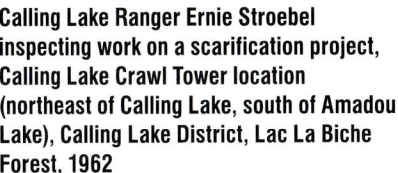

Calling Lake Ranger Ernie Stroebel inspecting work on a scarification project, Calling Lake Crawl Tower location (northeast of Calling Lake, south of Amadou Lake), Calling Lake District, Lac La Biche Forest, 1962
Alberta Government, AFHPC

Strip scarification in a deciduous stand, where spruce was planted or seeded along the scarified rows. Sulphur Lake area, Peace River Forest, 1965
Cliff Smith

Administrative control of the Rocky Mountains Forest Reserve returned entirely to AFS in 1959, when the area became once again fully integrated within the provincial AFS operations. The Eastern Rockies Forest Conservation Board continued in an advisory role until the agreement expired in 1962.

Reforestation

Support for silviculture and reforestation had been increasing gradually, spurred in part by the success of activities on the NWPP forest management area and encouragement from Reg Loomis who established a new posistion for a silviculture forester in 1959. Larry Kennedy, a

Larry Kennedy, first AFS Silviculture Forester in 1959
Alberta Government, AFHPC

UBC forestry graduate from Rocky Mountain House served 13 years before returning to his family ranch. When he arrived, as reported in the 1960-61 Annual Report, the AFS affirmed: "The objective of all cutting methods is to produce, if at all possible, natural regeneration on cutover lands."[3] Scarification was the predominant post-harvest site treatment. Scarification is the mechanical mixing of soils and organic matter after harvesting to create sites that will support natural seeding or planted seedlings. However, Kennedy recognized that successful

1948 - 1965 95

silviculture, the growing of trees, would involve a lot more than this. Working through and with field foresters he initiated programs for seed collection, set up seed collection zones and records, established seedling trials and improved the nursery operation at Oliver.

The increased need for planting by industry and government was met by seedlings grown by the AFS at an expanded Provincial Tree Nursery at Oliver. The AFS also established medium-scale containerized seedling production in Rocky Mountain House, at the Provincial Jail in Peace River, and with a private nursery contractor at Grande Prairie.

In 1963 Eric Huestis was promoted to the position of Deputy Minister of Lands and Forests. Robert G. Steele, one of the class of 1949 University of British Columbia forestry graduates recruited by Huestis, became Director of Forestry. Steele had been a leader in forest surveys and initiated the forest management planning process. He also served as Forest Superintendent in the Rocky-Clearwater Forest at Rocky Mountain House before returning to Edmonton.

Pine tree & river
Artist Lorna Bennett

Sketch of lookout cabin
Artist Lorna Bennett

Ranger Pat Foley reviewing forest cover type map in Slave Lake Forest, September, 1959
Cliff Smith

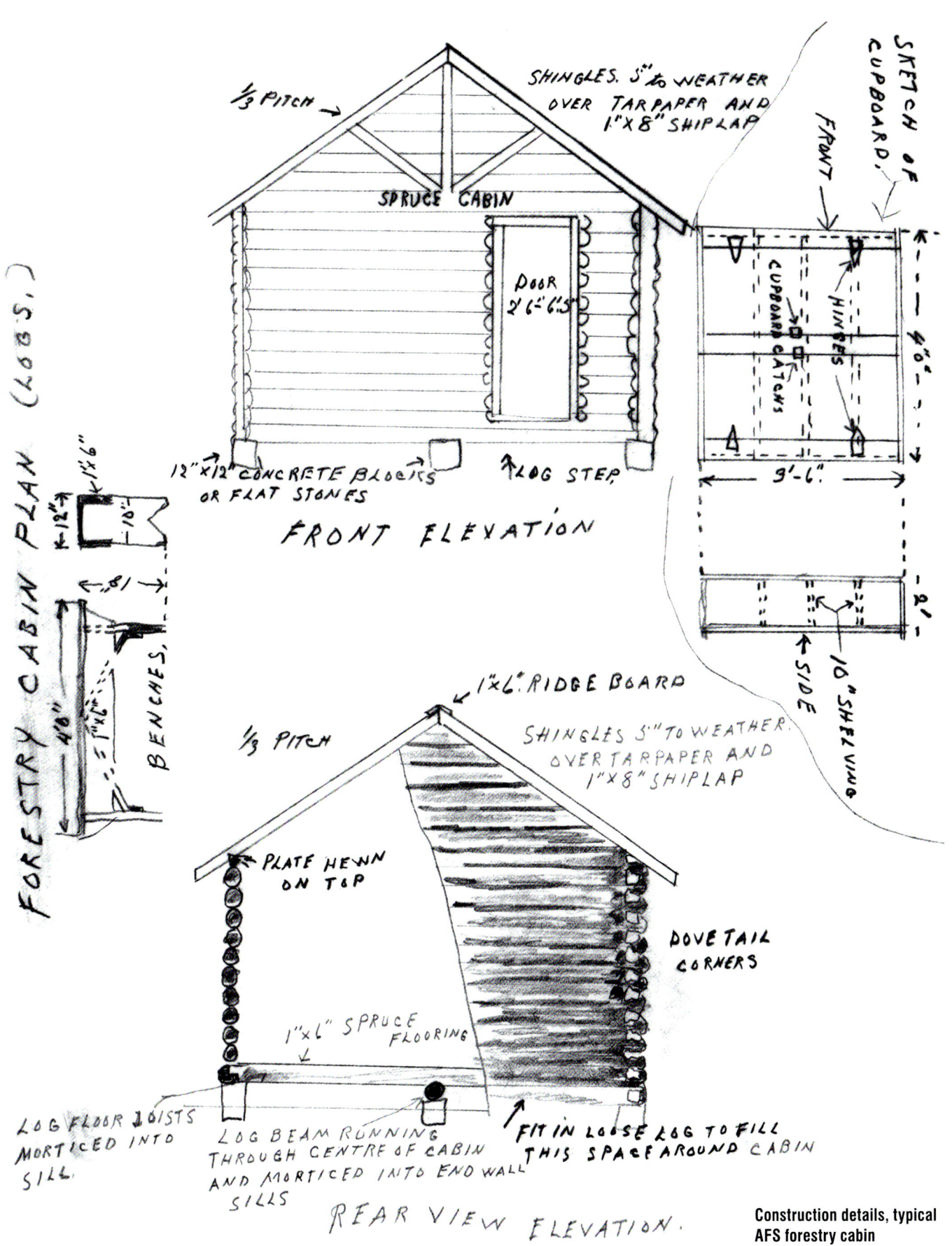

Construction details, typical AFS forestry cabin
Alberta Government, AFHPC

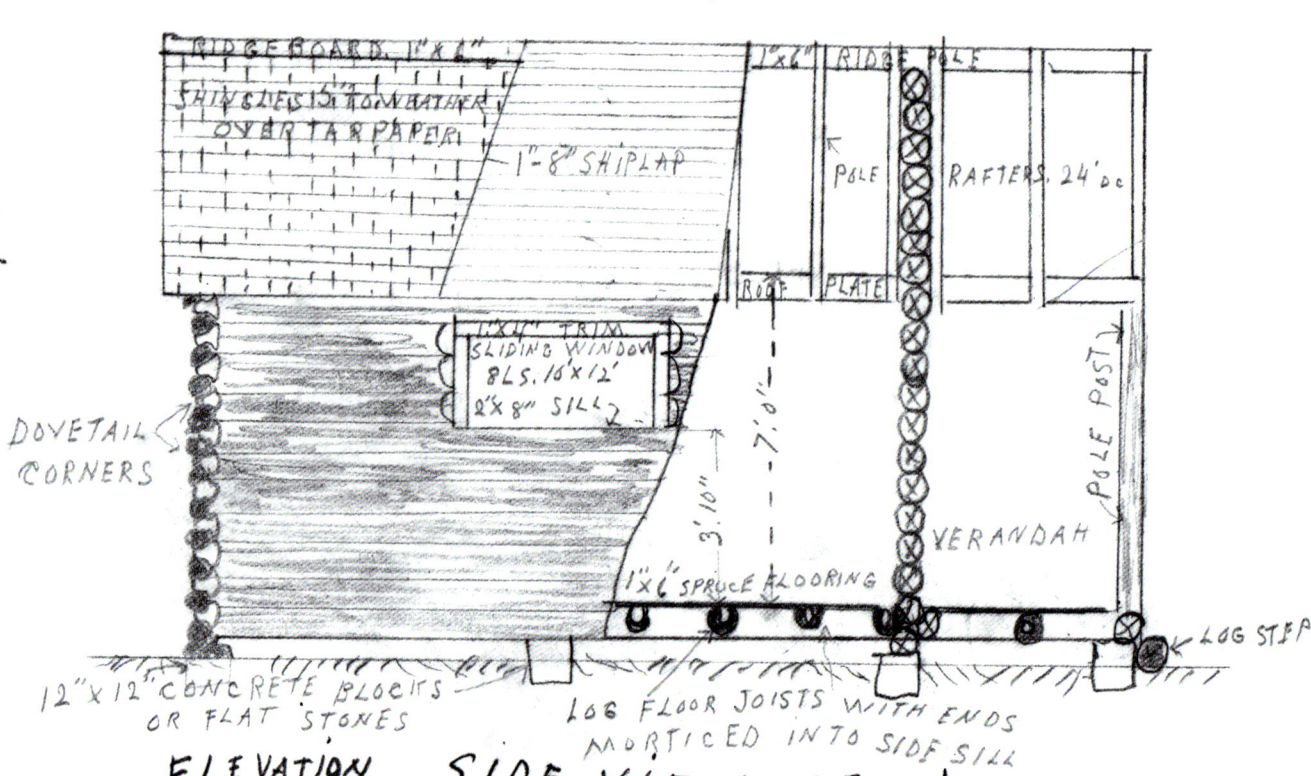

THE FORESTRY ROUND UP
By Harry Edgecombe, 1949

The Forestry had a round up,
Upon the River Tay,
The riders brought the camp outfit,
And Walker hauled the hay.

There was Morris from the Red Deer,
With his palomino steed,
Noted for its' endurance
And ever lasting speed.

Jack Walker rode old Nellie
The mare that liked to bolt.
He said he could not ride her
Unless the horn he had aholt.

Ronnie rode in on Tony.
They made a rugged pair.
He said he would bend those horses
And never turn a hair.

Dick came in on Corbit,
A prancing little dun.
He said he would catch a stallion
Before the day was done.

Jack and Dick from the Big Horn,
Came in with horses four.
They said if these were not enough
They would go back home for more.

The Brazeau sent Ben,
Who travelled all the way
To catch those wildies
Or help, was what he had to say.

From the Meadows came Harry.
His horse was known as Dan.
The meanest little horse
That ever carried a man.

The wheel of riders gradually grew
Until we were short one cog,
Then Bill Winters from Moose Creek
Came in on Spotted Dog.

For days we chased those wildies,
Through crusts of drifted snow.
We trailed them through the mountains,
Or where ever they did go.

After five long days of riding,
Only seventeen head were caught.
So the Forestry sent out Bloomberg
Who claimed to know a lot.

Next morning just at daylight,
Jack Browning called the crew.
We saddled up and headed west.
By ten o'clock Bloomberg was through.

For seven more days the chase went on.
Each day a band was caught,
Until thirty five head were corralled
On Bob Bugbee's feeding lot.

Now this is the story of the last round up.
The Forestry said it did not pay.
So ask Superintendent Hall
And see what he will say.

Forest Rangers were required to own at least two horses – a saddle horse and pack horse. Since most of the travel and work was done by horse, rangers were always on the lookout for more good animals. One source was to catch horses running wild. Harry Edgecombe, ranger at the Meadows, wrote this poem about what turned out to be the last big forestry horse roundup. It was held in the Clearwater Forest around 1949.

Names of the people mentioned are listed in order of appearance:

Morris Verhaeghe - Ranger
Jack Walker - Ranger
Ronnie Lyle - Ranger
Dick Knorr - Ranger
Jack and Dick Browning - Guides and Outfitters
Ben Shantz - Ranger
Harry Edgecombe - Ranger
Bill Winters - Trapper
Bill Bloomberg – Forester at Rocky
Jack Browning – Guide and Outfitter
Bob Bugbee – Guide and Outfitter
Herb Hall – Forest Superintendent at Rocky

First Ranger School – Banff School of Fine Arts, February, 1950
(L to R): Jack MacGregor, Larry Bunbury, Ben Shantz, Frank Jones, Ron Lyle, John Elliot, Robin Huth, Jack Roy, Rex Winn, Bert Coast, Ernie Ferguson, Russ Eckert, Wally Walton, Gordon Watt, Colonel Cormack
Alberta Government, AFHPC

1946 Forestry class. (L to R): Wally Harrison, Dick Radke, Stu Height, Bill Balmer. Front left corner - Vic Higgins
Alberta Government, AFHPC

Rangers Wally Harrison (L) and Sandy Brown walked from Entrance to the Wildhay River to bring in the horses to ready them for a summer on the Mountain Trail, Athabaska Forest, Rocky Mountains Forest Reserve, 1950. Neil Gilliat recalled "travel was primitive and shanks pony was always an alternative." In further discussion, Gilliat explained that 'Shanks Pony' was an old Scottish saying, at one time widely used, meaning you will have to walk. Shank was the leg; pony was on your own device. It was generally used in a statement indicating disappointment when some form of transport failed – "well I guess we will have to use Shanks Pony."
Neil Gilliat story and photo

Rangers Ron Lyle (L) and Ben Shantz attend the joint Banff National Park and Alberta Forest Service Forestry Training School, 1950
Jack Roy

Ranger Neil Gilliat haying meadows at Moberly Ranger Station, Athabaska Forest, Rocky Mountains Forest Reserve, 1950. Neil Gilliat recalled that upon employment "every Ranger was supposed to supply one riding horse and a packhorse. In return we were allowed 10 days to put up hay for the winter. The Athabaska Forest had a team of horses and a hay mower and horse powered hay rake." In this photo Gilliat is riding the hay sweep which was used to move piles of hay to stack inside the corrals protecting the hay from the elk and deer. "The haying was a cooperative effort of most of the rangers and we put up hay on the meadows around Gregg Lake and Winter Creek, both locations now in Switzer Provincial Park."
Neil Gilliat story and photo

Neil Gilliat breaking in a horse in preparation for summer trail use, Moberly Ranger Station, Athabaska Forest, Rocky Mountains Forest Reserve, 1950. Gilliat recalls that there "wasn't a horse in the west that couldn't throw me. My old bones stand witness to the fact to this day."
Neil Gilliat story and photo

First graduating class of the Forestry Training School outside on porch of the Colonel's Cabin, Kananaskis Forest Experiment Station, 1951
Back Row (L to R): Rusty Esson, Bill Forbes-King (jacket over shoulder), Joe McGrath, Jack Kilgore (behind Macnab), Jack Macnab (officer cap), Bert Varty, Bill Adams, Frank Theirault, Harry Edgecombe, Tad Garland, Bill Bulmer (at back in front of door), Ray Moss, Sandy Brown, Neil Gilliat. Front Row: Jim Hereford, Mike Reap, Des Crossley (hand on chin), Phil Nichols, Victor Heath (Instructor i/c), Buck Rogers, Jim Stewart (sitting very front), Bill Bloomberg (Instructor 2i/c), Johnny Doonanco
Neil Gilliat

First graduating class of the Forestry Training School inside classroom, Kananaskis Forest Experiment Station, 1951. Instructor with back to photo is Eric Huestis
(Left side of class, back to front): Ray Moss, Rusty Esson, Frank Theirault, Buck Rogers, Tad Garland, Bill Forbes-King, Mike Reap. (Right side of class, back to front): Jack Macnab, Jack Kilgore, Bill Bloomburg (standing – Instructor 2 i/c), Victor Heath (standing – Instructor i/c), Sandy Brown, Neil Gilliat, Bert Varty, Harry Edgecombe, Joe McGrath, Phil Nichols, Johnnie Doonanco
Alberta Government, AFHPC

Eric Huestis teaching students at the Forestry Training School, Kananaskis Forest Experiment Station, 1951
Alberta Government, AFHPC

Ranger Jack Grant, with horse and pack horse, Keg River Ranger Station, Peace River Division, 1952
Alberta Government, AFHPC

Students learning mechanics at the Forestry Training School, Kananaskis Forest Experiment Station, 1951
(L to R): Buck Rogers, Bill Forbes-King, Ray Moss
Alberta Government, AFHPC

Supper at Forestry Training School, Kananaskis Forest Experiment Station, 1951
(L to R): Face – Bill Bulmer; Flunkey's back; Face – Bill Adams, Bert Varty; Back – Jim Stewart, Phil Nichols, Mike Reap; Face – Bill Forbes-King, Johnny Doonanco, Neil Gilliat; Back – Jack Macnab
Alberta Government, AFHPC

Forestry Training School classroom building, Kananaskis Forest Experiment Station, 1950s. Building and site were the former location of an internment camp housing German prisoners during World War II
Alberta Government, AFHPC

Ranger Wally Harrison on horse patrol, Cathedral Mountain in the background, Athabasca Forest, Rocky Mountains Forest Reserve, 1951
Neil Gilliat

Ranger Wally Harrison in front of the Smoky Cabin with Neil Gilliat's outfit, Atahbasca Forest, Rocky Mountains Forest Reserve, 1951. Neil Gilliat recalled that the Smoky Cabin "was located at a place known as Clark's Crossing (named after former Ranger and Game Commissioner Stan Clark). The crossing was very dangerous and was located next to a big whirlpool. After at least one drowning (an outfitter from Jasper) the forestry built two small sheds to house canvas Peterborough canoes. We unpacked all the horses and put all saddlery and pack boxes in the canoes then herded the horses across the river. There was a fenced pasture on both sides."
Neil Gilliat story and photo

Athabasca Forest staff at the Rock Lake Ranger Station, 1951
(L to R): Chief Ranger Walt Richardson, Ranger Wally Harrison, Assistant Ranger Neil Gilliat, and Assistant Superintendent Charlie Jackson leaning against the International 4x4 Power Wagon assigned to Walt Richardson. This truck was to service both the Brazeau and the Athabasca Forests. Neil Gilliat recalls that "when I was Assistant at Entrance I think I spent half my life on the end of the winch of this vehicle. Everyone bragged about the ability of this machine to climb mountains and visit towers such as Adams Creek L.O. but very little was ever said about the amount of winching that had to be done across rivers, muskeg or up steep slopes. Unfortunately, I was the guy who would wade the river dragging the winch, crawl through the muskeg dragging the winch, and climb the rock face dragging the winch. Suffice to say I was not impressed with the capabilities of the Power Wagon but it was amazing where we went with it." On Walt Richardson, Gilliat recalls "there was a new generation of ranger staff taking over the Forest and Walt was a steadying influence and almost a father figure to all the young rangers."
Neil Gilliat story and photo

Alberta Forest Service Ranger meeting, 1950s
Photo includes John Currat, Pat Foley, Phil Nichols, Larry Gauthier, Bert Varty, Ted Hammer, Bert Prause, Bill Smith, Eric Dawson, Wally Harrison, Walt Richardson, Mike Reap, Sandy Brown, Alphonse Lawrence, Ray Smuland, Dexter Champion, Neil Gilliat, Rex Winn, Frank Platt, Phil Comeau, Morris Verhaeghe, John Elliott, Bob Lewis (Unable to identify all of the people in the picture)
Provincial Archives of Alberta, PA2798-1

Forest Rangers Banquet, Edmonton, March 13, 1952
Far Row Left (front to back) – Mrs. Rose Gauthier, Larry Gauthier, Mrs. Hazel Ryhanen, Hank Ryhanen, Mrs. Burke, Mike Burke
Second Row From Left (front to back) – Ben Sutherland, T.W. Farrell, Bill Adams, Wink Plews, Bert Prowse, G.L. Blackmore, Mike Gagnon, N.C. (Clarence) Peterson, August Gatzke, Johnny Booker, Frank LaFoy, Buck Rogers, R.G. (Tad) Garland, C.G. (Colin) Campbell, Bill Titley
Third Row from Left (front to back) – Joe McGrath, Alfonse Lawrence, Ernie Ferguson, Lou Babcock, John Holden, Pat Foley, Guy Randall, Jack Roy, Mrs. Jack Roy, Neil Gilliat, Mrs. Gilliat, Bert Coast, Ray Smuland, Mrs. M. Smuland, L.A. Reed, Mrs. Reed
Fourth Row from Left (front to back) – John Currat, Angus Crawford, T.D. (Dewy) Feland, John Sutter, Bob Lewis, J.G. Bryden, Phil Nichols, Mike Reap, Fred Smith, Vic Mitchell, Dexter Champion, Harold Parnell, Erwin Kellerman, H. Campbell, R. Grenier
Fifth Row from Left (front to back) – Stanley Smith, Peggy Clarke, H.T. (Bert) Varty, T.R. (Ted) Hammer, Jack Naylor, Harry Jeremy, Larry Bunbury, Ed Jackman, Ryan Krause, Mrs. Marie Macnab, Jack Macnab, Ron Linsdell, F.M. Pherrill, Wm. McPherson, Charlie Jackson
Back Row Standing (left to right) – Bill Bloomberg, Reg Loomis, Herb Hall, Mrs. Eric Huestis, Eric Huestis, Jack Janssen, Vic Heath, Norm Lind, Stan Hughes, Joe Bracke, Mrs. Jean Holmes, Jack Holmes, Don Keith, C.J. (Charlie) Curran, Jack MacGregor, L. (Donnie) Donovan, P.C. (Pete) Comeau, Dirdre Lizotte, Bill Lizotte, Sheilah Walker, Jack Grant, Gladys Clarke, Bill Brackman
Wells Photo

Moving camp gear, with a month's food order, along the Athabasca River, north of Fort McKay area, Lac La Biche Division, 1954. Peter Murphy in back, Bill McPhail sitting in middle, Jack Robson laying down. Peter Murphy recalls that "we were a 5-man forest survey crew – two cruisers (Paul Dworshak and me), two compassmen (Bill McPhail and Jack Robson) and a cook (Al Baisley). We drove to Lac La Biche, stayed overnight at the AFS bunkhouse and caught the Northern Alberta Railway Muskeg Flyer to Waterways, an overnight trip. We spent a couple of days at McMurray fixing up the canoes – three 18-foot Chestnut freighter canoes – and got a grub order. Ranger Mike Gagnon took us in the AFS Scow to Fort McKay and on to Bitumount where we started the inventory cruise. We got supplies once a month and would check with trappers along the way to get supplies for them while we were going. It was a neighbourly but spread out community along the river." The season generally went from the May long weekend until early September
Bill McPhail photo; Peter Murphy story

Forest survey camp on the Athabasca River, Lac La Biche Division, 1954. Camp was at Poplar Point - cruiser Peter Murphy in the foreground, cook Al Baisley in back. Peter recalls that the dog "just wandered into camp, we never found out where from." The dog was left with Ranger Lorne Lapp in Fort McKay
Bill McPhail photo; Peter Murphy story

Ranger August Gatske, in truck, with standby crew outside the Fort McMurray Ranger Station, Lac La Biche Division, 1956
Alberta Government, AFHPC

Firefighters and cat skinners enjoy breakfast at fire camp, Whitecourt Division, 1956. Crew was working on Fire 36-1-56
Alberta Government, AFHPC

Cat skinners taking a break from dozer guard construction on Fire 36-1-56, Whitecourt Division, 1956
Provincial Archives of Alberta, PA2786-3

(L to R): Rangers Bert Prowse, Angus Crawford and Charles Clark at Mile 10 Cabin, Brazeau-Athabasca Forest, February, 1958
Charles Clark

Lands and Forests fire prevention display at the Hudson's Bay Company store, Edmonton, June 20, 1957
Alberta Government, AFHPC

Bud Klumph (centre) crossing the Athabasca River, Fire 36-1-56, Whitecourt Division, 1956. Scotty Wagoner is operating the 34-foot forestry boat
Alberta Government, AFHPC

Forestry staff meeting in Edmonton, 1958
Side and Back Rows (L to R): Not Identified, Stella McCreedy, Alf Longworth, Peter Murphy (standing), Ted Keats, John Hogan, Lou Babcock, Bob Steele, Ray Smuland, Buck Rogers, Larry Gauthier (head down), Rex Winn, Reg Loomis (face covered), Gordon Smart, Nick Sosukiewicz, Jack Macnab (standing), Not Identified, Not Identified. Front two rows: Dick Radke (reaching for paper), John Schalkwyk, Mike Lalor, Doug Lyons, Neil Gilliat (putting pen in pocket), Jary Prokopchik, Jack MacGregor (over left shoulder of Jary), Not Identifed (talking to Neil), Art Lambeth, Not Identified, Not Identified
Alberta Government, AFHPC

Class of 1959, Forestry Training School. This was the last class held at KFES before the training school moved to Hinton

Back Row (L to R): Jim Andersen (Whitecourt), Jens Kristensen (Grande Prairie), Don Harvie (McMurray), Irv Allen (Manning), Gordon DeGrace (Edmonton – visiting instructor), Bugs Ross (Nordegg), Jack Macnab (FTS – Instructor 2 i/c), Vic Wilson (Hythe), Pete Murhpy (FTS – Instructor i/c), Sam Sinclair (Slave Lake), Ron Fytche (Edmonton – visiting instructor), Bill Mitchell (Medicine Lodge), Jack Brock (Steen River), Art Peter (Lodgepole), Phil Stoley (Grande Prairie). Front Row (kneeling L to R): Eric Seyl (Fort McKay), Al Werner (Slave Lake), Bruce Johnson (High Prairie), Don Lowe (Rock Lake), Hylo McDonald (McLennan), Maurice Verhaeghe (Red Deer RS), Jack Plews (Fort Chipewyan), Bud Sloan (Sunset House), Frank Jones (Canmore)

Alberta Government, AFHPC

First bridge constructed across Tony Creek on the Tony Tower forestry road, west of Fox Creek, Whitecourt Forest, 1959

Gary Giese

(L to R): Rangers Charles Clark, Angus Crawford and Dick Berghout relaxing at the Gregg Cabin, Brazeau-Athabasca Forest, January, 1958. The rangers were relaxing after a hard day of rolling up the old telephone line from Gregg Cabin to the Yellowhead Cabin

Charles Clark

Forestry pile driver being used for the construction of the first bridge over Tony Creek (west of Fox Creek) on the Tony Tower forestry road. Pile driver was built by AFS mechanic Ray Kimzey. Raymo Quaranta and Allan Wagnar operating the pile driver, Whitecourt Forest, 1959

Gary Giese

Ranger meeting in Slave Lake, late 1956
(L to R): Ted Greening, Pat Foley, Phil Nichols, Buck Rogers, Jack Macnab, Des Woodman, Arnold Lingberg (Radio Operator), Howard Morigeau, Elmer Johnson, Fred Riley

Lou Foley

Forest Survey crew member Cliff Smith receiving a haircut from crew mate John Toth, Slave Lake Forest, Flattop Tower, 1959
Cliff Smith

(L to R) Rangers Wayne Cole, Harry Kostyk and Pat Foley, Slave Lake Forest, December 10, 1959. Rangers were at a forest products check station at the Clyde Corner, east of Westlock
Cliff Smith

Rypien Brothers sawmill, south west of Calling Lake, Lac La Biche Forest, 1950s. One job of a ranger was to inspect the timber berths issued by the government to ensure that the operators were fully utilizing logged timber and managing the risk of fire from both the bush and mill operations
Rypien Family

Rangers being trained for lowering to the ground by sky-genie, a winch attached to a helicopter. The rangers were able to clear landing pads in forested areas for helicopter access – tower or lookout sites, fires, etc. Ranger has a power saw attached on a rope. Helicopter used in this training is the Alberta Forest Service Bell 47AJ, CF-AFK, early 1960s
Alberta Government, AFHPC

Forest Survey crew, Whitecourt Forest, summer, 1960
(L to R): Frank Hodge, Not Identified, Malcolm Broatch, Not Identified, Cliff Smith
Cliff Smith

Forest Survey crew, Whitecourt Forest, summer, 1960
(L to R): Ed MacDonald, Cliff Smith, Not Identified, Fermen L'Hirondelle, Gilles Lessard (possibly)
Cliff Smith

Alberta Forest Service Forest Superintendents, late 1950s or early 1960s
Back Row (L to R): Bert Coast (Lac La Biche), Hank Ryhanen (Edson), Bob Steele (Clearwater), Larry Gauthier (Peace River). Front Row: Rein Krause (Whitecourt), Ray Smuland (Grande Prairie), Jack MacGregor (Slave Lake)
Alberta Government, AFHPC

Doug Lyons and Gordon DeGrace at their forest management survey camp, early 1960s
Alberta Government, AFHPC

Staff members of the Public Lands Division of the Alberta Department of Lands and Forests were brought together in Edmonton to discuss general and particular aspects of their duties, 1960. Chaired by A.D. (Art) Paul, supervisor of land classification, the three-day meeting was attended by V.A. Wood, the Department's Director of Public Lands, and E.P Shaver, his assistant
Seated (L to R): C.W. Harke, N. Kufel, Mr. Paul, Dr. Wood and Mr. Shaver. Standing: C.E. Paquin, J.A. Campbell, R.A. Wroe, D.I. Peters, R. Nieberding, S.E. Carter, R.N. Ireland, J.B. Milne, T.J. Gorman, F. Breyenton, W.T. Galliver, L.S. Yule, S.G. (Bud) Klumph, H.A. Brick, L.M. Forbes and C.D. Sawyer
Alberta Government, AFHPC

Forestry Training School Ranger Class – with the Fish and Wildlife component, Forestry Training School, Hinton, 1960
Back Row (L to R): Len Allen, Ken Wheat, Murray McDonald, Dick Russell, Bill Sanregret, Ernie Stroebel, Ray Sloan, Cyril Lanctot, Maurice Mitchell (Clerk). Front Row: Jack Gosney, Doug Jackson (F&W Lethbridge), Frank DeWindt, Dan Jenkins, Bob Woodward (F&W Edson), Chuck Rattliff, Wayne Markle, Chuck Geale (Valleyview), Garry Giese, Ed Larsen (F&W Athabasca), Jerry Kleinhout, Carl Leary, Jerry Tranter, Steve Zacharuk, Carl Roscovich (F&W St. Paul), Don Caldwell (F&W Pincher Creek), Terry Whiteley, Dennis Howells
Alberta Government, AFHPC

Forestry Training School Ranger Class, with Foresters – taking the Fish and Wildlife component, Forestry Training School, Hinton, 1961

Back Row (L to R): Don Fregren, Edson; Frank Brown, Edson; Joe Rickert, Slave Lake; Jim Young, Bow River; Vic Hume, Clearwater; Carl Ducommun, Bow River; Horst Rohde, Peace River; Jim Monroe, Edson. Front Row Standing: George Ontkean, Clearwater; Laurence Johns, Peace River; Fred Schroeder, Crow; Max Stanchfield, Lac La Biche; Ken Hennig, Grande Prairie; Frank Smeele, Slave Lake; Ken Olson, Lac La Biche; Edo Nyland, Whitecourt; Ogden Cole, Clearwater. Seated in Chairs (L to R): Neil Rutt, Slave Lake; Jack Robson, Grande Prairie; Harold Ganske, Crow; Stan Clarke, Grande Prairie; Fred Neumann, Edson. Seated on Floor: Norm Rodseth, Grande Prairie; Bill Davies, Whitecourt; George Trachul, Crow; Murray Doherty, Peace River

Alberta Government, AFHPC

Lac La Biche Forest Superintendent W.E. (Bert) Coast patrolling the Athabasca River downstream from the Calling River, Lac La Biche Forest, 1961. The boat was called the M.V. Calling River, or "Inchworm"

Ernie Stroebel

Honourable Norman Willmore, Minister of Lands and Forests, accepts the provincial Junior Forest Warden charter at the official presentation ceremony, March, 1962

(L to R): W.H. Myring, Vancouver, Secretary-Manager of the Canadian Forestry Association and Chief Warden; Dr. G.B. Griffith, National President presenting the charter; Terry Whiteley, Alberta Chief Warden, and Minister Willmore. Presentation of the provincial charter marked the affiliation of the Alberta JFW with the national Junior Warden association.

Alberta Government, AFHPC

Joe 'Lucky' Lieskovsky inspecting one of the signs he made in the Forestry paint shop, Edmonton, early 1960s

Provincial Archives of Alberta, PA3467

Joe 'Lucky' Lieskovsky in the Forestry paint shop where he was responsible for creating and making signs for the department, early 1960s
Provincial Archives of Alberta, PA2728-1

Forestry Weather Course, Forestry Training School, Hinton, February, 1962
Staff were Chief Rangers in their respective Districts. Back Row (L to R): Charles Clark, McMurray; Bill Kostiuk, Edson; Jim Hereford, Grande Prairie; Joe Kirkpatrick, Slave Lake; Art Lambeth, Edson; Del Hereford, Blairmore; Harry Edgecombe, High Level. Front Row: Jack Macnab – FTS (Instructor 2i/c); August Gatzke, Lac La Biche; Bernie Brouwer, Peace River; Dick Mackie, Calgary; Frank Jones, Rocky Mountain House; Ben Shantz, Rocky Mountain House; Peter Murphy – FTS (Instructor i/c). Front Row: Mike Burke, Whitecourt; Ted Boodle, Slave Lake. Missing: Aircraft Dispatchers Pat Donnely and Jack Grant
Alberta Government, AFHPC

Forestry Training School Ranger Class, Hinton, 1962
(L to R): Jack Macnab – FTS (Instructor 2i/c); Joe Burritt, Bow River; Francis Schenk, Grande Prairie; Melvin Tessmer, Rocky Mountain House; Laverne Larson, Slave Lake; Andy Kostiuk, Lac La Biche; Wayne Cole, Slave Lake; Larry Derbyshire, Whitecourt; Wayne Robinson, Grande Prairie; Bill Wuth, Lac La Biche; Lou Blasius, Lac La Biche; Ernie Duchesne, Peace River; Stan Fischer, Peace River; Gordon Japp, Edson; Albert Cauchie, Crowsnest; Peter Klymchuk, Peace River; Herb Walker, Slave Lake; Gordon Matthews, Bow River; Bob Richmond, Edson; Conrad Bello, Clearwater; Larry Kennedy – FTS (Acting Head); Roman Bizon, Whitecourt
Alberta Government, AFHPC

Basic Ranger Course, Forestry Training School, Hinton, 1963
Back Row (L to R): Frank Nuspel, Leonard Kennedy, Maurice Bolduc, Ken South, Len Westhaver, Mike Dubina, Larry Huberdeau. Front Row: Morris Mitchell, Owen Bolster, Archie Miller, Dave Dodds, John Stepaniuk, Gordon Bossenberry, Lorne Goff, Roger Olson, Art Giroux, Ray Dubak, Dave Brown, Al Needham, Harry Sondergard
Alberta Government, AFHPC

Launch of a forestry tug boat into the Clearwater River, Fort McMurray, Athabasca Forest, mid-1960s. This boat was built in Edmonton and transported by truck to Fort McMurray. Ranger watching could possibly be Lou Babcock, Superintendent Athabasca Forest
Alberta Government, AFHPC

Rangers John Stepaniuk (L) and Owen Bolster with Instructor Rocky Hales at right, Forestry Training School, Hinton, 1963
Alberta Government, AFHPC

Forestry tugs and barges were the main mode of transportation for hauling materials and supplies from Fort McMurray north and south. Photo is the SS Athabascan and two barges on the Clearwater River, Fort McMurray, Athabasca Forest, mid-1960s
Neil Gilliat

Refresher Course, Forestry Training School, Hinton, 1963
Back Row (L to R): John Elliott, Colin Campbell, Gordon Campbell, Ed Beebe, Harry Jeremy, Bert Hadley, Don Dawson. Middle Row: Gerry Stuart, Mike Gagnon, Dick Girardi, Phil Nichols, Des Woodman, Bert Varty, Howard Morigeau, Eric Dawson, George Deans. Front Row: Joe Passamare, Ken Janigo, John Holden, Don Crawford, Hy Baker
Alberta Government, AFHPC

Refresher Course, Forestry Training School, Hinton, 1963
Edson Rangers (L to R): Gordon Campbell, Eric Dawson, Joe Passamare, Don Crawford
Alberta Government, AFHPC

Refresher Course, Forestry Training School, Hinton, 1963
Rocky Mountain House Rangers (L to R): Ken Janigo, John Elliott, Gerry Stuart, Des Woodman
Alberta Government, AFHPC

Forest Officers Art Peter (L) and Harold Enfield, fall, 1963
Alberta Government, AFHPC

Refresher Course, Forestry Training School, Hinton, 1963
Slave Lake Rangers (L to R): Harry Jeremy, John Holden, Howard Morigeau, Phil Nichols
Alberta Government, AFHPC

Western Land Directors' meeting, Natural Resources Building (now the Bowker Building), Edmonton, 1963
Back Row (L to R): A. D. (Art) Paul, Alberta Public Lands (field staff); Cyril B. Kenway, Registrar, Alberta Lands and Forests; Nick Kufel, Head, Alberta's Lands Branch Field Staff; S. G. (Bud) Klumph, Grazing Reserve Supervisor, Alberta Lands Branch; and C. E. (Charlie) Paquin, Kufel's assistant (Lands), later to become ADM, Public Lands (1973-1981). Centre Row: Vi A. Wood, Alberta's Director of Lands; J. A. (Scotty) Campbell, Alberta's Grazing Appraiser; Gordon M. Smart and R. D. (Reg) Loomis, Alberta Forest Service. Front Row: A. J. LeBlanc, Saskatchewan Lands Branch; A.M. (Art) Thomson, Saskatchewan Director of Lands; J. "Arni" Barr, Manitoba Lands Branch; David Borthwick, B.C. Superintendent of Lands; and R.(Bert) Torrance, B.C. Ass't Deputy Minister of Lands and Forests
Provincial Archives of Alberta, PA2751

Refresher Course, Forestry Training School, Hinton, 1964
Back Row (L to R): Bob Lewis, Stan Carlson, Vic Fischer, Ben Abel, Keith Thompson, Johnny Johnson, Oliver Glanfield. Middle Row: Al Walker, Harold Enfield, Ron Lyle, Karl Altschwager, Dave Schenk, Glen Sloan, Wilf (Jock) Kay, Vic Schneidmiller. Front Row: Doug Allen, Ray Hill, Ted Loblaw, Fred Facco, Mike Burke, Bill McPhail. Instructors at Front: Jack Macnab, Peter Murphy
Alberta Government, AFHPC

Lac La Biche Forest Division Spring Ranger Meeting, 1964
Back Row (L to R): Jack Kreutzer (Anzac), Bill Wuth, Ken Kolodychuk (LLB Warehouse), Not Identified, Pat Leibel (Fort McMurray), Not Identified, Gary Pollock (LLB Fish & Game), Owen Bolster, Ed Johnson (Fort MacKay), Len Blasius (Wandering River). Middle Row: Len Swatsky (LLB Warehouse), Al Needham (Anzac), Lawrence Johns (Fort McMurray), Not Identified, Ken Olson (Fort McMurray), Stan Olzowka (Fort Chip), Frank Nuspel, Oliver Glanfield (Lac La Biche), Ernie Stroebel (Calling Lake). Front Row: Lawrence Yanik (Fort Chip), August Gatske (LLB Chief Ranger), Jack Williams (LLB Administration), John Booker (LLB Assistant Superintendent), Bert Coast (Superintendent), Bert Varty (Assistant Superintendent), Not Identified, Len Allen (Fort McMurray Chief Ranger), Eric Syles (Fort McKay)
Alberta Government, AFHPC

Arsenault Sawmill in High Level, December, 1964. Note grain elevators in picture
Cliff Smith

Basic Ranger Course, Forestry Training School, Hinton, September to November, 1964

Front Row (sitting L to R): Jurgen Moll, Keg River; Harold Dunlop, Swan Hills; Gerald Armfelt, Little Red River; Ken Porter, Calling Lake; Ron Sears, Meander River. Middle Row: Ross Purves, Steen River; Mag Steiestol, Grovedale; Barry Nelson, Hines Creek; Ken Kolodychuk, Embarras Portage; Fred Thorn, Salt Prairie; Jack Kreutzer, Anzac; Gordon Baron, Wandering River; Nick Galon, Conklin. Back Row: Ken Gatzke, Slave Lake; Ray Howarth, Lac La Biche; Bill Francis, Spirit River; Sandy Donaldson, Kinuso; Ed Pulleyblank, Fort McKay; Ed Johnson, Slave Lake; Joe Smith, Muskeg

Alberta Government, AFHPC

Basic Ranger Course, Forestry Training School, Hinton, 1965
Back Row (L to R): Denis Loiseau, Don Law, Frank Lightbound, Ray Kover, Dave Chabillon, Don Dawson, Harvey Megli, Russell Verhaeghe, Ed Dechant.
Front Row: Pete Nortcliffe, Ian Methuen, Francis Zboya, Adolph Porcina, Denis Sanregret, Al Gehman, Dick Seaman, Gerald Labrie, Harold Evanson, Hank Louwerse, Bob Yates

Alberta Government, AFHPC

Ted Hammer (standing) and Frank Platt (in uniform leaning over map) with Beaver aircraft, Slave Lake Forest, 1965
Neil Gilliat

1948 - 1965

Athabasca Forest Spring Ranger Meeting, 1965
Back Row (L to R): Bruno Farro (Boat Cook), Joe Brodoski (Standby Crew), Alvin Scott (Deckhand on Boats), Armando Tedesco (Standby Crew Foreman), Carl Brakstad (Forest Carpenter), Dan Law (Standby Crew), Cliff Henderson (1st Athabasca Forester) (hidden), Not Identified, Mickey Patterson (Standby Crew). Front Row: Bert Varty (Assistant Forest Superintendent), Carol Henson (Secretary), Lou Babcock (Forest Superintendent – 1st), Bob Steele (Director of Forestry), Stan Hughes (Chief Timber Inspector), Len Allen (FO III), Jerry Karasinski (Boat Captain)
Alberta Government, AFHPC

Forestry water truck with hose and equipment boxes attached to side, Edson Forest, 1965
Alberta Government, AFHPC

Forestry Trunk Road near the Willow Creek Ranger Station, Crow Forest, mid-1960s
Alberta Government, AFHPC

Reg Loomis, Assistant Director of Forestry, inspecting seedlings planted at the Three Creeks Arboretum, Peace River Forest, 1965
Cliff Smith

Canadian Institute of Forestry host team meeting, Banff, 1966
(L to R): Bob Steele, AFS Director of Forestry; Peter Murphy, AFS Head Training Branch; Doug Lyons, AFS Forest Survey Branch; Pat Duffy, CFS Soil Scientist, Chair of Host Committee; Mike Drinkwater, CFS Program Manager, Rocky Mountain Section Chair; Iris Steele, Chair Women's events; Stan Hughes, AFS Head Forest Protection; Bob Stevenson, CFS Entomologist; John Hogan, AFS Head Forest Surveys; Phil Thomas, CFS Head Northern Forestry Centre; Ron Fytche, AFS Forest Surveys; Joe Baranyay, CFS Forest Pathologist

Phil Debnam, CFS Photographer

Ranger John Holden showing homemade snowshoes and birch bark canoe, two important means of conveyance, Wabasca, Slave Lake Forest, 1965
Alberta Government, AFHPC

Ranger Don Lowe changing tire on Nodwell, Eagles Nest area, Edson Forest, c. 1961. Billy Magee, Predator Control Officer and Johnny Bader, Assistant Ranger fixing bait trap for wolves for the rabies control program
Don Lowe

Sawmill in Salt Prairie District, north of High Prairie, Slave Lake Forest, 1963. A ranger's duties included inspecting sawmills and bush operations to ensure operations met harvesting and utilization standards, wood was sawn to specifications and all operations met fire prevention and control requirements
Don Lowe

Ranger Phil Nichols (plaid jacket) on mill inspection, Salt Prairie District, Slave Lake Forest, 1963
Don Lowe

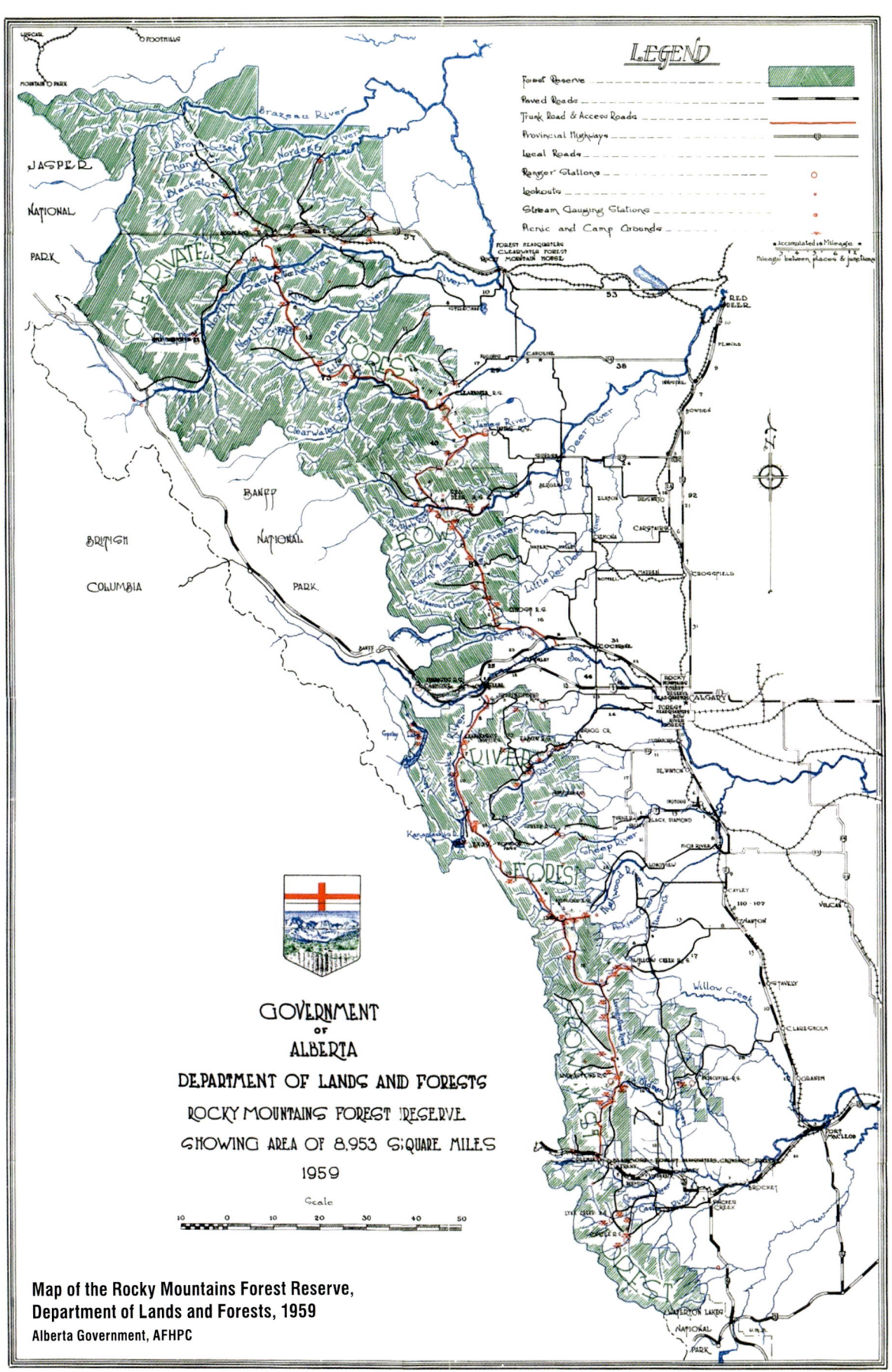

Map of the Rocky Mountains Forest Reserve, Department of Lands and Forests, 1959
Alberta Government, AFHPC

CHAPTER 5

Lookouts and Communications

Forester Harvey Reginald Macmillan, who went on to found the successful Macmillan-Bloedel forest company, began his career with the Dominion Forestry Branch (DFB) in 1908. After studying fire problems on the forest reserves in Alberta, he neatly summarized the essence of a fire control system: "The measures adopted to protect the forests from fire are now generally understood. They are the removal by education or legislation adequately enforced of the causes of fires, the organization of a patrol to find and extinguish such fires which will inevitably start, and the improvement and organization of the forest areas so as to render most efficient the efforts of firefighters and to minimize the chances of any fires getting beyond control."[1]

Given the limitations of prevention, early detection was a priority in the quest to combat forest fires from the beginning of both the DFB and Alberta Forest Service (AFS). Patrols on horseback were initiated even before the Forest Reserves were set up. The intent was first to teach travellers about fire prevention and to fight any fires that came to their attention. There was no early attempt to build lookouts, since there was no way to communicate and report any fires that might be seen.

In northern Alberta, detection started out as little more than reporting of a fire by the public to a Fire Ranger, or the discovery, investigation and suppression of a fire by the Fire Ranger himself. As a result, only fires along the routes of travel, either horse and wagon trails, rivers and later railways, were reported and actioned. Usually these were human-caused fires. Lightning-caused fires were rarely actioned unless they threatened life or property. The structured detection system thus began its existence along regular routes of travel, punctuated by any high points in the terrain that provided a good view of the surrounding countryside. Ingenious rangers increased their capacity for smoke detection in high incidence areas with the construction of 'crawl tree' structures, ladder rungs or steps made from poles attached to two trees in close proximity.

A lack of communications technology and problems of geographical access posed some serious frustrations for early forest officers. The Annual Report for 1932-33 stated "In the northern district . . . there is a distinct lack of those improvements or developments so essential to successful forest protection. There are no trails, roads, telephone or telegraph lines built specially to answer the purposes of forest protection and administration. Fire detection except by the old method of patrol is non-existent and equipment is insufficient."[2] The "northern district" was described as all those areas outside the Forest Reserves.

Eric Huestis described the limitations of detection and initial attack during the 1920s and

Goose Mountain Tower (built 1939-40), Slave Lake Forest. The stumps are from trees felled to increase visibility for the observer
Mel Willis

early 1930s by quoting the words of the timber inspector at Carrot Creek: "I had a team of horses and a democrat buggy. I would get word by CN telegram that there was a fire down at Breton or Winfield. I would hook up the team and away we'd go and four days later I'd arrive. Either the fire was all over hell's half acre or it was out. So then I'd get on the phone or go to the nearest town and phone in and they would tell me there's another fire up near Edson. So I'd head back and the same thing would happen - either it was all over hell's half acre or it was out, depending on conditions. It was a hopeless proposition to do anything as far as fire protection was concerned, with only about a dozen rangers in the whole north country."[3]

trees. Wooden towers made from local logs and rough sawn timber later served the purpose. The first lookouts were initially located close to the ranger station or close to fire trails cut by those who patrolled the districts. They were linked to the ranger station by telephone line, the wires usually strung along the road or trail. It was not until the simple single-wire ground-return telephone became available in the early 1910s that these lookouts were built. Then, over time, high hills or mountain tops in suitable locations and with a good view of frequently-traveled routes, settlements or ranching areas were selected as sites for the towers. A deciding factor in the choice of location was the feasibility of constructing pack trails for construction access, maintenance and regular servicing trips.

The ranger was expected to put in considerable effort and time to establish these horse trails. He was also expected to patrol and maintain the telephone lines that linked the lookouts to the ranger station. When fire hazards allowed, lookout men also helped to patrol and maintain the telephone lines.

John Currat talked about building the first Heart Lake Tower in the early 1940s with fellow ranger Peter Comeau. This was a wintertime project. They were both skilled axemen. After they cleared the site and built a log cabin, they felled four of the best spruce they could find so they could build

Trail below summit of Burke Mountain-Cameron Lookout, Bow River Forest, Rocky Mountains Forest Reserve, 1929
Dominion Forestry Branch, AFHPC

Good visibility and keen eyesight were essential: Pigeon Lookout, Barrier Mountain. This structure previously was a guard tower at the prisoner of war camp at nearby Barrier Lake, Kananaskis Forest Experiment Station
Alberta Government, AFHPC

First sites and structures

A system of facilities that could be used by observers to detect fire evolved into two kinds of structures – a lookout on a high open point, built on the ground or on short stilts, or a tower designed to place the observer above obstacles to vision. The first constructed lookouts were simple ladder-like structures fastened to two

Lookouts and Communications

a sixty-foot tower and skidded them up behind a team and sleigh. They hewed them square and tapered the logs to about a five-inch top. John lamented that they were putting up the fourth one when the rope slipped, the log dropped and it broke. So they had to go back down and do it all over again. However, what really pleased him was when they put the side rounds on and got them bolted into place – he said it looked like the Eiffel Tower because of the way the legs swept in to the top.

As with all new endeavours, some poor decisions in creating the lookout system had to be corrected. For example the highest mountaintop did not necessarily provide the best visibility as frequent cloud cover or haze could interfere with visibility. Detecting or confirming a smoke by providing a triangulating "cross shot" to pinpoint its location was impossible in such circumstances.

As settlements extended into the northern portions of the province, so did the forest rangers. There were high hills in key locations but these were covered with trees, some over 80 feet in height. This prompted the introduction of high steel towers to enable the observer to look over the trees.

Initially these sites were chosen by rangers who knew their district and the hills that provided maximum visibility. Rangers climbed a lot of trees trying to find which hills provided the best coverage. Once aerial photography became common, contour maps were drawn for proposed sites. From these maps a "visible area" around each vantage point was calculated and optimum sites were chosen to provide the most complete detection network possible.

Helicopters provided another useful means of choosing and verifying potential sites, especially in the northern forests. Frank Platt developed an aerial survey process in the 1950s and 1960s. Joe Niederleitner, head of the Planning Section with Forest Protection, later played a key role in checking these sites with helicopters to evaluate and compare visible area mapping with actual on-site inspection. The helicopter hovered 100 feet above the ground to simulate the height of the tower and panoramic photographs were taken in a 360-degree arc. These photos would be checked against the visible area map for errors in coverage calculations and the best of the possible sites in the locality was chosen. [4]

A vertical photo was also taken from 10,000 feet. For anyone who had never ridden in a small helicopter such as a Bell 47AJ2 this was a memorable experience. Often the District Ranger, in whose district the proposed new tower was to be built, was taken along. A helicopter ride at tree level was old hat but one at 10,000 feet quickly revealed those with an anxiety for heights. References to the horizon are much more difficult than at lower levels and thus the pilot had to be more alert in keeping the helicopter level. Some said this was akin to balancing oneself atop a flagpole.

The visible area of a lookout or tower varied with the terrain but as a standard it was set to encompass an area with a radius of 25 miles. This 50-mile wide circle, encompassing 1,941 sq miles, is only slightly less than the area of the province of Prince Edward Island with 2,184 sq miles. Alberta's lookout network originally called for a lookout

McLennan Crawl Tower, 1965. This native timber structure enabled the ranger to 'crawl' up for a look around his area
Ken McCrae

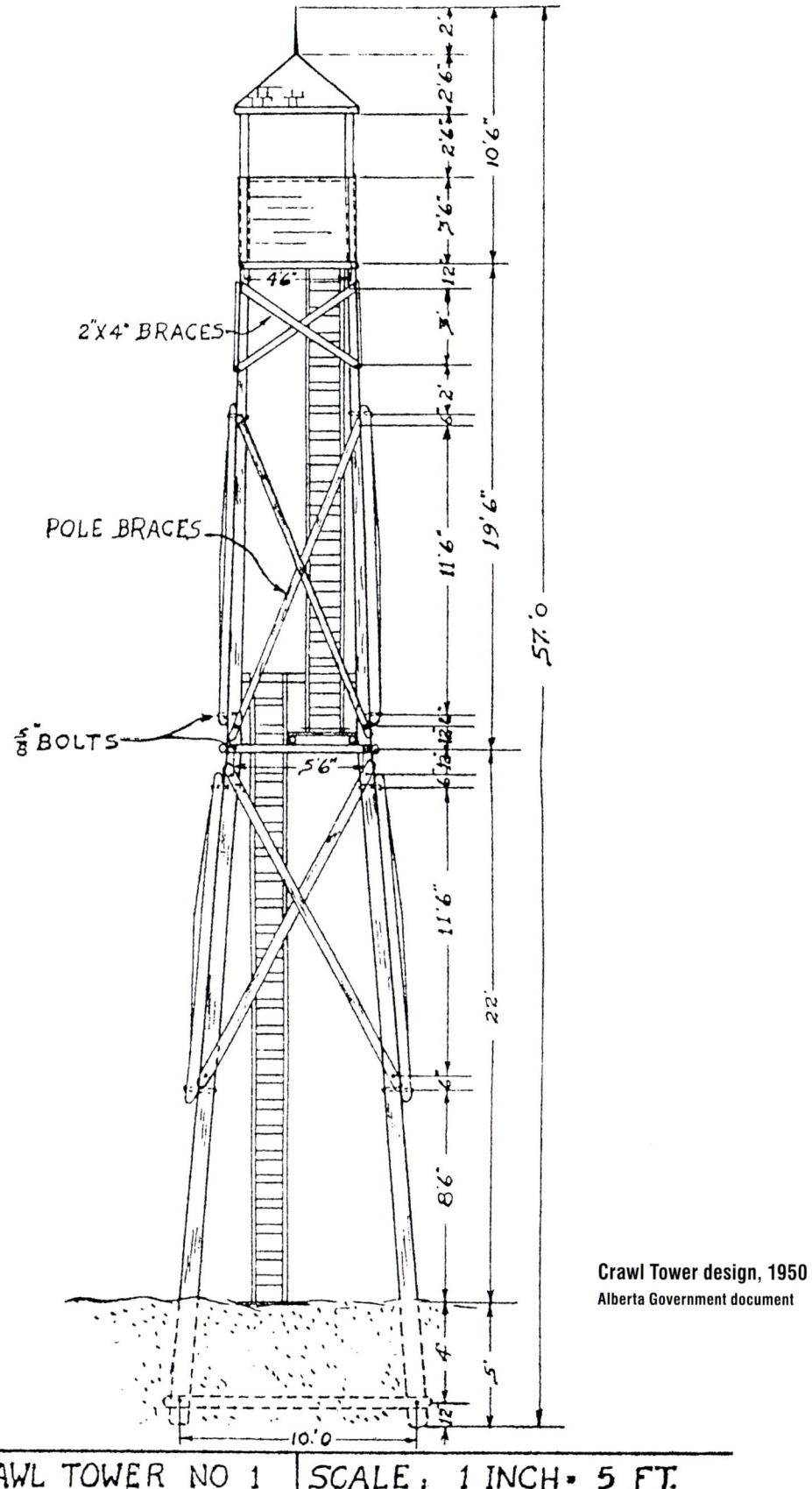

Crawl Tower design, 1950
Alberta Government document

spacing of 20 miles in alpine and sub-alpine regions and 40 miles in the northern boreal based on studies by Americans George M. Byram and G.M. Jamieson. Joe Niederleitner outlined in his 1984 Fire Detection Study that the decision made in the late 1970s was to use a 25-mile spacing. The distance of 25 miles was considered the maximum distance an observer could reasonably be expected to spot a fire the size of a campfire with a smoke column rising approximately 100 feet into the air. This of course varied depending upon wind speed dispersing the smoke, colour of the smoke relative to the background and whether sighting with or against the sun. The objective of the lookout network was to have all forest areas covered within 25 miles of at least one lookout.

Detection

The towers and lookouts spaced throughout the province are known collectively as Alberta's fixed detection assets. All other fire detection systems not permanently attached to one specific site are classed as mobile detection assets.

The first mandate of fixed detection was, and still is, the detection and reporting of fires.

Only people with good vision were hired. A simple eye test was introduced around 1956 for use during job interviews and training. The test involved a small black dot on one quadrant of a white piece of paper. The paper was spun and after it had stopped spinning, the applicant had to identify the quadrant position of the dot from a certain distance. Failure to do this meant that the candidate was asked to get a proper eye examination and glasses if necessary. Studies in the 1940s and again in the 1960s led to a variety of different tests for lookout personnel. Some tested the ability to spot objects in various light conditions, while others tested the maximum distance that a smoke could be detected. Extensive tests were run using candidates from various backgrounds such as trapping, the military, academia and forestry, with the results showing that having a trained observer was more important than perfect eyesight. The eye exam was eventually phased out.

Some observers did indeed have excellent sight. On one occasion a lookout person on the old Battle River Tower site west of Manning reported an intermittent smoke in the late

Slave Lake Chief Ranger Len Allen coordinating wildfire resources in Lac La Biche during the May, 1968 fires. Under the column *State of Control*, O.C. means *Out of Control*
Alberta Government, AFHPC

afternoon. The ranger plotted the sighting on the office map and then proceeded to check the smoke out. The lookout person had the bearing and distance right on and after considerable searching and with darkness only an hour away, the ranger sat beside some construction equipment that was working on building the Mackenzie Highway. It was then that he noticed that each time the "cat skinner" opened the throttle a cloud of diesel smoke billowed out because the diesel injectors of a crawler tractor were slightly out of adjustment. By talking on the radio the ranger could confirm, before the lookout person could see it, when each intermittent puff of smoke would occur.

Clyde Ulm set a record in the late 1950s when he spotted a smoke 110 miles from Marten Mountain and located it within a quarter section. Lookouts prided themselves on their knowledge of the country and ability to tell the ranger where the smokes were. Needless to say these were the types of "eyes" the AFS depended upon.

Not all first-time observers were suited to the loneliness of a lookout person's life. It was often difficult to determine from an interview who would be susceptible to loneliness and who would enjoy the solitude. One unorthodox method was to look at the hands of the person being interviewed. If the person wore a lot of rings or other jewellery, the thinking at the time was that individual would not last long on a tower – because there'd be no one there to admire the decoration.

Trappers were among the best observers. They were accustomed to living alone and they enjoyed the lifestyle. Usually there was just enough time for them to move off the trapline, sell their furs, then report to the ranger station for the fire season each spring. Trappers often looked out over the very area where they trapped in winter. Their livelihood depended in more ways than one on the quick detection and suppression of forest fires.

In recent years this has changed and today artists and writers are often found on lookouts and towers, taking advantage of quiet times to paint, write or study in their chosen disciplines. One of them noted that he learned to play the bagpipes without disturbing a soul.

After smoke detection, communication was the second priority of the detection network. In the early days the telephone line was the sole and critical means of communication. Later the telephone was replaced by single side band (SSB) radios, and the monitoring and relaying of messages, especially during times of poor reception, were constant chores for the observer. This duty became much less important as VHF (very high frequency) radios came into use and radio repeater sites were installed. Communication was then relegated to third priority, with weather observation becoming second priority.

During the Second World War, lookouts were required to employ a code when submitting their

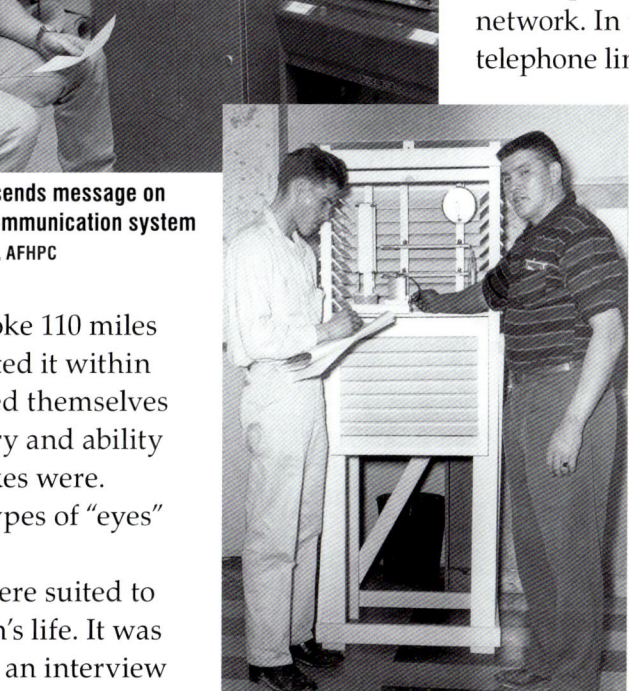

John McQueen sends message on headquarters communication system
Alberta Government, AFHPC

A training session on the Stevenson Screen and weather instruments, 1964. Doug Doerkson on left
Alberta Government, AFHPC

Medal given to tower personnel for reporting aircraft movements during the Cold War and into the 1960s
Tim Klein

weather readings, lest the information be used by enemy agents. The code was fairly simple, but its use required extra time for the sender and the receiver.

Provincial Detection Technician Tim Klein says that during the Second World War, when the Allies were assisting Russia, radio relay sites were set up to monitor and assist aircraft being moved from North America to Russia through Alberta and along the Northwest Staging Route.

"Some of our lookouts were used for these purposes. One was Whitecourt, which had a staff of up to five people on a 24-hour basis. I do not have documentation on the duration of this operation, although one would assume that it was only for that period when the aircraft were being moved."

Klein says that an agreement was later struck between the federal government and the AFS to involve the lookout observer network in watching for hostile aircraft as part of the RCAF Ground Observer Corps (GObC). The GObC came to prominence with the Cold War in the early 1950s when the threat was that long-range Soviet bombers might head over the polar regions to hit targets in the U.S. "The RCAF supplied our staff with posters and recording pads (type and number of aircraft, heading, etc.) depicting silhouettes of Canadian, U.S. and Soviet bloc aircraft."[5]

The idea was that lookout personnel would report all low-flying aircraft heading in a southerly direction (90 to 270 degrees). The AFS and its lookout personnel took the job seriously. GObC instruction became an integral part of the AFS lookout training courses. The RCAF sent their own staff to teach this topic and supplied a movie called "The Dangerous Mile," referring to the mile of atmosphere above the earth through which enemy aircraft might fly to escape radar detection. Ground Observer Corps badges were handed to observers and at the end of the program small GObC medals and certificates were given for services rendered.

Ted Blefgen, Director of Forestry, referred in his Annual Report for 1942-43 to several fires starting in remote areas without evidence of lightning or human activity. "The great number of aircraft flying in the northern part of the province was, perhaps, the cause of at least some of the fires fought in the outlying districts. Of nine fires that occurred in the

Construction of the Cline Lookout on the Upper Saskatchewan River west of Nordegg. This was the first time a helicopter was used to haul in all the materials needed. It saved many weeks of packing materials in by horse
Alberta Government, AFHPC

Radio operator Don Hutchins at Whitecourt, late 1940s
Alberta Government, AFHPC

Athabaska Valley, north and west of Whitecourt, seven were directly on the route between Edmonton and Fort St. John, British Columbia, and two of these were only a short distance on either side." Blefgen and staff postulated that these might have been started by cigar butts thrown out of the cockpit windows of those propeller-driven planes. The reasoning was that only a cigar butt would be big and heavy enough to stay ignited all the way down and drive through the canopy of trees, and primarily only U.S. military people smoked cigars."[6]

With the replacement of the telephone by radio, a power source was needed. At first, cumbersome batteries were used with the aid of wind-powered chargers to help extend battery life. Then, for dependability, gas-driven electrical generators were installed in an engine house at lookouts and towers. This was a boon to the observer as he now also had electrical lighting as opposed to the simple gas lantern. The system, however, was not 110-volt, so electrical appliances could still not be used. The AFS has continued to experiment with both wind and solar power.

Construction of steel tower. Prefabricated cupola to be placed on top
Alberta Government, AFHPC

Among the regular visitors to towers and lookouts were the radio technicians, another vital group within the AFS that tried to ensure the best possible communications. It was an ongoing challenge, given the great distances, topographical variation, the pioneering nature of radio in the early days and the inherent impulse of towermen to try to fix things themselves. Tony Earnshaw was the first radio superintendent. Starting in 1938, he experimented, tested and installed the prototype radios and developed the radio system. In 1962 he transferred to Alberta Government Telephones when AGT assumed responsibility for the system.

In the meantime, the technical work and installations were largely done by the radio technicians; people such as Gordon Fowlie who operated

Lookout Instructor Course, Forest Technology School, Hinton, 1969
Back Row (L to R): Myron Sterr, Harry Freeman, Len Allen, Colin Campbell, Rolly Jourdain, Floyd Schamber. Middle Row: Lou Boulet, Tom Stewart, Bert Varty, Jack Naylor, Lee Watson, Len Smith. Front Row: Bruce Byron, Emanuel Doll, Irv Frew, Cliff Henson, Not Identified
Alberta Government, AFHPC

a tower in 1941 and stayed on as a technician until he retired in 1976. One of Fowlie's first installations was at Heart Lake tower after John Currat and Pete Comeau built it in the mid-1940s. He recalled the long gruelling trip in, part of the way by team and wagon then back-packing to the top of Heart Mountain since the tower had been built in winter with only a sleigh track for access. Then there were the clouds of bugs during the month he was there to cut and raise antenna poles and complete the wiring. He recalled some of the pioneers: Bill Norton and Les Braunberger out of Calgary and Doonie Donovan and Ron Linsdell in Edmonton.[7]

For many years all weather information was sent to a central tower in Edmonton at 10322-146 Street, then on the western edge of the city. In 1950-51, the system was changed so that information was sent to Cooking Lake tower, for better reception. From there the information was relayed to Edmonton. These facilities were later dismantled and moved to 120 Avenue and 109 Street, and are now located in the Provincial Forest Fire Centre in Edmonton's downtown Forest Protection Division offices.

Lookouts also became important in keeping track of AFS aircraft activities in their observation area. The recording of the time for helicopters' "time up" in the morning and "time down" for the night were noted. Also, field crews in the tower area would use the lookout to relay and receive messages.

Training of lookout personnel began in 1956 when Frank Platt demonstrated the Osborne Firefinder and Doonie Donovan talked about radio and weather. Until then the ranger showed new lookouts the ropes when they packed in at spring opening. However, the ranger often had to hurry back to fight fires or seal beaver so the training was uneven.

Starting in 1957, Training Branch staff Peter Murphy and Jack Macnab developed a three-day course that they took to forest headquarters in March. Radio staff Donovan and Ron Linsdell instructed on radio and weather. The spring training was centralized at the Forestry Training School at Hinton in 1960, attracting 71 candidates from all over Alberta including Cypress Hills and Jasper National Park. The Hinton course was decentralized by the mid-1980s and held in local districts. Ten years later the course was extensively revamped and returned to Hinton. Local areas continue to run supplementary courses. The provincial course at Hinton is now mandatory for all new tower staff.

Luscar Lookout, Edson Forest – first tower built on stilts by the AFS, late 1940s
Alberta Government, AFHPC

Blue Hill Lookout on wooden stilts, west of Sundre. Note wiring for lightning diffusion
Alberta Government, AFHPC

Weather

Weather observations had long been a regular part of the lookout observer's duties, but correct procedures and accurate recordings grew in importance in the mid-1950s when the Fire Weather Index was developed by the Canadian Forest Service.

Recordings were taken at 0730, 1230 and 1830 hours, seven days a week. The noon reading was for fire danger rating, the morning and evening for the climate record. Readings included temperature, relative humidity, maximum and minimum temperatures, wind direction and speed and precipitation. By 1940, some towers had very basic weather instruments. Between 1947 and 1955, more weather equipment was purchased for towers and ranger stations.

When the collection of weather data first began, fuel moisture sticks were used. These were a set of four round half-inch diameter wooden dowels connected side by side and adjusted to weigh exactly 3.53 ounces when bone dry. These fuel moisture sticks were mounted on a wire frame about one foot above ground in a shaded area. There they would absorb moisture from the air or else dry out in low humidities. Any weight over 3.53 ounces represented the per cent of fuel moisture – the lower the weight, the dryer the fuels. The "Stevenson Screen" was a rectangular louvered "house" painted white to provide a shaded, ventilated and protected enclosure in which weather instruments were housed. The federal Meteorological Service supplied most of the weather instruments to support its climatological studies.

Some individuals, once the weather observations were calculated and recorded and they were waiting for the weather "Sked" to be called in, just had to be doing something with their hands. There's a humorous story that one of these pastimes would be to "whittle" on the fuel moisture sticks. Needless to say the "whittler" was quickly discovered by the Forest Headquarters' check of the data, prior to forwarding by teletype to Edmonton. A sudden loss of weight indicated one of only two things. Either the relative humidity at the site was very low and all the forest fuel was drying at a very rapid rate or there was a sudden change in the weight of the fuel moisture sticks.

Lookout observers learning to weigh fuel moisture sticks
Alberta Government, AFHPC

Recruits learn about the Osborne firefinder at training school, instructed by Jack Macnab (L), early 1960s
Alberta Government, AFHPC

Lookouts and Communications

Another story recounted the activities of a "keep it clean" person. This individual was always "sprucing" things up and the fuel moisture stick ended up getting painted. The Forest Headquarters' daily check of the weather data uncovered a sudden increase in the weight (by the added paint) of the fuel moisture sticks. If it was not pouring rain and the "sticks" (and the forest itself) taking on moisture then what was happening? A quick "QSO" (discussion on the radio) occurred for all to hear and learn from (much to the embarrassment of the perpetrator) and a new set of fuel moisture sticks was sent out pronto.

In the mountainous regions of the province the siting and construction of many of the early

saw 14 foot square cabins and the addition of a square second storey for observation. Later they incorporated the same cupola as is used on steel towers. The cupola is an octagonal structure with windows in the top half of all eight sides. These could be opened from the inside to provide unobstructed visibility, for cleaning or to allow increased air flow on hot days. The majority were made of fibreglass,[8] though some aluminum cupolas were also installed. During the late 1950s Edmonton Transit was developing fibreglass panels for buses to reduce costs of maintenance and repair. Frank Platt worked closely with them to successfully apply those techniques to cupola construction, later setting up a full-scale section in the AFS Fire Centre.

A crucial piece of equipment in the lookout was and is the Osborne firefinder. This is a sighting instrument mounted on a stand at the centre of the cupola. It is oriented to true north and provides azimuth directions for reporting either wildfires or unauthorized smokes to the Ranger District headquarters. The firefinder aids in identifying the location of permanent points within the visible

Towers and their cabins can be remote and isolated
Alberta Government, AFHPC

lookouts required considerable ingenuity and sheer determination on the part of both man and beast. They were required to haul all the building materials to remote and challenging rocky ridges and mountain peaks. The materials list included lumber, windows, doors and other building supplies, not to mention barrels for rain water for cooking and drinking, firewood for heating and cooking and kerosene for lighting.

The mountain lookouts were situated at or above the treeline and were cabins approximately 12 feet by 12 feet. Some of these cabins were mounted on short 12-foot legs. Later designs

Sam Fomuk typing reports to submit to headquarters, late 1970s
Dave Brown

area or other pertinent information that may aid the observer in estimating locations and distances when reporting smokes within his or her area of responsibility. Often mirrors were used, from fixed and known landmarks, to provide triangulation in orienting the firefinder to true north. The Annual Report of 1945-46 said: "Osborne firefinders are now standard equipment at all lookout towers, and their use has made fire locating considerably more accurate than before."

Lookout observer Sam Fomuk, who retired in 1994 said that before 1945-46; "Firefinders were rather crude – a plastic circle on a pin and a rotary metallic sighting slot on it – the Osbornes were a big advance. I have a note on file that says: 'A few years ago the Kootenai National Forest in Montana rounded up 20 Osborne firefinders in the west and shipped them off to British Honduras as this Central American country was establishing a lookout system. In 1995, neighbouring Nicaragua placed a similar request with the Washington office of the United States Forest Service and the call went out across the nation for more Osborne firefinders to make them happy, too. A commendable international gesture wouldn't you say? The trouble is that the last known supplier of new Osborne firefinders in the world was Forestry Equipment Suppliers of Jackson, Mississippi, and they had dropped them from their catalogue in 1992 after selling their last ones for $3,700.' We should have hidden away a few!"[9]

The use of a telescopic sight mounted on the firefinder is unique to Alberta and was the brainchild of tower observers Steve Quaranta (Whitecourt) and Jay Sumner (Brazeau). Provincial Detection Technician Tim Klein said the pair started experimenting in 1958 with two types of home-made mounts. "These designs Steve then drew out and submitted to Edmonton. I have drafted drawings by 'A.S.' for the Weaver Scope Attachment for Osborne Firefinder Ring (AFS Edmonton), one type dated August 18, 1959 and the other type November 2, 1959. By the mid-1960s both had been modified and combined to the present system, although as with any change it took a few years for those upgrades to make it

Lookout observer Bob Tough using scope and firefinder to identify location of smoke, Moose Mountain Lookout, Bow Crow Forest, 1987
Roger Meyer

to every tower."

The cabin located near the base of the lookout or tower was initially a one-room structure serving as kitchen, bedroom, office and living room. Not only did the individual deal with times of loneliness but also with cramped quarters, especially when confined to the cabin due to inclement weather or fierce mountain-top winds.

Collecting and conserving water was a task in itself as the only readily available source was rainwater. Seldom was there a spring or small stream of water close enough to the mountain-top to make it practical to carry water, one or two pails at a time, uphill. In very dry times the ranger would haul a supply of water by packhorse. Bathing was relegated to times of plenty when all the rain barrels were full. The northern towers were slightly more fortunate as a small creek, low area or even a hole dug into nearby muskeg provided more accessible water. Without the modern amenity of refrigeration, food storage was an issue. It demanded the use of dried and canned foods or else digging a "root cellar" in a shaded north-facing hillside for short-term storage of perishables.

Before gas-fuelled stoves were supplied at the lookout sites another task of the ranger and/or lookout person was to skid firewood by horse to the site. The trees or logs were then cut into stove-length blocks with a bucksaw, split and stacked. A minimum of one cord (stacked wood measuring four feet wide, four feet high and eight feet long) of firewood was expected to be in place by the time the site was closed at the end of the fire season, ready for the next spring's opening. During the height of the lightning season was no time to be cutting firewood, when the prime purpose of the lookout person's job was the detection of fires. Once again the lookout personnel in the north had a much easier time keeping up with the wood supply as standing timber was closer, or timber cleared for construction of the site was still available.

Lightning is the cause of many forest fires, and it also was a cause of anxiety for the observer, whether on a lookout or a tower. Slave Lake lookout Joe Decoigne was seriously injured by lightning on Flat Top Mountain in the 1950s and had to be rescued. The lookout structures were eventually, of necessity, well grounded. At one mountain lookout a first-timer on the job, a keen individual, thought he would do the Forest Service a favour and clean up the site. There was a network of wire and cables laid out around the building in an attempt to provide a route to earth for the lightning strikes and allow the charge to disperse harmlessly into the surrounding mountainside. In this case the "keener" picked up all the wire and cables and neatly stacked the debris for the ranger to haul out the next time he was up. Needless to say the ranger was fit to be tied over that incident.

Lookout person Sam Fomuk recounts an experience from Nose Mountain in mid-July, 1962. "The day began calm and with heavy fog which did not lift at all. After passing my weather report on the noon sked, I turned around to walk out of the radio room. Just then a deafening blast rocked the cabin. I distinctly saw what appeared to be a fireball in front of me and a strong smell of ozone. Apparently directly overhead a thunderhead cloud had been building up for some time, its presence unknown to me. The lightning charge came down the antenna wire, shattering a heavy copper wall switch to fragments, some of which were embedded in the wallboard of the cabin. Outside, the charge came down one of the tower guy cables, blasting out a pail-sized hole in the ground. Most of the transistorized radio equipment, both the government's and my own, was out of commission. There was just an emergency tube-type portable radio left. After improvising an antennae of sorts, I was able to advise HQ what happened.

"Now the incredible sequel to the story - next day there was an encore performance. Being somewhat jittery, I exited the radio room in haste after passing the noon weather report, and just in time! The second strike knocked out what radio equipment was still left. Eventually a technician arrived in a helicopter with replacements. So lightning does strike in the same place twice, but when it strikes almost at the same time on two consecutive days, that is a hard act to follow! Over the years we found that transistorized VHF equipment was very susceptible to damage from nearby lightning hits.

"There was one instance of a cupola destroyed during a severe night-time storm. In another case, the occupant on my neighbouring tower was knocked unconscious in the cupola for a short

High cupolas gave good visibility, but also could put tower staff in the thick of the storm
Alberta Government, AFHPC

time when a powerful charge hit the tower. In another severe case the firefinder itself became welded to its moorings.

"It's a wonder that we didn't have a few fatalities over the years - because it was so exposed. You can just imagine me hanging back by the table in the radio room on the first blast when all those copper fragments went flying in all directions. A person could have been blinded there easy as nothing. So maybe there is a guardian angel somewhere who was keeping half an eye out on us so-called lookout persons!"

The steel tower came into being in the 1920s. The metal structure for the province's second steel tower, erected in 1928, is abandoned but later also installed on mountain-top lookouts.

The original invoice for the steel used for the 1952 construction of the Sweathouse tower shows a cost of $1,370 – not bad value considering the tower was not replaced until 2003. The replacement cost for the Sweathouse tower and cabin in 2003 was about $70,000.

The observers on the steel towers were warned, with emphasis, to never climb the steel whenever there was an electrical storm within five miles. "If you are down, stay down – if you are up, stay up." One foot on the ground and the other on the first rung of the ladder was definitely not a healthy position to be in with an electrical storm overhead. There are no plumbing facilities in an eight-foot cupola at the top of a 100-foot steel tower. Needless to say there were some anxious times during a long afternoon of constant thunder and lightning activity.

Care had to be taken in what materials were used in the construction of the cabin and other outbuildings on the site. At Keg Tower (and others) a peak rabbit

Fairchild aircraft (1928) were commonly used on northern Alberta fire patrols; some were based at Cooking Lake near Edmonton
Dominion Forestry Branch, AFHPC

still at Ministik Bird Sanctuary southeast of Edmonton. Steel towers ranged from 20 feet (Blue Hill) to 120 feet (Puskwaskau, Berland) in height, depending upon the site and the height of the surrounding timber. Two companies, Ajax Engineering and Manitoba Bridge and Steel supplied most of the tower steel used in construction. Ajax supplied the blueprints for the first wooden octagonal cupola that saw service on many towers from 1938 until replaced by one of the fibreglass designs in the 1960s. The cupola was mounted at the top of the steel structure with the firefinder at its centre. These cupolas were

Air patrol from High River flies fire detection route along the Rockies, 1921. Flights were at 13,000 feet mostly in open cockpits
Dominion Forestry Branch, AFHPC

population occurred one winter. The siding used on the buildings looked nice and was durable as far as the elements were concerned, but it was no match for the rabbits. Some type of preservative used in the treatment of the siding had a special attraction for the rabbits and in the spring when

Lookouts and Communications

opening the tower for the start of the fire season, staff found the first three boards of the siding were chewed off. Only the protruding nails were left behind. It was a good indication of how deep the snow had been around the cabin. A fast repair job was needed before the driving rain would leak in and rot the base of the cabin walls.

The once-monthly servicing of the towers was looked forward to, particularly by the lookout personnel. For the first hour one had only to nod one's head in agreement until the observer had "talked himself out." After an hour or so things would get back to normal and a good conversation would be had, with both parties brought up to date on happenings of the past month.

If the lookout person was a trapper it was expected one would sit down and enjoy some rabbit stew with him. It was sheer entertainment if the ranger brought along a new assistant ranger for orientation, especially if he was "city raised." Originally the towers were manned for the fire season from April 1 to October 31. Then in the 1990s, the records of all lookouts and towers in the province were evaluated, and an appropriate "manning" season was set for each site. The length of time for each site varied depending upon whether the visible area was a total green (forested) area or overlooked settlements and other sources of man-caused fires. Priorities also played a deciding role. This made good sense as lightning season did not normally start until June and this meant the months of April and May could be uneventful, especially if no weather observations were taken until "snow gone" had occurred.

Originally only men were employed, but as time went by women started to make their appearance. Safety and security was always important but with women living alone on a remote site additional concerns arose. In one instance the ranger had set up a code for security known only to the two of them. This paid off when an individual of doubtful character arrived in the middle of the afternoon causing some anxiety for the female observer. She feigned having to send a report that was expected, and sent a "coded" weather report. In less than 20 minutes the ranger was on site and the situation resolved without incident.

Mobile Detection

The mobile detection system, particularly the use of fixed-wing aircraft, was intended to supplement the fixed detection program by providing designated patrol routes that covered the blind areas not visible from lookouts and towers. These were primarily river valleys that were in the "blind" portion or areas beyond the 25-mile radius of one or more towers. Extra attention was also given to areas that were getting into the high and extreme categories of fire danger.

Meteorologist Wilbur Sly (C) discusses the 'convective index' with tower staff Ike Doerkson (L) and Sam Fomuk
Alberta Government, AFHPC

Fixed-wing aircraft owned by the AFS were often used to patrol the forests, but private contracts for fire patrol and casual hire also formed a part of the aerial detection force.

In more recent times "loaded" patrols (helicopters with highly trained seasonal firefighters on board) were used to follow lightning storms, looking for fire starts from these strikes and attacking them within minutes, thus keeping the fires small in the majority of instances.

The Observers

Many devoted and talented people have been associated with the lookout operations of the AFS, and more than a few have service records dating back 20 years or more. In southern areas, many were cowboys and wranglers from neighbouring ranches. Others were "upstart" ranger hopefuls interested in the outdoors and forestry work. All of the early lookout people learned on the job, but after 1960 the training was delivered through the Forest Technology School in Hinton. Many lookout people worked their way up through the ranks to become rangers and foresters. Some, like Sam Fomuk, have become legends in the AFS.

An article written by Jeff Henricks, Forest Protection Technician at Fort Vermilion in August, 1994, said Sam's family moved to Canada from the Ukraine in 1932, settling on a homestead in the Lac La Biche area. "He started with the Alberta Forest Service on April 19, 1945, working the Brazeau Tower for a wage of $3.50/day."[10] Fomuk retired in 1994 after 49 years of paid service, plus four or five weeks of voluntary work in his 50th year when he returned to clean up his cabin. Fomuk was instrumental in helping develop lookout operations and offering valuable hands-on help to ensure advances in radio and electronics were adapted to Alberta conditions.

The lookout observer and his lonely perch
Alberta Government, AFHPC

Typical construction and layout of tower and cabin
Alberta Government, AFHPC

Staff members have many stories to tell about encounters with wildlife. Lookout personnel are "on their own" and must be vigilant. Wildlife, especially bears, were often present. The smells of garbage, food and cooking were great attractions for these animals. Many sites included excellent blueberry or raspberry patches in the late summer or early fall. On one occasion the lookout person at Chinchaga was cooking supper on a hot summer evening. The windows were open to allow some airflow. Hearing a noise in the radio room the man, cast-iron frying pan in hand, looked over to see a young black bear with its head coming through the window. A stiff crack with the frying pan over the head of the bear resolved that incident. Observers quickly learned never to climb the tower while leaving the cabin door open and the rifle inside the cabin. A rifle shot from the tower was often all that was required to scare off a bear, especially if the garbage pit was kept clean by burning and liberal doses of lime were used on the remaining waste.

On another occasion at the old Naylor Hills Tower, since torn down and relocated to the Kimiwan Tower site, Frank Vogel, of German ancestry, met a bear just outside his cabin door. Startled, his first reaction was to holler at the bear in German. The bear immediately ran off. On reciting the story later he said the bear was a good one - it understood German.

Other situations involved lookout personnel running low on "grub." When Sam Fomuk was at Nose Mountain Tower in 1951, a very remote

Lookouts and Communications 135

location south of Grande Prairie, the autumn supply pack outfit couldn't reach him. He was almost out of food. He took the opportunity to correct this when, he says, "one day a young bull moose visited the lookout site. One shot from my 35 Remington at 50 yards hit the chest area and the animal wandered 75 yards and dropped dead." Sam recalls that with local huckleberries and his remaining rice he managed nicely until the first snow, when he closed the tower and met the district ranger for the horse trip back to Grovedale.

During part of Sam's career he assisted with the task of recruiting new lookout staff. Much of the turnover resulted from marriage and family demands. "When observers married they found very few of their wives were able to put up with the remote lookout life, so they usually left," he told interviewer Peter Murphy. "What determined the end of a married lookout's service was when their first kid had to start attending school. In Ike's case (Ike Doerkson from Grande Prairie, on Economy Creek Tower) his wife

Portable radios helped the ranger coordinate firefighting tactics
Alberta Government, AFHPC

knew enough about it to at least teach or supervise the first couple seasons of a child's training via correspondence course. Some parents couldn't be bothered with that or they didn't have enough skills themselves. Usually you could bank on it. A married man lasted on a tower only until the first child started approaching school age and then they were gone. They either took some assignment in town or just left the Service.

"Those were the days when anybody that wanted to stay on, stayed. So after a year's probation I got on the permanent staff and of course on a tower we were expected to work, if conditions required, from sun up until after sundown. Even when helicopters came around they could legally fly only until close to sundown, so you had to stick around until they were down. It could mean a long, long day in those days. If it was so foggy you couldn't see to the end of your yard or a couple days of pouring rain, there was no point in going into the "dog house" [cupola], so you had to look at it philosophically. Sometimes you really had a long hard day and other days as long as you sent those weather reports and checked in a couple times a day there was nobody breathing down your neck or giving a damn what you were doing.

"You could read. Or you could 'press blankets.' That was a comical catch phrase in the Service. In the early times, we had to keep a little diary – a Ranger's Diary. There were spaces to enter how many miles you travelled today by horse, by speeder, by railway, etc. Not many of those were options for the tower staff. So, on these so-called easy days what could you put down? Maybe you did some catch-up reading, maybe had a darn nice snooze after noon, etc. So just as a gag we entered, aside from routine

Early photo of isolated Baseline Lookout, Clearwater Forest, 1928
Alberta Government, AFHPC

136 Alberta Forest Service

duties, 'pressing blankets.' Eventually we got a circular memo: 'Henceforth, lookout men shall not use the entry of pressing blankets.' It wasn't that lookout staff were shirking their jobs, but if you couldn't see as far as your engine house or it was raining cats and dogs all day long or maybe a couple days in succession, what extra-curricular activities could you put down?

"Lookout staff had to be available on the job every day during the fire season, requiring them to work on weekends and holidays. They were not paid extra for these days. It was 20 years before they received an allowance for the extra days. They could then get cash for extra days worked at time-and-a-half or could take time off in lieu at the end of the season. Then it was agreed that permanent observers should also have an extra week for annual leave – which was subsequently extended to three and then four weeks for long service. If the accumulated working days were added to vacation time they would support a long break in the season during the winter."

On one occasion the observer went for a walk in the evening and was never heard of again. The story about Ben Knutson remains a mystery. Despite the railway lines up the Coal Branch, as well as towns, mining and logging activity, this area was still remote. Access to the Grave Flats Lookout, located east of Mountain Park and overlooking the Cardinal River valley, was by pack horse trail. On August 31, 1946, experienced lookout observer Ben Knutson did not respond on the scheduled daily morning round of calls among lookouts and ranger stations. As Ranger Angus Crawford's daughter Doris Gosney recalled:

"Dad was at Grave Flats with my sister Glad when Ben Knutson went missing. They went up to the lookout to check things out. Ben wasn't there, both dogs and their packs were gone and his rifle, so Dad assumed he had gone hunting. Dad went on into Mountain Park and when he still couldn't contact Ben he notified headquarters at Edson that Ben was missing. This was August 31, 1946. A search party of about 30 volunteers from Mountain Park including myself took part in a Labour Day search but nothing was found.

"Some time shortly after, on another trip back to Mountain Park from Grave Flats, Dad's assistant whom Doris' brother Don thinks was Evert Stanley, went up to the lookout and Dad and Don took the bottom trail. [Ben's] dog Paddy was at the lookout and still had his pack on. They took the dog on another search hoping he would lead them to Ben, but no luck again. About a month later the second dog, Blackie, showed up at the Red Cap cabin. His hair was worn off and his stomach was all raw from the pack he had been carrying. Dad kept this dog and used him in the dog team."[11]

Berland Tower 110-foot single pylon construction, 1988
Alberta Government, AFHPC

There was great speculation as to what happened to Ben Knutson. His bed was apparently found made-up for the day and beans were soaking in a pot, which indicates he planned to return. The search lasted for over a month before it was called off. No trace of Ben has yet been found.

Chuck Rattliff, a forest management forester at Grande Prairie, added more stories in interviews.

"One time Foggy Tower was trying to get hold of Sam Fomuk. Foggy is located 60 or 80 miles north of where Sam was at the time in Wadlin Tower. They couldn't get hold of him on the regular FM radio system, but there was a smoke near Foggy Tower and they wanted to get a cross shot from Sam. So finally we got hold of Sam on the VHF system from headquarters and said, 'Sam, Foggy thinks he sees a smoke near your tower. Do you have anything that you see there?' He said, 'The engine house is on fire.' That was why his FM radio wasn't working.

"There's another story about a tower just north-west of Hinton (Athabasca Tower). I don't remember the lookout person's name but the superintendent at the time was Hank Ryhanen. It was a Sunday morning when Hank and his wife went for a drive and visited the tower. The cabin door was open. They couldn't see the lookout person anywhere but they did see a grizzly bear that took off out of the yard. The observer had gone out to the toilet first thing in the morning without a stitch of clothing on. By the time he was ready to go back to the cabin, the grizzly was out there. He couldn't get back to the cabin so he stayed in the toilet until Ryhanen came along - and even then he couldn't come out because of his nakedness!"[12]

In 2005, lookouts formed an observation network over most of Alberta's forest landscape. It is the biggest such network in Canada and probably the world – the largest such system in the U.S., with 95 towers, is in Wisconsin. With 129 active locations providing vital lookout coverage, the Alberta network affords prompt reports of smoke for quick response by initial attack crews. Sophisticated technology continues to enhance their effectiveness and ensure their place in modern forest firefighting, but lookouts still depend upon people. Living and working conditions along with better salaries have helped make the isolated duties more pleasant, and for many people, there is still the allure of isolation in the wilderness.

Editor's note: Thanks to Tim and Hope Klein, who helped ensure the accuracy of information in this chapter.

All photos on the following pages, unless otherwise identified, are from the Dominion Forestry Branch or AFS, and in the possession of the Alberta Forest History Photo Collection (AFHPC).

Construction of the Marten Mountain Lookout, Lesser Slave Forest Reserve, 1923

Flat Top Tower construction, raising 20 foot ladder section, Slave Lake Forest

Nose Mountain Tower, Grande Prairie Division, Northern Alberta Forest District, 1949
Ron Linsdell

Baldy Lookout, north of Nordegg, Clearwater Forest, 1966

Buck Mountain Tower, Clearwater Forest, 1938

Radio Superintendent Tony Earnshaw at Moose Mountain Lookout, mid-1950s

Lookouts and Communications

Chisholm Tower, 1959
Cliff Smith

Narrow Lake, southwest of Athabasca - spring fire season lookout

Power supply system, Mockingbird Hill Lookout, Bow Forest, mid-1950s

Sam Fomuk (R) receives certificate of recognition for 50 years of tower service from Fort Vermilion Chief Ranger Bill Lesiuk (L) and Forest Officer Jeff Henricks, 1995
Jeff Henricks

Heart Lake Tower, built by Rangers John Currat and Pete Comeau, Lac La Biche Forest, early 1940s

Livingstone Lookout, Crowsnest Forest, 1974

Puskwaskau single pylon tower construction, Grande Prairie Forest, 1990s
Cliff Smith

Cupola exchange on Pass Creek Tower, 2004, Whitecourt Forest

Lookouts and Communications 141

Cupola exchange completed by A-Star helicopter at Pinto Tower, Grande Prairie Forest, 1997
Tim Klein

Moose Mountain Lookout, Bow River Forest, Rocky Mountains Forest Reserve, August 29, 1932. Dexter Champion, who spent three and a half seasons as lookoutman on Moose Mountain, said that in dangerously dry weather he had been as long as two months without leaving, and had gone 26 days without seeing a living soul. Champion married Miss Louise Clarke and took his bride to the cabin on the peak for several months' honeymoon
Jay Champion photo; Calgary Herald 1937 article

Radio Technician Ron Linsdell standing in front of Lovett Lookout, Athabasca Forest, Rocky Mountains Forest Reserve, 1939. Ron Linsdell began with the Alberta Forest Service in 1939 and retired after 35 years of service, the last 12 with Alberta Government Telephones
Ron Linsdell

Ranger Harold Parnell and Radio Technician Ron Linsdell at Lovett Lookout, Athabasca Forest, Rocky Mountains Forest Reserve, 1939
Ron Linsdell

Lookoutman Dexter Champion hiking up to Moose Mountain Lookout, Bow River Forest, Rocky Mountains Forest Reserve, May, 1934
Jay Champion

Mayberne Tower, Athabasca Forest, Rocky Mountains Forest Reserve, 1939
Ron Linsdell

New steel Brazeau Tower, Brazeau Forest, Rocky Mountains Forest Reserve, 1950s
Ron Linsdell

Repairing yoke on stone boat during trip to Brazeau Tower, Brazeau Forest, Rocky Mountains Forest Reserve, 1940s
Ron Linsdell

Brazeau Tower, Brazeau Forest, Rocky Mountains Forest Reserve, 1939
Ron Linsdell

Ranger Gordon Ramstead tying load to pack horse for trip to Brazeau Tower, Brazeau Forest, Rocky Mountains Forest Reserve, 1940s
Ron Linsdell

Lookouts and Communications 143

Helen Ledingham and Joe Wuetherick at the central radio dispatch centre in Edmonton connect to forestry field staff, c. 1940

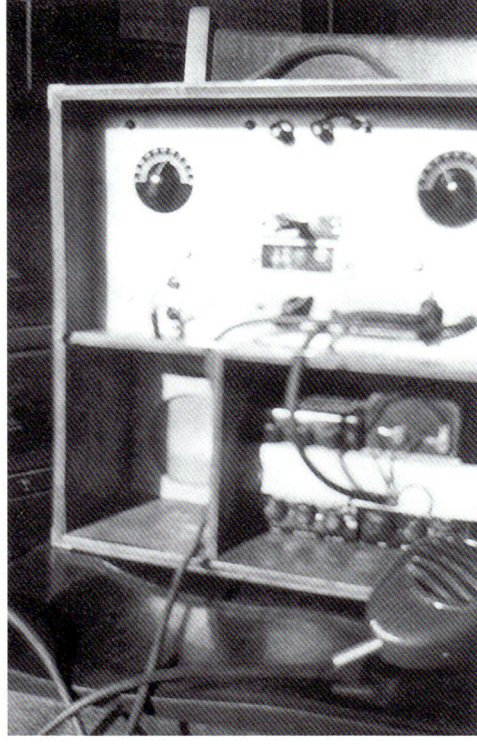

3 Volt Ultra High Frequency (UHF) Radio used for communications between Lookouts, Towers and Ranger Stations, early 1940s
Ron Linsdell

Wooden Whitecourt Tower, Whitecourt Division, Northern Alberta Forest District, early 1940s
Ron Linsdell

Whitecourt Tower and cabin, Bill Norton towerman, Whitecourt Division, Northern Alberta Forest District, early 1940s
Ron Linsdell

Towerman Al Schultz launching a weather balloon at Whitecourt Tower, Whitecourt Division, Northern Alberta Forest District, 1940s
Ron Linsdell

Lovett Lookout, Athabasca Forest, Rocky Mountains Forest Reserve, 1942
Ron Linsdell

Radio room set-up with paperwork at Goose Mountain Tower, Slave Lake Division, Northern Alberta Forest District, 1940s
Ron Linsdell

Goose Mountain Tower ready for radios and operation, Norm Smith first towerman. Slave Lake Division, Northern Alberta Forest District, 1941
Ron Linsdell

Nearing completion of construction on the Goose Mountain Tower, Slave Lake Division, Northern Alberta Forest District, April, 1941

Lookouts and Communications

Ranger Jim Randall packing construction supplies to Goose Mountain Tower, Slave Lake Division, Northern Alberta Forest District, 1941

Fickle Lake Crawl Tower, Brazeau-Athabasca Forest, 1940s

House Mountain Tower, Slave Lake Division, Northern Alberta Forest District, 1940s
Lou Foley

Flattop Tower, Slave Lake Division, Northern Alberta Forest District, 1940s
Lou Foley

High Prairie Ranger Monte Alford and Mrs. Alford standing at the base of Goose Mountian Tower, Slave Lake Division, Northern Alberta Forest District, early 1940s
Ron Linsdell

Athabasca Lookout, shed and radio mast (wooden pole and guy wires), Athabasca Forest, Rocky Mountains Forest Reserve, May, 1942. Fred Hendrickson was lookout observer
Ron Linsdell

Buck Mountain Tower with cabin and shed, Clearwater Forest, Rocky Mountains Forest Reserve, 1942
Ron Linsdell

(L to R): Don Bruce (first went to Buck Mountain at 16 years of age), Not Identified and local farmer Mr. McKenzie, Buck Mountain Tower, Clearwater Forest, Rocky Mountains Forest Reserve, 1944
Ron Linsdell

Towerman Don Bruce (lead) and Radio Technician Ron Linsdell (on white horse to right) taking pack train to Goose Mountain Tower from the High Prairie Ranger Station, Slave Lake Division, Northern Alberta Forest District, 1940s
Ron Linsdell

Lookouts and Communications

Pat Donnelly standing in doorframe at Edmonton Radio Office, June, 1940. Sign says "Government of Alberta, Department of Lands and Mines, Forestry Service Radio"
Ron Linsdell

Calgary Radio Office, 1944
Ron Linsdell

Pat Donnelly standing outside the Lac La Biche Radio Office, Lac La Biche Division, Northern Alberta Forest District, 1944
Ron Linsdell

Radio Technician's truck fording the Pembina River near Lodgepole, Brazeau Forest, Rocky Mountains Forest Reserve, early 1940s
Ron Linsdell

Radio Technicians at Little Red River, east of High Level, January, 1945 (minus 31° F)
Ron Linsdell

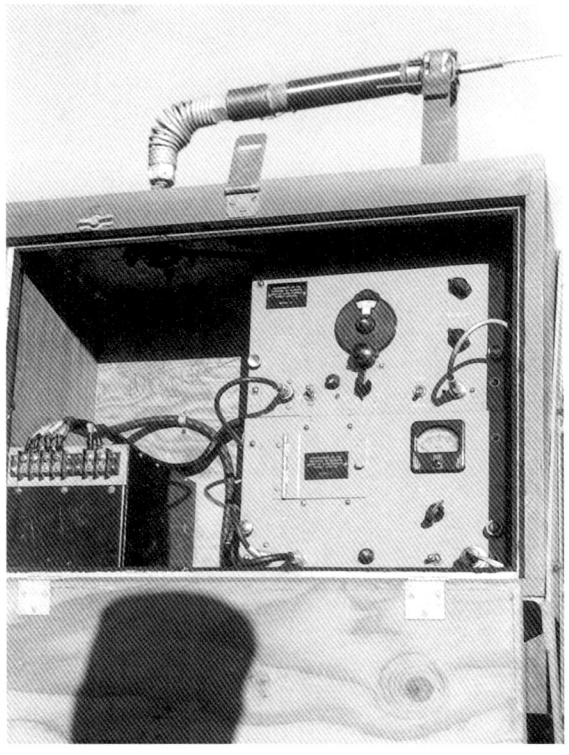

View of battery and portable radio, truck box mounted, 1940s
Ron Linsdell

Radio dispatch centre with 3 NC 100 National Receivers, Morse Code pad and typewriter, 1940s
Ron Linsdell

Portable radio mounted on the side of Radio Technicians truck, 1940s
Ron Linsdell

Lookouts and Communications 149

Construction of the Lenarthur Tower using winches fastened to standing trees to raise the 70 foot poles into standing position, Lac La Biche Division, Northern Alberta Forest District, 1946
Gunner Brauti

Lenarthur Tower - wooden crawl tower on Northern Alberta Railway near Anzac. Access was by rail or along the railway right-of-way

Fording Pinto Creek on return trip from Pinto Tower, Grande Prairie Division, Northern Alberta Forest District, 1949
Ron Linsdell

Transporting tower construction supplies to Lenarthur Tower along the Northern Alberta Railway, near Anzac, Lac La Biche Division, Northern Alberta Forest District, 1946
Gunner Brauti

Snuff Mountain Tower construction, Grande Prairie Division, 1953. Ted Hammer on top of cupola, Neil Gilliat one of people climbing up ladder
Neil Gilliat

Snuff Mountain Tower, Grande Prairie Division, early 1950s
(L to R): Ranger George Sebastian (Valleyview), Timber Inspector Ted Hammer (Grande Prairie), Mary Sebastian (cook), Assistant Inspector Campbell, Ranger Jack Grant (Debolt), Ranger Neil Gilliat (Hythe), Ranger Pete Comeau (Grovedale)
Neil Gilliat

Luscar Lookout, Athabasca Forest, Rocky Mountains Forest Reserve, July 11, 1950
Neil Gilliat

Wooden Crawl Tower, Brazeau Forest, Rocky Mountains Forest Reserve, 1940s
Neil Gilliat

Trout Mountain Tower, Slave Lake Division, early 1950s

Hauling supplies to Cowpar Tower, north of Lac La Biche, Lac La Biche Division, 1950s
Tim Klein

Norseman aircraft helped freight supplies to northern observers, 1951

Lookouts and Communications

Ranger Rex Winn completing telephone line maintenance, Athabasca Forest, Rocky Mountains Forest Reserve, 1951
Neil Gilliat

Norm Smith limbing tree for use as radio mast, Economy Tower, Grande Prairie Division, 1953. Norm agreed to limb the tree if Ron Linsdell topped it
Ike Doerkson

Ron Linsdell topping tree for use as a radio mast, Economy Tower, Grande Prairie Division, 1953
Ike Doerkson

Thickwood Tower, Fort McMurray, Lac La Biche Division, February, 1955
Dennis Howells

Radio Operator Joe Wuetherick in the Radio Room, top floor of the Natural Resources Building (now the Bowker Building), c. 1955. In front of Joe is a bank of teletype machines – the technological advance of the day. Other radio operators at the time were Helen Ledingham, Doonie Donovan and Ron Linsdell
Provincial Archives of Alberta, PA2793-1

Rocky Mountain House AFS office and Rocky Tower, Clearwater Forest, Rocky Mountains Forest Reserve, 1940s
Ron Linsdell

The original Rocky Mountain House Tower was located at the top end of the Main street, Rocky Mountain House. It was across from the Hotel and next to the Imperial Bank. Ben Shantz and Neil Gilliat took it down in 1959. The idea was to move it to a new location north of town; however, the steel was so twisted out of shape it could not be used again. This tower had been a landmark in Rocky Mountain House for 30 years
Neil Gilliat

Lookouts and Communications 153

Flat Top Tower, Slave Lake Division, July, 1959. Forest Survey crew camp to the right of the tower road
Cliff Smith

Chungo Lookout, Rocky Clearwater Forest, 1960s
Bill McPhail

Ranger Bill Sanregret on Nodwell used to travel the Tony Tower forestry road. Sanregret and Gary Giese were opening Tony Tower for the 1960 season. Whitecourt Division
Gary Giese

Buck Mountain Lookout, Clearwater Forest, 1960s
Bill McPhail

Falls Lookout, Clearwater Forest, 1960s
Bill McPhail

Knut Hansen Pass Creek towerman, Whitecourt Division, July, 1962. In this picture Knut Hansen was 64 years old - he served on Pass Creek from 1957 to 1972
Vere Paré

La Corey Crawl Tower, La Corey District, Lac La Biche Forest, 1971

Pass Creek towerman Knut Hansen (L) and pilot Paul Kristapovich, Whitecourt Division, September, 1965
Vere Paré

Road problems while opening Goose Mountain Tower, Slave Lake Forest, April 28, 1971
Jamie McQuarrie

Lookouts and Communications

Construction of La Crete Tower, Footner Lake Forest, fall, 1972. Logs were harvested by Swanson Lumber near the confluence of the Wabasca and Peace Rivers
Dave Brown

Bell 206B slinging cupola into place on Deadwood Tower, Peace River Forest, 1970s
Terry Van Nest

Securing cupola on Deadwood Tower, Peace River Forest, 1970s
Terry Van Nest

Forget Me-Not Lookout, Bow River Forest, 1972

Advanced Communications Course, Forest Technology School, Hinton, March 5-9, 1973. This course was designed for lookout observers and radio communications personnel

Back Row (L to R): Larry Huberdeau, Doug Doerkson, George Schultz, Jerome Marsh, Len Smith, Gary Pulleyblank, Steve Quaranta. Third Row: Clyde Ulm, John McKinney, Gerald Bursey, Bill Norton, Paul Campbell, Martin Johnson. Second Row: Bob Sheets, Ken Cheesman, Dooney Donovan, Len Larson, John Morrison. Front Row: Harry Edgecombe (Instructor), Bob Thomas, Gordon McDonald, Roy Mitchell, Dennis Mah, Lee Watson

Fire Detection Officers Course, Forest Technology School, Hinton, March 19-23, 1973

Back Row (L to R): Pat Hendrigan, Egan Isaksson, Floyd Schamber. Third Row: Martin Johnson, Ike Doerkson, Paul Campbell. Second Row: Bob Thomas, Myron Sterr, Jay Sumner, Tony Stuart, Jim Sartorious. Front Row: Chuck Rattliff (Instructor), Nick Paulovits, Bruce Bryan, Cliff Anderson, R.A. Mitchell, John Morrison (Instructor)

Lookout Observer course, May, 1976, Forest Technology School, Hinton

Lookout Observer course #1 April, 1981, Forest Technology School, Hinton

Lookout Observer course #2 April, 1981, Forest Technology School, Hinton

Lookouts and Communications

Alberta government Bell 222 C-GFSI slinging tower section to workers building single pylon tower, 1980s

Alberta government Bell 222 C-GFSI hovering with tower section while workers bolt the sections together, 1980s

Darryl Johnson, Forest Protection Technician, climbing down outside Pass Creek Tower checking tightness of bolts on tower legs and cross pieces. Generally a crew of four to five Forest Officers would work on the tower. Fox Creek District, Whitecourt Forest, 1991

Alberta government pilot Roger Tessier (L) and Mike Dubina completing a lookout helipad inspection at Torrens Lookout, Grande Prairie Forest, 1991

One of the luxuries of a lookout observer – a bathtub! Algar Tower, Athabasca Forest, 1985

Construction of footings for the new Flattop Tower, Lesser Slave Wildfire Management Area, 2003
Tim Klein

Ernie Basaraba – trapper and towerman, 37 years of service in 2004. Standing at the base of the new Simonette Tower, Smoky Wildfire Management Area, 2004
Tim Klein

Class of 2004 Lookout Observers, Hinton Training Centre, Hinton
Tim Klein

Lookouts and Communications

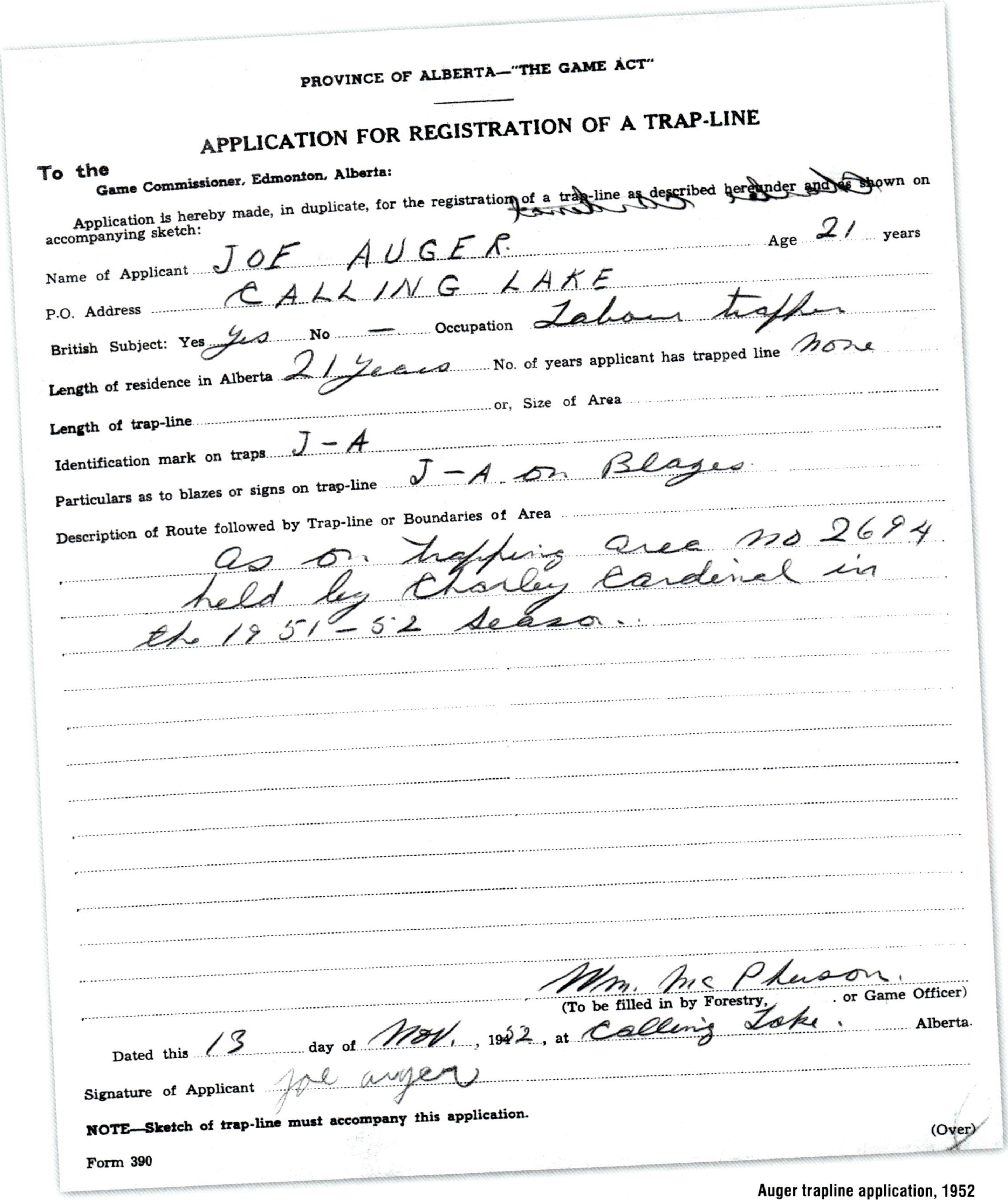

PROVINCE OF ALBERTA—"THE GAME ACT"

APPLICATION FOR REGISTRATION OF A TRAP-LINE

To the Game Commissioner, Edmonton, Alberta:

Application is hereby made, in duplicate, for the registration of a trap-line as described hereunder and as shown on accompanying sketch:

Name of Applicant __JOE AUGER__ Age __21__ years

P.O. Address __CALLING LAKE__

British Subject: Yes __Yes__ No __—__ Occupation __Labour trapper__

Length of residence in Alberta __21 years__ No. of years applicant has trapped line __none__

Length of trap-line _____ or, Size of Area _____

Identification mark on traps __J-A__

Particulars as to blazes or signs on trap-line __J-A on Blazes__

Description of Route followed by Trap-line or Boundaries of Area __As on trapping area no 2694 held by Charley Cardinal in the 1951-52 season.__

__Wm. McPherson__
(To be filled in by Forestry, or Game Officer)

Dated this __13__ day of __Nov__, 19__52__, at __Calling Lake__, Alberta.

Signature of Applicant __Joe Auger__

NOTE—Sketch of trap-line must accompany this application.

Form 390

(Over)

Auger trapline application, 1952
Alberta Government documents

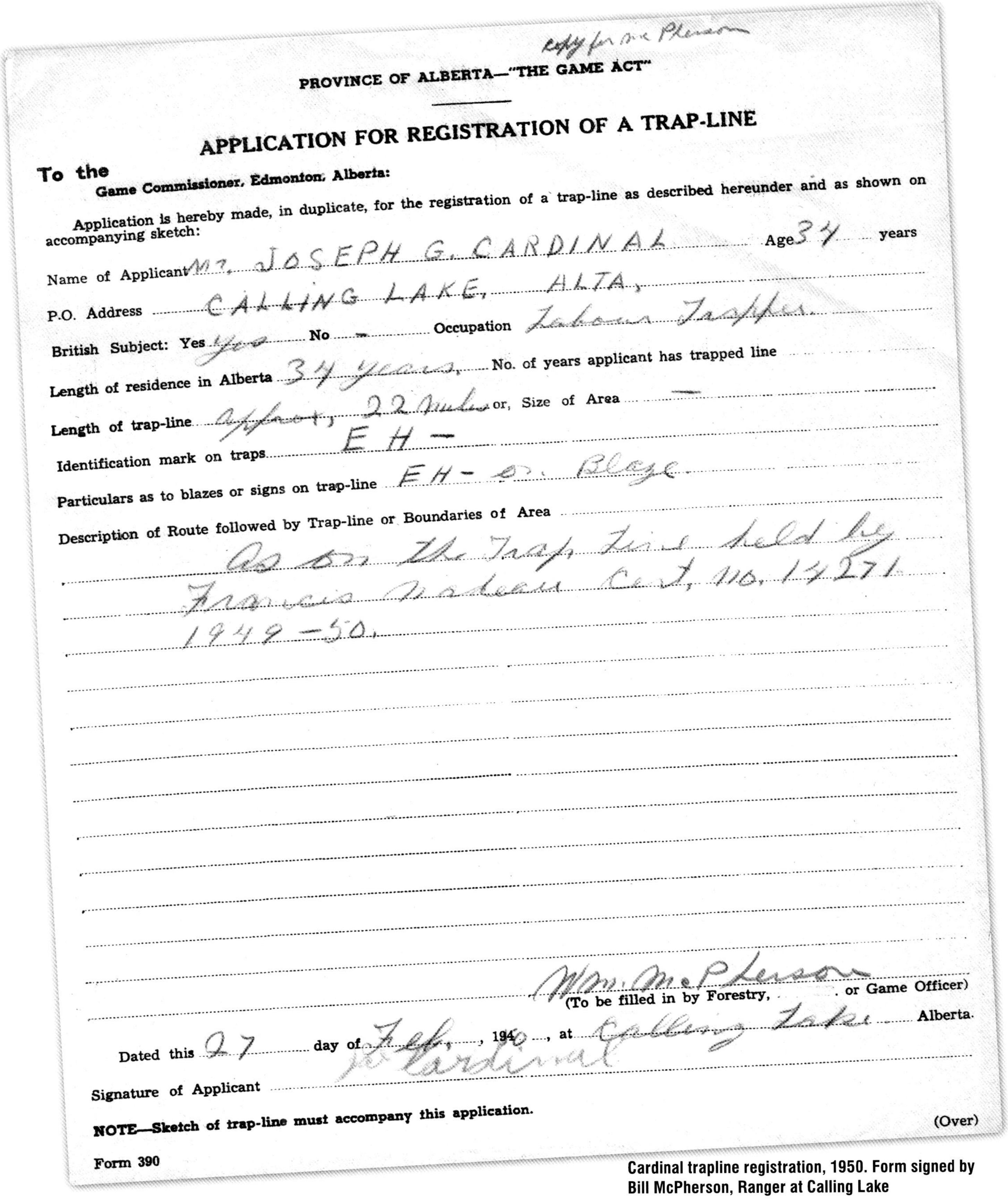

copy for McPherson

PROVINCE OF ALBERTA—"THE GAME ACT"

APPLICATION FOR REGISTRATION OF A TRAP-LINE

To the
Game Commissioner, Edmonton, Alberta:

Application is hereby made, in duplicate, for the registration of a trap-line as described hereunder and as shown on accompanying sketch:

Name of Applicant _Mr. JOSEPH G. CARDINAL_ Age _34_ years

P.O. Address _CALLING LAKE, ALTA._

British Subject: Yes _yes_ No ___ Occupation _Labour Trapper_

Length of residence in Alberta _34 years_, No. of years applicant has trapped line ___

Length of trap-line _approx. 22 miles_ or, Size of Area ___

Identification mark on traps _EH—_

Particulars as to blazes or signs on trap-line _EH— on Blaze_

Description of Route followed by Trap-line or Boundaries of Area
As on the Trap Line held by Francis Nadeau Cert. No. 14271 1949-50.

Wm. McPherson
(To be filled in by Forestry, or Game Officer)

Dated this _27_ day of _Feb._, 194_9_, at _Calling Lake_, Alberta.

Signature of Applicant _Jos. Cardinal_

NOTE—Sketch of trap-line must accompany this application.

Form 390 (Over)

Cardinal trapline registration, 1950. Form signed by Bill McPherson, Ranger at Calling Lake
Alberta Government documents

CHAPTER 6

1966 - 1984

The years preceding this period saw completion of the first forest inventory, creation of a silviculture unit within the Alberta Forest Service (AFS) to support seed collection, site preparation and planting, and the development of a sustained yield policy based on long-term tenure within agreements for sustainable forest management. These developments set the stage for a new era of forest management in Alberta.

The Quota System

The experience of having the pulp industry participate in forest management via the Forest Management Agreement (FMA) was a success. The first FMA at Hinton was well underway with its state-of-the-art planning and reforestation programs, and another possible agreement at Grande Prairie was being discussed in the late 1960s. As a result, head of forest management Reginald Loomis decided to work with the Alberta Forest Products Association to explore the possibilities of extending the participatory concept to the lumber industry. Different mechanisms and divisions of responsibility were applicable to this sector, so a somewhat different form of tenure was devised. What emerged in 1966 was the Timber Quota System.

The Quota System had its roots in the sustained yield policy, balancing harvests with forest growth and ensuring that

Reg Loomis in retirement
Alberta Government, AFHPC

Cruising timber in the Athabasca Forest, 1967
(L to R): Peter Nortcliffe, Cliff Henderson, Dale Huberdeau, Ted Cofer and Larry McKechern
Don Lalonde

all harvested areas were reforested. This was supported by the forest industry's strong desire to obtain a stable, long-term timber supply. This stability enabled industry to invest in permanent wood processing facilities (primarily sawmills), and contribute to economic and community development. The system, which continues today, allows a company to harvest a share of the annual allowable cut (AAC) in a forest management unit, in return for payment of dues and completion of regeneration. The company has to comply with the *Forests Act* and the provincial operating ground rules. Like the Forest Management Agreements, holders of timber quotas were made

Staff attending Forest Management meeting in Peace River, mid-1960s. In this photo they are at Footner Airstrip on a side trip to look at the new administrative Forest being established at Footner Lake
(L to R): Don Bunbury, Bob Gray (at back), Charlie Jackson, Not Identified (Hidden), Mike Lalor, Alf Longworth, Reg Loomis (holding coat), Not Identified (Hidden), Not Identified (Hidden), Not Identified (Hidden), Ray Smuland (dark glasses), Lou Babcock (uniform and tie), Les Harding (in back in plaid), Larry Gauthier (buttoned up dress coat), Rein Krause, Con Dermott, Hank Ryhanen. Edo Nyland is in the door of the DC3
Alberta Government, AFHPC

Timber Management cruising party, working mainly in the High Level area, 1974. Many of these individuals faced the waters of F8 (wettest FMU in Alberta) on a daily basis for six weeks, where their pants were literally eaten away by the acidic waters
Back Row (L to R): Margarete Hee, Harry Archibald, Maurice Freehill, Jim Steele, Deb Pelchat, John MacGarva, Randy Panko, Gord Cook, Hugh Boyd, Dale Gobin, Bill Chadderington (47J helicopter pilot). Front Row: Steve Harrison, Mike Cassidy, Kevin Freehill (Asst Party Chief), Mel White, Paul Folkmann, Buck Dryer, Cordy Nordal. Missing: Henry Desjarlais, Party Chief; John Best, T/M Technician and Rick Keller, Forester
Margarete Hee

full partners in the business of sustained yield forest management.

The initial volume and growth estimates were based on the first broad forest inventory called Phase I which was started in 1949 and completed in 1956. This inventory served as the information base for forest management planning. It included all publicly-owned forestlands in Alberta except Indian Reserves, National Parks and the Rocky Mountains Forest Reserve.

The inventory area was divided into zones, each with different growth conditions, species and elevation. Timber volume tables, by cover type, were developed for each zone. The Phase I inventory was used for defining lands suitable for agriculture, to determine allowable timber harvest levels, and to plan forest protection and timber and industrial developments.

A more detailed forest inventory, named Phase II, was started in 1956 and completed in 1966. It included areas not covered by Phase I, particularly those areas with timber suitable for manufacture of lumber. The data were also used for AAC calculations for Forest Management Units where timber quotas were to be issued.

Detailed information collected through the Phase II inventory included species composition, stand density, stand height, site class and stand age. Minimum stand size identified was 16 hectares, compared to 65 hectares under Phase I.

Between 1965 and 1967 the detailed Phase II forest inventory was augmented by a major ground-truthing project called the Quota Reconnaissance Survey. This survey was used to increase the accuracy of the forest inventory and to identify areas for new Quotas, and for Quota planning.

The field work was organized through each of the Forests, using AFS staff and hired crews. By this time helicopters were also available to move crews around, so productivity was high.

Typically a two-member team was set down in a sampling location in the morning and picked up at a different spot in the afternoon, after a rigorous day of measuring and walking through the bush.

Members of these "cruising" teams had their share of encounters with bears. Chuck Rattliff, a timber management

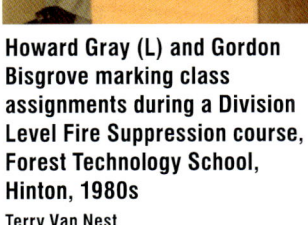

Howard Gray (L) and Gordon Bisgrove marking class assignments during a Division Level Fire Suppression course, Forest Technology School, Hinton, 1980s
Terry Van Nest

Lou Boulet (front) and Don Law working on an overhead team exercise, Division Level Fire Suppression Course, Forest Technology School, Hinton, 1980s
Terry Van Nest

1966 - 1984   163

forester stationed at Grande Prairie, recounts the following story:

"Bears were quite rambunctious once in a while, and put the occasional cruiser up a tree. One fellow who worked in Peace River had his boots pulled off and his legs torn up by a bear chasing him up a tree. I think it happened to him twice. We worked alone with no rifle so we were taking a mighty risk and we had a few close shaves with the bears.

"Otto Barrett was a timber cruiser who worked for me. We went up to House Mountain and were going to let Otto out on an open ridge, but two or three grizzly bears were standing there that morning. So nobody was getting out at that location. So we took Otto down into the valley and let him out, saying we would pick him up at a wellsite that evening. It was in the spring and very dry and the grass was tall and dry. Otto always carried a big roll of wax paper and kitchen matches. His plan was to start a fire if a bear threatened him. Otto came out on this wellsite about 3 pm ready to be picked up. There was a grizzly bear sleeping in the tall grass on the site. Otto walked over to the tall grass and all of a sudden there is this grizzly bear. The grizzly didn't know he was there but Otto got so excited he screamed. The bear jumped up and down and started to roar and Otto got excited and set the grass on fire. The blaze chased the bear off, but we spent the rest of the evening fighting the fire."[1]

The Phase II inventory was considerably more detailed than the first broad-scale inventory. Both were focused primarily on coniferous forest stands so the information on deciduous trees was much less detailed. This limited the value of the inventory for industry's emerging interest in poplar species, although it did get the Quota

Ranger Gary Mandrusiak inspects lumber in mill yard, mid-1970s
Alberta Government, AFHPC

Small sawmill and beehive burner
Alberta Government, AFHPC

system off to a good start.

The Quota system proved to be a catalyst for industry's investment in larger, more efficient mills. It also resulted in the design of more rational road systems, and greatly expanded silviculture and reforestation activities. Between 1966 and 1976, for example, the number of seedlings planted by the AFS and the forest industry ballooned from 490,000 to 7.3 million a year and, by 1984, had expanded further to 17.1 million a year.

Under the Quota system, 220 Timber Quota Certificates were issued in 1966 for an AAC of 439.7 million board feet. A board foot is the amount of wood in an unfinished board measuring 12 inches by 12 inches, and one inch thick. The 439.7 million fbm (foot board measure) is equivalent to one million m^3 in present-day terms. Of the 220 certificates, 185 were granted and 35 were sold through auction. Additional quotas for 29 million board feet were made available for sale "at such time as the lumbering industry has consolidated operations in the existing quotas."[2]

While the Quota certificates established the timber industry's right to a long-term (20 years) timber supply, they did not actually authorize logging. Logging authority was conveyed through the issuance of timber licences that generally provided the licensee with the quota volume for about five years. The actual volume harvested was compared with the volume authorized by the Quota certificate every five years, which became known as the Quota Quadrant period. Both under-production and over-production were discouraged by the government in order to achieve "even-flow" harvests over multiple years.

Forester Cliff Smith in the Fox Creek area, 1970
Cliff Smith

Under the Quota system, the holder of the certificate was responsible for reforestation of the logged areas. Holders could complete the reforestation work themselves or transfer the work to the AFS through payment of a reforestation levy. Initially, all Quota holders could transfer the reforestation work to the AFS. Later on, and up to the early 1990s, only the small Quota holders could do this, while medium and large Quota holders had to do the work, generally at their own cost – with the AFS providing free tree seedlings for planting programs.

Another significant element of the Quota system was an appraised rate of dues to replace the previous competitive bidding rates for sawmill timber. This meant the value of timber was now set by formula rather than by actual auctions and sales. Base rates were based on studies of actual costs and returns and then adjusted according to timber quality and accessibility in the various regions. Four factors were used in the timber appraisal - cull volume, average tree size, gross volume per acre and haul distance.

Companies paid dues according to the amount of timber they reported that was brought out of the bush. This amount was calculated on measurement of the diameters and lengths of logs (a process known as scaling, from which board-foot or cordwood volumes were determined) or else on the end product of manufacture. In 1979 Alberta began to conduct timber scaling to metric standards and compiled scaled volumes in cubic metres. In those situations where dues were previously assessed on the manufactured foot-board measure (fbm) product volume, the fbm product volume was now converted back to the cubic metre roundwood volume deemed to have been used to produce the product.

Scalers, both from AFS and industry, had to pass a rigorous examination to qualify for their licence. By 1966, about 20 per cent of the wood was being scaled by weight. This procedure used average conversion data and was much faster. Today, volumes (in cubic metres) are mostly determined by conversion from weight.

Lumberman Merrill Muttart (L) and Minister Norman Willmore sign Tree Farm Licence
Alberta Government, AFHPC

Timber Permits

The Permit Program was one more tier of timber allocation for short-term cutting rights, typically only for one year, and for small volumes. Permits were initially issued for quick salvage of dead or damaged timber or timber that would otherwise be cleared, for example, for roads or pipelines. They were occasionally used to provide short-term winter employment to individuals along the forest-settlement border areas. The AFS managed the permit program and carried out all reforestation with funds collected from the permittees through their reforestation levies. Today, three more-specific types of permit are used: Commercial Timber Permit (CTP), Deciduous Timber Permit (DTP) and Local Timber Permit (LTP). LTPs can be issued directly to applicants or by draw if demand exceeds supply.

A unique allocation was initiated about 1955. This was the Tree Farm Licence in the Cold Creek-Nojack area. It was issued to long-time lumberman Merrill Muttart. As Eric Huestis recounted, Muttart told him that forestry had been good for him as a businessman and he wanted to give something back. His vision was to set up a tree farm that was open to the public. This project would show sustained yield forest management and small-scale manufacturing of forest products. They negotiated an area around Nojack and Muttart invested in roads, buildings and a small sawmill. He hired forester Herman Oosterhuis to run the operation. The tree farm lasted two or three years until Muttart passed away. The tree farm reverted to the AFS, and the other assets were sold. The Muttart family later shifted its focus to support the Muttart Conservatory in Edmonton, among other initiatives.

The 1960s also saw development of electronic data processing through the use of punch-cards and the beginning of the application of main-frame computers. The systems were used to track harvest and planting information required by regulation. Data to support growth-and-yield and timber supply were also transferred to computer programs, but the personal computer had not yet arrived.

Peace River Forester Fred McDougall on woods inspection in the Clear Hills (Doig River) area, 1964
Cliff Smith

Industry Development

There was an absence of investment in pulp mill development for 13 years after the Hinton pulp mill was built. Then, in 1968, renewed interest in mill development resulted in public hearings for a proposal in the Grande Prairie area. These hearings lead to the negotiation of a Forest Management Agreement with Procter and Gamble Cellulose to construct Alberta's second pulp mill. This project was completed in 1973.

An amendment to the *Forests Act* enabled disposition of deciduous timber on a long-term basis, similar to the coniferous quotas introduced in 1966. This led to Deciduous Timber Allocations (DTAs) for plywood and sawmill operations in the Lesser Slave Lake area, marking the first significant industrial utilization of aspen and poplar. Also about this time, new technology was being developed that would make it economically possible to use deciduous species for pulp and waferboard. This generated additional investment interest in the forests of northern Alberta. The response also heightened public awareness about the potential impact of development opportunities associated with the forest. Groups such as the Alberta Wilderness Association began campaigns of concern respecting coal exploration in subalpine and alpine areas, some of which were adjacent to national parks.

Minister John Zaozirny
Bob Stevenson

A Forest Products Development section was established within the Department of Energy and Natural Resources in 1975. It began simply with a report on the utility pole and fencepost industry, and continued in a modest capacity until 1978. It was the start of a proactive approach by the Alberta government to encourage expansion of the forest products industry. Its first initiative in 1978 was to invite proposals for a pulp mill and other timber processing facilities for the Berland-Fox Creek Timber Development Area (TDA). In November, 1979, after a round of public hearings on the proposals received, the Minister awarded the FMA to BC Forest Products Company Ltd. (BCFP). The company had proposed a $230-million sawmill at Grande Cache, a second sawmill at Knight near Whitecourt, and a thermo-mechanical pulpmill and newsprint mill at Hurdy, 25 km west of Whitecourt. BCFP started construction of the Grande Cache sawmill in 1980. As part of the Berland-Fox Creek TDA, 665,000 m³ of annual conifer cut was offered to existing quota holders in the Fox Creek area to enable expansion of their mills. Additional volume was offered to other established timber operators without quotas in the Edson and Whitecourt areas.

Al Brennan, Executive Director, FIDD, 1986
Bob Stevenson

Cliff Smith, ADM AFS, 1986
Bob Stevenson

Millar Western's Eugene Leighton with the commemorative piece of lumber that marked Alberta's first annual production of one billion board feet of lumber. Leighton was at the controls in 1983 when the landmark event occurred
Millar Western

AFS planning staff had also identified other possible TDAs that might support forest products industries on a sustained-yield basis. A 1979 Timber Management Branch report, *Alberta: Location of Future Forest Industry* was prepared for distribution to the forest industry in western Canada. This set the stage for an active program to encourage further investment in the forest products industry.

The early 1980s were characterized by a major economic slump that resulted from falling prices in both the energy and forestry sectors. This considerably slowed the pace of investment and, in the case of BCFP, contributed to it having to abandon its commitments beyond the Grande Cache sawmill. This resulted in the loss of the company's FMA except for the area supporting that sawmill.

Despite these economic setbacks, the Alberta government announced the availability of the Brazeau TDA and received 12 proposals by June, 1982. By the end of 1983 there were signs of recovery in the forest industry. Annual production for the sawmilling sector had reached one billion board feet – a record to that point. The ceremonial "billionth board foot" was produced by Millar Western at Whitecourt. By this time, timber volumes were being measured in cubic metres (m³) – the billion board feet represented 4.29 million m³. To put forest harvest volumes in perspective, total annual production, primarily of lumber and pulp, had climbed from three million m³ in 1966 to 7.3 million m³ in 1984.

The ongoing softwood lumber dispute first emerged in October, 1982, when a coalition of U.S. forest companies filed a petition with the U.S. Department of Commerce alleging government subsidies on Canada's exported softwood lumber. The Alberta government and industry collaboratively prepared a response that on preliminary determination in March, 1983, resulted in a judgment that ruled no significant subsidy existed. However, this was to become only the first of a long series of legal interventions by U.S. forest industries that continued into the 21st century.

A particularly significant event occurred in October, 1983, with the opening of an oriented strand board (OSB) plant in Edson. This mill, using aspen as a fibre source, was the initiative of entrepreneur Al Owen of Pelican Spruce Mills. It was the first such mill in Canada and was the visionary forerunner of major investment in this product. As John Zaozirny, Minister of Energy and Natural Resources, noted, "The harvesting of this under-utilized resource and its conversion into a highly acceptable product is of major significance to the future of Alberta's forest industry." He also noted: "With less than 50 per cent of our total timber resources committed to development at the present time, Alberta's long-term opportunities to expand to meet future world demand are highly favourable."[3] These were prophetic remarks.

The year 1984 saw the introduction of another major 'turning-point' in forest policy in Alberta. The *White Paper on Economic Development*, generated in the aftermath of the 1982 economic recession, had a tremendous impact on the security and well-being of the province. Out of a realization that the economy was vulnerable to boom-and-bust cycles in traditional resource markets, the paper promoted economic diversification. There was a call for direct government intervention in stimulating the economy. Alberta's forest resources were identified as relatively untapped and ripe for new investment. The next 15 years saw unprecedented levels of investment with the introduction of world-scale greenfield pulp, panel and lumber mills. This resulted in a near-complete allocation of the province's sustainable softwood and hardwood resources.

Administration

Bob Steele was named Deputy Minister of Renewable Resources in 1973-74, replacing V.A. (Vi) Wood on his retirement. Steele moved to Alberta Utilities and Telephones in 1978, and Fred McDougall replaced Steele as Deputy Minister. Fred was raised in Calgary, received a forestry degree at the University of New Brunswick and became the first field forester at Peace River in 1959. He worked for a while with the Swanson Lumber Co., and then resumed his AFS career in staff positions in Forest Management. He was serving as head of the Forest Management Branch when he was appointed Director of Forestry, and later became Deputy Minister.

J.A. (Al) Brennan was hired as

Weldwood of Canada pulp mill, Hinton
Bob Stevenson

Three Directors of Forestry who would later carry the title of Deputy Minister (L to R): Fred McDougall, Eric Huestis and Bob Steele, c. 1979
Alberta Government, AFHPC

Director of Forestry in 1979, the position at that time renamed Assistant Deputy Minister for the AFS. Brennan was another University of New Brunswick forestry graduate, who was Chief Forester for his home province of Newfoundland before moving to Alberta.

In 1984 Brennan was invited to participate in an intensive 18-month senior-level management course through the Royal Military College at Kingston. C.B. (Cliff) Smith was appointed as Acting Assistant Deputy Minister for the AFS in his place. Smith had worked for the AFS after high school and during the summers of his forestry program at the University of Montana. He began work with the B.C. Forest Service in Prince George in 1963, and moved to Peace River as a forester that fall. His career took him to two section-head positions in the Forest Management Branch. He became Superintendent in Grande Prairie, and, in January 1981 Head of the Forest Protection Branch. When Brennan returned to Alberta from his management course in 1986, he became Executive Director of the newly created Forest Industry Development Division. Smith then became Assistant Deputy Minister for the AFS. He finished his career in the position of Deputy Minister in 1993.

The 1966-67 fiscal year was noteworthy for several other developments besides introduction of the Quota system.

The Forest Protection Division had become more encompassing. It now comprised six specialized sections: Fire Control, Fire Weather, Communications, Fire Research, Mechanical and Equipment Development, and Construction and Maintenance. The AFS owned three helicopters and three fixed-wing aircraft, and leased 14 others. The number of lookouts had increased to 146. Single sideband radios were introduced to provide better communications, especially in the northern areas where VHF showed weaknesses over the great distances and rough terrain.

The Forest Surveys and Planning Division expanded its program to meet growing demands from both the Department and the public. The new Land Use Section handled upwards of 1,500 land surface applications annually, with most of them coming from the oil and gas industry.

In the provincial Forests, many ranger districts were combined into fewer but larger areas "to provide improved working conditions and balanced workloads for ranger staff. Responsibility for the supervision of field activities gradually shifted from head office in Edmonton to Forest and District offices. Major increases in land uses and reforestation practices required staff additions, particularly in specialized fields."[4]

V.A. (Vi) Wood, Deputy Minister from 1966-67 to 1973-74
Alberta Government, AFHPC

A new, more geographically comprehensive, Phase III forest inventory was initiated in 1970 and completed in 1984 (see Chapter Three for details of Phase I and II inventories). The inventory covered the majority of forested land in the province under provincial ownership (395,399 km^2 or 60 per cent of Alberta's land area), outside the national parks. This inventory again focused more on coniferous stands, although the deciduous sample sized was increased. The Phase III inventory was used to support programs that encouraged more investment by forest industries. It was also used in the review and revision of quota allocations in 1986, the first renewals since the Quota system was introduced in 1966. These renewals were based on new utilization standards that obliged Quota holders to use smaller trees and lower-diameter logs. Besides the computer-based Phase III inventory, a program called Timber Revenue Enumeration and Evaluation System (TREES) was completed in January, 1985, to provide for better accounting and statistical reporting in support of forest management.

The Phase III inventory was kept up to date for forest depletions, such as burned areas, timber harvesting, land clearings, etc., and geo-administrative changes such as the Green and White Areas, Forest Management Agreements and Forest Management Units.

Two developments highlighted deficiencies in the Phase III inventory during the mid-1980s. First was the rapid growth of interest in the commercial potential of Alberta's deciduous species, primarily trembling aspen and balsam poplar. It became apparent that the deciduous inventory was weak. Secondly, the need for inventories of other forest attributes as a basis for multiple-use planning was recognized. Consequently a committee of government forestry, fish & wildlife and public lands staff, and holders of Forest Management Agreements, was established in 1984 to design a more comprehensive forest inventory that would provide resource information on timber and other forest-based resources such as wildlife habitat.

Junior Forest Wardens at Forestry Youth exhibit in Edmonton
Bob Stevenson

The new inventory was named the Alberta Vegetation Inventory (AVI). AVI is a computer-based enhanced digital inventory that includes such additional details as understorey and soil/site information. Forest Management Agreement holders agreed to do the bulk of the work to complete and maintain AVI on their respective areas. The company information is shared with the government through data-sharing agreements.

A new Progressive Conservative government was elected under the leadership of Peter Lougheed in 1971. Most AFS programs were initially continued, but all were given intense scrutiny over the next few years.

In 1972, the government commissioned a study into the environmental effects of timber harvesting in Alberta, prompted in part by public criticisms of natural resource management. C.D. Schultz (SHULCO) was hired to do a study in the fall of 1972 and reported in September, 1973. The report made 75 recommendations, but began by noting that: "The Forest Service in its relatively short life has set worthwhile objectives for Crown forest lands, and has successfully advocated legislation, regulations and tenure systems for attainment of these objectives. Related legislation of other departments in government is generally in harmony. Much of this legislation and regulation is too recent for observation of results on the ground."$_5$

Among the conclusions was a statement that: "Timber harvesting can remain as a principal and highly legitimate use of the project area. In many cases trade-offs between this use and competing uses are feasible and would avoid conflicts without major sacrifice."

The report noted that environmental damage associated with roads, both during and after construction, was far greater than that associated with all other phases of harvesting. However, Schultz suggested that much of the damage could be reduced in the future simply by better planning, supervision and attention to detail on the ground. Overall it recommended strengthening AFS capabilities in forest planning and management.

After about four years of study and evaluation, the Alberta government brought about a major reorganization among government departments. On April 2, 1975, the AFS became part of the new Department of Energy and Natural Resources, combining former Lands and Forests with Mines and Minerals. Two Deputy Ministers were named within this new Department, with Bob Steele as Deputy Minister of Renewable Resources. The mandate of the AFS was defined as responsibility for the management of Alberta's forest lands "to ensure a supply of timber products while maintaining a high-quality forest environment."₆ Parks and Wildlife were moved to a new Department of Recreation, Parks and Wildlife.

As part of this reorganization, all government-owned aircraft and the four pilots on AFS staff were transferred to the Department of Government Services. The Construction and Maintenance Branch of AFS was dissolved, and road and airstrip construction responsibilities transferred to the Department of Transportation.

In 1976, the Junior Forest Wardens were returned to the AFS from the former Department of Youth. The organization grew to 106 clubs with over 2,000 members by 1985. The Junior Forest Ranger work program continued with 120 JFRs in 10 field camps in 1976.

The changing nature of the work of the AFS resulted in another reorganization within the Renewable Resources section of the new Department of Energy and Natural Resources in 1977. A new Resource Evaluation and Planning (REAP) Division was established, reporting to the Deputy Minister, with inventory and planning staff transferring from Forest Land Use and the Timber Management Branches. The mandate of REAP was to focus on resource evaluation and planning to try to further reduce conflicts in land use. The AFS was reconsolidated under five branches: Forest Protection, Timber Management, Reforestation and Reclamation, Forest Land Use and Program Support.

The final report of the Environment Conservation Authority (ECA) - *Forest Management in Alberta: Report of the Expert Review Panel* - was issued in February 1979. Bruce Dancik of the University of Alberta chaired this forestry panel. Other members included Des Crossley, Chief Forester for North Western Pulp and Power Ltd., J.F. Reynolds from Grande Prairie and A.D. Crerar from the ECA. The panel's first community meeting was held in September, 1977, and meetings continued through 1978. Their report stimulated a lot of discussion and prompted a reassessment of programs within the AFS. However, it also provided reassurance about many aspects of forest management. For example, the preamble to the report's recommendations on timber harvesting included these remarks:

"Many of the potential environmental problems associated with timber harvesting have been or can be resolved with an intelligent application of the current Ground Rules. The ground rule system works well, and, with continuing input from appropriate personnel in other departments and industry, it can continue to function well and give adequate protection to environmental and other resource values while maintaining viable forestry operations. Continued diligence by personnel of the AFS and other interested departments is needed to ensure that the letter

Professor Bruce Dancik chaired the ECA panel on forest management, 1979
David Holehouse

Lodgepole pine regeneration, south of Hinton. White spruce regen along roadway in the foreground
Bob Stevenson

and the spirit of the rules protect important environmental values. This is not to say the system should become static; it has changed and should continue to change to reflect new information and shifting values."

The metric system was formally adopted by the AFS and other public bodies on November 1, 1979. This decision to conform to national standards was reached after extensive consultation with the forest industry.

In 1980 the AFS celebrated its 50th anniversary. It was a milestone event. Robin Huth's *Horses to Helicopters* was published as an anniversary book. As well, the Junior Forest Wardens (JFW) celebrated the 50th anniversary of their establishment in British Columbia and their 25th year in Alberta – an anniversary celebration was held at the new Alberta Long Lake Camp. The JFW in Alberta by this time included about 1,120 members in 50 clubs.

A new Forest Research Branch was established in 1980, led by Joe Soos. As the Annual Report stated: "Serious reductions in forest research efforts were implemented by the federal government in the past 10 years. It has been recognized by the Alberta government that research results are lacking in many areas of forest resource management which may hinder progress toward more intensive forest management in the province. As a result, a new branch was created in April, 1980, to carry out forest research in seven major areas: site classification, reforestation, growth and yield, genetics, tree physiology, land reclamation and range management, and watershed management."[7]

Sign at the entrance to Pine Ridge Forest Nursery
Bob Stevenson

Advanced Fire Behaviour class, 1984. Education was important to support advances in the science and technology of fire management
Back Row (L to R): Dahl Harvey (PFFC Weather Section), Clayton Burke (NWT), Jamie McQuarrie, Norm Olsen, Frank Lewis, Bob Yates, John Brewer, Rick Lanoville (NWT). Middle Row: Andy Gesner, Darryl Rollings, Herb Walker, Harold Dunlop, Mansel Davis, Bruce MacGregor, Larry Warren, Don Sarafinchin. Front Row: Mike Dubina, Stan Clarke, Maurice Lavallee, Don Dawson, Not Identified (Yukon), Gordon Dumas (Yukon), John Redburn (NWT), Bob Lenton
Alberta Government, AFHPC

These results were to be achieved through mission-oriented research and by ensuring research results were transferred into practice. The Branch was also expected to coordinate work among the other research agencies in Alberta.

A separate Reforestation and Reclamation Branch was set up in 1977 to focus on forest renewal, under the leadership of Con Dermott. Its mandate was to establish, maintain and improve forest stands for timber production and for recreation, grazing, watershed and wildlife benefits. It was responsible for reforestation, for the Pine Ridge Forest Nursery at Smoky Lake and the Genetics and Tree Improvement Program. Construction of Pine Ridge Forest Nursery in 1980 was made possible through a $9-million grant from the Heritage Trust Fund, and it became a flagship nursery operation in Canada.

The elevation of these responsibilities to a

Branch status reflected the AFS commitment to sustained yield, including forest renewal. It was also a tribute to the pioneering silviculturists, including Larry Kennedy and Kare Hellum. Hellum was a UBC forester who led during a time of great experimentation with container-grown seedlings and scarification techniques. He led the search for the new nursery site near Smoky Lake and, with forester Ed Ritcey, developed a computer-based silviculture record system. Hellum left the AFS to teach at the University of Alberta.

The Reforestation and Reclamation Branch also addressed the need for preventative measures and restoration on disturbed sites as a result of resource exploration and extraction. This need was highlighted in the 1979 ECA report on forest management. $_8$ Comparing the relative influence of the forest products industry and exploration by petroleum companies for the 20-year period from 1956 to 1976, the report estimated 25,600 square kilometres of disturbance by forest industry and 23,400 square kilometres by the petroleum industry – almost equal areas. The ECA recommendations emphasized both prevention and restoration.

With increasing industry involvement in reforestation, the demand for planting stock increased greatly. This led to trials of bareroot stock in nursery beds and a variety of containers in greenhouses to develop greater efficiencies in seedling production. The AFS nursery at Oliver had been expanded to increase bareroot stock, but increasing problems with soils and weeds led to a search for a new location with more suitable soils and room for expansion. A new site was selected near the town of Smoky Lake on which the new Pine Ridge Forest Nursery was constructed. It was completed in 1980 with 13.2 million seedlings being grown in two nursery buildings and additional bareroot stock started in outdoor beds. By 1984 the facility was shipping 9.3 million trees and, in that same year, a special tree-planting ceremony was held to celebrate the shipping of the 100 millionth tree. Also in 1980, a Silviculture Improvement Advisory Committee was established, combining industry and AFS members in a cooperative approach to research and field trials. In 1982 the AFS purchased land in the Grande Prairie area for a seed orchard complex that was jointly developed with forest industry.

Seedlings delivered for field planting project, Tony Tower area, southwest of Fox Creek, late 1980s
Alberta Government, AFHPC

Kare Hellum, former AFS Silviculturist and U of A Professor, 1993
Bob Stevenson

The first "satisfactory reforestation" (SR) deadline under the Quota System was reached in 1976. All logged areas had to be surveyed for regeneration success by the end of the seventh year following harvest. Not sufficiently regenerated (NSR) areas had to be planted in the year following the survey, and surveyed again in the tenth year. Significant increases in industy reforestation led to greatly increased planting programs. The total number of seedlings planted by the AFS and industry passed the seven million mark in 1976, according to the AFS statistical record. This collaborative approach resulted in increases to over 10 million seedlings in 1980 and 17 million in 1984.

Plans Course participants, 1972. The course prepared personnel for the position of Plans Chief, one of the key members of a wildfire overhead team
Back row (L to R): Harold Enfield, Bob Plankenhorn, Tom Oliver, Lou Foley, Keith Whyte, Ed Stashko, Dave Cox. Second row from back: Bob Miller, Ken McCrae, Bob Petite, Clyde Ulm, Ken South, Gordon Bisgrove, Cliff Henderson. Third row from back: Norm Woody, Art Peter, Gordie Masson, Oliver Glanfield, Ed Pichota. Front row: Bill Fisher, Brian Carnell, Leon Graham, Bill Francis, A. Forbes, Rick Manwaring and Harry Jeremy
Alberta Government, AFHPC

A Maintaining Our Forests (MOF) project was approved under the Alberta Heritage Savings Trust Fund, and initiated in April, 1979. It was proposed as a seven-year $25-million program to establish and maintain coniferous timber stands. This infusion of money resulted in significant additions to regeneration surveys, site preparation, planting and stand spacing. The program came out of a review conducted by the AFS in 1978-79 into the depletion of the coniferous resource by activities other than authorized timber harvesting.

The MOF program was designed to stop the erosion of the coniferous landbase. It included four themes: reforestation & afforestation of potentially productive lands, wetland drainage and tree-growth improvement trials, tree improvement (genetics), and stand improvement (spacing and thinning of wild stands). The program was approved for $25 million, starting in 1979-80. For logistical reasons the reforestation sub-program ran for eight years. A major proportion of the reforestation program was spent on conversion of young, very dense aspen stands to white spruce, which reflected the conifer bias that prevailed at the time.

The closing report for MOF provides the following information on expenditures and program achievements to 1987:

Reforestation **($22,933,774):** stocking surveys, stand conversion, 64 million seedlings planted, cone collection for seeds, aerial seeding, stand thinning.

Tree Improvement **($962,895):** superior trees identified, two seed banks, clone banks and one research arboretum established, planting stock for genetic research and 6,564 grafts, more than 700 seedlots added to the genetic seed bank.

Stand Improvement **($865,978):** stand thinning.

Wetland Tree-growth Improvement **($200,275):** drainage ditches on 35 ha of treed muskeg.

Evolution of fire management

Although progress had been made in building fire control capability, the system was severely tested in the spring of 1968. As Bob Steele, Director of Forestry, reported: "The many improvements during the year were marred by the worst forest fire year in Alberta's history. Fires destroyed almost one million acres of forestland, with most of the damage occurring during one week of extremely unfavourable weather in May. Settler fires became uncontrollable and

Type III firefighter training, south of Fox Creek, 1986. Mike Dubina giving instruction
Alberta Government, AFHPC

swept into the forested areas in central Alberta with unprecedented vengeance. Only a small per centage of the fire-killed timber could be salvaged. This problem further demonstrates the necessity for greater control of settlers' burning practices and a much improved weather

Trainee firefighters take morning fitness walk, south of Fox Creek, 1986
Alberta Government, AFHPC

forecasting system."₉

The strong connection between weather and fire behaviour had long been recognized. The problems were first to obtain reliable weather data over the full geographic region, second to try to quantify what the data meant in terms of fire hazard, and then to interpret the data and forecast fire weather and behaviour. The Canadian Forest Service fire research staff had developed tables based on weather readings that gave a "Fire Danger Index" to indicate relative ease of ignition and fire intensity. These were adapted to Alberta in the 1950s. These tables illustrated the importance of weather readings and, with support of Canada's Meteorological Service, properly equipped weather stations were extended to all lookouts and most ranger stations. Radio Branch handled the collection and compilation of readings, which were also sent to federal forecasters and climatologists. Forest Protection sent out daily generalized assessments of fire danger, with the Training Branch including fire weather in its programs for lookoutmen and fire management staff. During this time, federal forecasters also began to focus more on the warning factors such as low relative humidity, cold fronts and instability.

The appointment of W.J. (Jock) McLean as fire weather officer in 1963 was the first major advance in applying weather to fire management. This appointment was a result of the urgings of Frank Platt. Then, in 1967 Jock McLean recommended that Alberta hire its own meteorologist on an experimental basis to test the idea of a specifically tailored forecast and closer liaison with both the federal weather service and AFS fire control staff through to the firefighter. Meteorologist Ed Stashko was hired to fill this position.

B-26 drops water mixed with fire retardant
Alberta Government, AFHPC

Dale Huberdeau, Forest Protection Officer, on river patrol
Alberta Government, AFHPC

Although reaction was positive, tentative plans in 1968 for continuation of such service fell through when operating funds were slashed. Ironically, the spring of 1968 was also the year of the disastrous fires in which more than a million acres burned. This included the Vega fire that spread nearly 40 miles in 10 hours, threatening the town of Slave Lake. However, the experience highlighted the importance of fire weather forecasting.

In 1969 the AFS Fire Weather Office commenced full-time routine operation with Blane Coulcher as permanent meteorologist and Ed Stashko added for the summer. For the 1970 and 1971 seasons Ed was the permanent meteorologist. Two additional meteorologists, Bill Meheriuk and Joe Eley, were on staff and available and, for the first time, a field forecaster was posted at Footner Lake Forest for two months. Under the able and innovative supervision of these meteorologists and Dahl Harvey, Ben Janz and Nick Nimchuk at AFS headquarters, fire weather and fire behaviour forecasting became integral components of Alberta's fire management system.[10]

As fire weather forecasting improved through the 1960s, the Canadian Forest Service was introducing the Canadian Forest Fire Danger Rating System (CFFDRS). It was a product of over 50 years of research with data that included test fire observations from Alberta and across Canada. The CFFDRS was based on observations of experimental fires and definitive measurements of fuel moisture. The system predicts fire spread and fire intensity through weather, fuel and topography inputs. Dave Kiil was the senior fire scientist with the Canadian Forest Service in Edmonton who, along with Dennis Quintilio, Dennis Dube and George Chrosciewicz, conducted experimental burns to improve the data on fire behaviour elements of the new danger rating system. An impressive series of over 60 of these carefully planned and recorded burns was done during the 10-year period from 1965 to 1975 in cooperation with AFS. The aim was to develop and quantify the fire behaviour indices and to assess the potential of fire as a silvicultural tool. The fires also served as training exercises. Some of the burns were also monitored to assess post-fire vegetation and stand succession. Fires were ignited in study areas at Hinton, Kananaskis Forest Experiment Station, Slave Lake, Darwin Lake and Steen River.

The Darwin Lake trial in northeastern Alberta, for example, was conducted in 1974 by AFS and CFS staff. Seven research fires were set in jack pine stands and measured to provide reliable data to strengthen the components of the CFFDRS. Photographs clearly illustrated the changes in fire intensity and have been widely used in training. Participants included Dale Huberdeau from AFS, Dennis Quintilio from CFS and Charlie Van Wagner from CFS Petawawa along with firefighters from Fort Chipewyan.

In 2005, nearly 200 weather monitoring sites were used to provide field staff and managers with up-to-date fire behaviour and weather forecasting information. This information has been shared with the federal government for the past 30 years and represents the only forestry data used in the national climate archives.[11]

Fire management was given a boost in 1969 when the new Provincial Forest Fire Centre – the "Fire Depot" – was opened in Edmonton on the east side of the City Centre Airport. It brought the fire control elements together including the warehousing, equipment development and construction activities. This provided greater coordination and collaboration. A DC-3 transport aircraft was also purchased to enhance transport of fire crews, and in 1973 the first provincial firefighting competition was held.

Refinement of fire management capabilities continued into the 1970s. Three new airtanker bases were established in 1977 and infrared sensing devices were introduced. The system was again severely tested in 1980 with 1,300 fires and 670,000 hectares burned. It was a

Lt. Gov. Ralph Steinhauer (standing centre back) with Sandy Lake's winning squad (Alexander First Nation) in a firefighter competition, 1976. Ranger front left is Maurice Lavallee
Alberta Government, AFHPC

$26-million year for fire costs, and was a precursor to what was described as the worst fire season to date in 1981, with 1,556 fires and over 1.3 million hectares burned. Firefighting costs that year exceeded $60 million. Lightning caused most of these fires and one of them, the Moosehorn fire, forced the evacuation of the town of Swan Hills in early August. More than $5 million was spent controlling this fire in a period marked by no rain.

The 1981 season was a catalyst for reassessing the fire management system and for instituting additional major changes. While these major changes were being evaluated, 1,285 fires occurred in 1982, costing $75 million to fight and burning 680,000 hectares. The fires of 1980 and 1981 prompted the Timber Salvage Incentive and Forest Employment Bridging programs, designed to help the forest industry salvage burned timber and to provide an employment opportunity for woods workers.

As Dale Huberdeau, Fire Protection Officer, commented, "… the Protection Branch made the most significant changes from the disastrous year of 1981. To about 1984 we pretty well shook the organization upside down and built a new one in terms of forest protection. There were some very significant changes and advances made."[12]

One improvement was the introduction of rappel training, using a new tower constructed at the Forest Technology School at Hinton. Graduates were employed as helitack teams, rappelling down onto early-stage fires from helicopters. By 1983 a new Presuppression and Preparedness Resource System (PPRS) was implemented, with the objective of allowing only a 15-minute initial attack response time for all fires during high and extreme hazard periods. To help make this possible, eight helitack and three 25-person university student crews, together with air support, were employed for the fire season. Also that year, the AFS completed negotiations to participate in a federal-provincial agreement to buy two CL-215s each, for a total of four to be supplied in 1987. A DC-6-B retardant airtanker was also added to the provincial fleet of airtankers – bringing the total to 11 B-26s, three DC-6-Bs and three Cansos.

The system was severely tested again in 1984 with the second-highest incidence of fires in

First NAIT graduating class at Hinton, 1966
Back Row (L to R): George Nemeth, Brian Carnell, Gordon Bisgrove, Horst Rohde, Stan Lux, Blaine Dahl, Brent Simmonds, Revie Lieskovsky, Ron Gordey, Rod Gustafson. Middle Row: Jack Susut, Don Campbell, Bruce Cameron, Emanuel Doll, Bart Presley. Front Row: Larry Huberdeau, Louis Kilarski, Archie Smith, Dennis Cox, Bill Kovach, John Edwards, Francis Donnelly, Ken Paulson, Bruce Robson. Missing: Arnold Mogdan. Instructors were Dick Altmann, Stan Lockard, Jack Macnab, John Wagar and Peter Murphy
Alberta Government, AFHPC

First graduating class of the Forestry Crew Worker program 1981, Alberta Vocational College, Lac La Biche
Back Row (L to R): Bobby Cardinal, Rene Noulta, Robert Harrison, Marcel Bruneau, Leroy Goodeagle, Bert Varty (Program Head), Alvin Cardinal, Lou Bougie, Douglas Smith, Jeff Reynolds, Toby Desjarlais, Ken Tyler. Front Row: Evelyn Calliou, Leona Buckler, Cathy Butler, Louise Babin. Both Evelyn Calliou and Doug Smith became Forest Officers with the Alberta Forest Service. Margarete Hee became program head in 1982
Alberta Vocational College

history. The total rose to 1,368, with a record 654 fire starts during 16 days in mid-summer.

Early detection and good preparation resulted in only nine fires escaping early containment, although 70,998 hectares burned. By this time an automatic lightning detection system had been installed and was working well. Alberta's capability in fire detection was recognized in 1984 when the AFS was asked to participate in a Canadian International Development Agency (CIDA) project on the Jiagedaqi model forest management project in the northeast Chinese province of Heilongjiang. Fire detection specialist Joe Niederleitner was seconded to the project to develop the fire detection system.

First Forestry graduating class at the University of Alberta
University of Alberta

Forest Health

The mandate of the Forest Protection Branch was redefined to: "protect Alberta's forests from damage and destruction by fire, insects or diseases and to administer the *Forest and Prairie Protection Act* and Regulations." The Branch was also to provide meteorological, emergency communication and survival services as well as statistics and analysis on forest fires. Also during the 1977-78 year, a "minor outbreak" of mountain pine beetle was discovered in the Crowsnest Forest, the first occurrence since the early 1960s. This was to evolve quickly into a major sanitation, salvage and burn operation assisted by the Canadian Forest Service. By 1984, the epidemic appeared to have stopped spreading.

Education and Research

In 1970, the University of Alberta approved a new BSc degree program in Forestry within the newly named Faculty of Agriculture and Forestry. This provided a local opportunity to obtain a forestry degree, and marked the beginning of an additional forest research capability for Alberta.

Dr. Fred Bentley, noted soil scientist and Dean of Agriculture at the university, had earlier discussed the idea of a forestry program with leaders such as Bob Steele, local Canadian Forest Service Director (Dr. G.P.) Phil Thomas and Arden Rytz, Manager of the Alberta Forest Products Association. Dr. Fenton MacHardy succeeded Bentley as Dean in 1970. He led the new program, along with Dr. Steve Pawluk, another soils scientist who had conducted research into the identification of lands that were most suitable for forestry.

The first forestry class enrolled in the fall of 1970. The program was directed largely by Steve Pawluk with Dr. Bill Corns, a plant scientist familiar with forested ecosystems. The Forest Technology School in Hinton organized a first-year camp for University of Alberta students in 1971, to provide more specific orientation before summer work. Two new teaching positions in the program were filled by Dr. John D. Schultz (program chair) and Jim Beck, who was still working on his PhD thesis but came well recommended. Jack Schultz resigned as chair in January, 1973 and P.J. Murphy took that position in the spring of 1975.

The first class graduated in the spring of 1974, with Dr. Allan Warrack, Minister of Lands and Forests, as guest speaker. Eight of those graduates later worked with the AFS or related provincial resource agencies: Keith Branter, Ryerson (Morley)

Christie, Bill Gladstone, Fred Moffat, Rod Simpson, Doug Sklar, Brydon Ward and Mel White.

The Forest Development Research Trust Fund was established in the early 1970s, through the AFS, to encourage greater collaboration among forest research organizations in Alberta and to provide additional funding for forest research.

Teaching an awareness of and appreciation for forests in Alberta was a particular responsibility of the Alberta Forestry Association, successor to the Prairie Provinces Forestry Association of the 1930s and the Canadian Forestry Association formed in 1900. From 1962 Greg and Gladys Stevens toured schools annually throughout Alberta with their small travel trailer to spread their messages of forest and wildlife conservation, as well as fire prevention. This genial couple influenced thousands of students before retiring in 1983, after 21 years of outreach work.

The Forest Technology School's Forest Fire Simulator won the Alberta Forestry Innovation award in 1990 and has enjoyed a long and very successful track record.

The simulator's beginnings stem back to 1967 when the institution's first fire simulator was purchased from Decision Systems Inc. of Paramus, New Jersey. The simulator was employed as a command and control decision-making training system for fire supervisors who had undergone basic theory training in fire behaviour, suppression resources, tactics and safety. It was a good machine in its time. The equipment consisted of a large console housing 5,000 watts of light power bounced through a series of mirrors, prisms, rotating discs, carbon paper and a single 35 mm slide depicting fuel and topography. The simulator created the illusion of smoke, flame and char moving across the topography through a slide projected on a screen.

A combination of radio/telephone communications and a multi-track tape sound effects system helped trainees make informed fire suppression decisions based on what they saw on the screen and what they heard.

When Rob Thorburn, the new fire management instructor, arrived at the school in 1986, he recognized that after almost 20 years, the machine was outdated. Wires were bare, sockets were corroded, and it seemed that there was more smoke being generated in the control room than on the screen. To add to this dilemma, the technology was no longer catching the interest of the students who, by this time, were familiar with the IMAX theatre, 360-degree movies with surround-sound and computer-based virtual reality. All this combined with the fact that set-up time amounted to several hours caused Thorburn to realize that with the technology of the day, there might be an easier and more effective way to deliver simulation.

Planning began in 1987 to consider application of laser disc technology to support a new fire simulator system, and it was this work that ultimately led to the award-winning interactive CD-ROM training program. Dennis Quintilio, Marty Alexander, and Terry Van Nest had spent two weeks at the University of Idaho scripting a laser disc fire behaviour course. When they returned with the final product Thorburn quickly

Greg and Gladys Stevens, well known throughout Alberta for leading the Alberta Forestry Association school tours for over 20 years
Alberta Government, AFHPC

(L to R): Butch Shenfield, Don Law, Ed Johnson Participants are roleplaying Overhead fire team positions in a FTS simulator exercise. This tool allowed people to learn and make mistakes in a 'friendly' environment, 1980s
Terry Van Nest

began adapting it to simulation training in Hinton. Though still in its infancy, laser disc technology offered exciting potential for 60 minutes of motion video and 108,000 colour slides accessible within two-tenths of a second. Thorburn conceptualized a design for a control software package and contracted a private company to do the programming.

Interface electronic hardware to connect to a computer was all hand wired by Thorburn at the school. By 1989 a working prototype had been evaluated and it led to a more permanent system by 1990. Word soon got out nationally and internationally, and Thorburn was hosted by agencies around the world to demonstrate the Alberta simulator. A complete fire simulator was built by Thorburn for the British Columbia Institute of Technology in 1992 and is still in use today. The California State Fire Marshall was the first non-Canadian agency to take delivery of another system and today many agencies around the world utilize this simulator for both training and certification purposes.

The simulator has undergone many transformations over the last 15 years. Laser discs have been replaced by digital video stored in a computer. Audio tapes have been replaced by digital audio. Software includes accommodation for computer-based fire growth modelling, simulated time clocks and real-time capture of a student's performance. The simulator is the best hands-on training that can be provided in a classroom setting and provides a safe environment for learning invaluable lessons.

Land Use Planning

The first major land-use issue in Alberta emerged in the 19th century as agriculture and settlement began to put pressure on forested areas.

On one hand, clearing forests on the better soils made sense and resulted in successful farms. On the other, clearing on poor soils did little to produce good farm livelihoods. In addition, there were associated challenges of out-of-control wildfires started by settlers clearing their land, loss of wildlife and natural habitat, and a demand for remote roads and schools that the province could ill afford.

In 1961 a new Land Use Section led by Gordon Smart was set up in the Forest Management Branch, largely to handle the

Gordon Smart, (R) Administrator of the new Land Use Section, 1965 with Senior Assistant Chuck Geale
Alberta Government, AFHPC

Coring samples for coal, Panther River area, Bow River Forest, 1969
Bob Stevenson

growing impacts of the energy sector. This was a different kind of "multiple-use" activity – one that disturbed the forest but did not require any of the forest resources to produce its product, nor was it amenable to management through land classification. Prior to the establishment of this new Section, field staff were operationally focused and did not have the tools or expertise to do effective land use planning. Their role revolved mostly around identifying sensitive sites before activities were approved, but the review periods were short and development pressures high.

Coal exploration trail on Dog Rib Ridge south of Ya Ha Tinda, Red Deer River area, Bow Forest, 1969
Bob Stevenson

Forest Surveys staff, as early as the mid-1950s, began trial land classification systems. Foresters Ron Fytche and Gordon DeGrace interpreted landforms on air photos, in an attempt to rationalize other forest-based activities such as agriculture, grazing, oil and gas. In 1969, Land Use was made a full branch under the direction of Gordon Smart with responsibility for managing forest land uses (also including recreation, watershed, erosion and reclamation) and planning for them. Land Use Branch staff also became involved in coordinating the Canada Land Inventory (CLI) Forestry Sector mapping. This was a federal-provincial program under the *Agricultural and Rural Development Act* (ARDA), through which estimates were made of the land's suitability and productivity for a number of potential uses, including forestry, recreation, wildlife and fisheries.

Brush piles from construction of the Big Horn Dam reservoir, west of Nordegg, Clearwater Forest, 1969
Bob Stevenson

During this time, the variety and intensity of land use impacts on the eastern slopes began to highlight conflicts among users and uses such as grazing, recreation, fishing and hunting, oil and gas and logging. Through a combination of these uses, damage to many individual sites occurred. Two inter-agency projects, the Foothills Resource Allocation Study and the Hinton-Yellowhead Regional Land Use Study were launched in 1970 to identify important and sensitive sites and to try to prevent and mitigate land-use conflicts through zoning. These exercises built on the study-team approach to evaluation and allocation of land identified by the CLI mapping.

Around the late 1960s, the AFS became aware of growing public concerns about industrial impacts through volumes of calls to field staff and articles in the press. One response to this pressure was intensive reclamation work to recontour and eliminate roads, trails and pits caused by coal exploration machinery in alpine areas.

Public lands are also managed for grazing
Alberta Government, AFHPC

In 1969 Bob Steele, Director of Forestry, reported that AFS head office was reorganized "because of increased interest in other uses of forest land and the necessity to implement greater administrative control of forestry policies and practices."[13] The resulting six branches took responsibility for Administration, Construction and Maintenance, Forest Land Use, Forest Protection, Timber Management and Training. Activity in the Forest Land Use Section was a reflection of growing land use concerns. Steele reported in 1971: "The public has become much more concerned in recent years with the manner in which the forested areas of the province are being managed. This has resulted in careful and frequent reviews of forest land policies and practices and in greater complexities in administration."[14]

In 1971 an erosion research project was initiated to study soil loss and reclamation in the Swan Hills area. It was undertaken by the AFS with staff from the Eastern Rockies Forest Conservation Board. The study led to recommendations that the Canadian Petroleum Association (now Canadian Association of Petroleum Producers) initiate corrective action and has led to increasing collaboration between the two sectors.

The mandate of the Forest Land Use Branch was to manage Alberta's public forestlands for watershed, grazing and recreational benefits, as well as reviewing applications for surface disturbances. This included looking after the 130 AFS campgrounds throughout the Green Zone; coordination of emergency action on oil and chemical spills; review of applications for surface rights such as seismic operations, sand and gravel deposits, coal mines, wellsites, pipelines and roadways; and administration of grazing of domestic livestock in the Green Zone.

The Environment Conservation Authority (ECA), an independent government-funded agency, conducted hearings into *Land Use and Resources Development in the Eastern Slopes* to identify public interests in and concerns about this foothills area. The ECA recommendations in 1974 identified watershed and public recreation priorities and called for an integrated resource policy and land use planning.

In response to the ECA recommendations, and encouraged by the work of the Foothills Resource Allocation Studies, a separate Resource Evaluation & Planning (REAP) division was set

Rider and packhorse cross the Clearwater river
Bob Stevenson

Oil leaked from wellhead in the Red Earth District, Slave Lake Forest, June, 1971. One duty of a ranger was to do land use inspections where environmental problems such as this were discovered. The ranger would then contact the company and ensure that the site was cleaned up and restored
Jamie McQuarrie

up in 1976 under the direction of Les Cooke. Cooke implemented the pioneering land-use zoning method advocated in the Foothills Resource Allocation Study, with the support of Deputy Minister Fred McDougall. It was decided at this time that the forest inventory function should also be centralized so that all agencies could have input and access to it. The inventory unit took over delivery of the Phase III forest inventory and all mapping functions. It also initiated physical and ecological land classification in support of the integrated resource planning effort.

In 1977 the use of forested lands by petroleum and natural gas industry resulted in a 25 per cent increase in applications that year to 2,688, and seismic line approvals increased to 6,325 miles from 3,643. In addition, AFS was directly involved with the administration of five active and two proposed coal mines.

The REAP evaluation and planning within the foothills culminated in 1977 in the approval of a *Policy for Resource Management of the Eastern Slopes*. This policy document described eight land use zones, listed permitted and prohibited uses within each, and delineated the boundaries of each zone on a series of maps. The area essentially encompassed the five original Forests of the Rocky Mountains Forest Reserve from the Crowsnest in the south to the Athabasca in the north. It included the headwaters of the Saskatchewan, Athabasca and Smoky River systems. The eastern slopes policy addressed oil and gas exploration and development using a compatibility matrix. For the first time these activities were prohibited in those zones designated as Prime Protection, and constrained within Critical Wildlife zones. Other trial Integrated Resource Plans (IRPs) were prepared for agricultural land expansion and oilsands development areas.

In 1980 enabling legislation permitted the Department to establish Forest Land Use Zones, or areas of land to which controls may be applied to resolve specific land-use problems. They were largely designed to manage recreational motorized vehicles, for example, to protect ecologically sensitive sites and critical wildlife habitats. This authority made it possible to extend aspects of eastern slopes-type zoning throughout the rest of the forested area, as well as addressing specific problems in the eastern slopes.

The eastern slopes policy was revised in 1984 "to provide for the maximum delivery of the full range of values and opportunities in this important region." It was a policy that prevailed into the new millennium. REAP continued its land use studies in an attempt to help rationalize and mitigate land-use conflicts. It achieved some successes within the renewable resource sectors, including forestry, wildlife, fish, watershed, grazing and recreation. However, the energy sector had become a major influence on land uses and, given the time-dependent nature of its exploration and development, it remained largely outside land use planning efforts.

Lodgepole Blowout

Forest Officer Lloyd Seedhouse of Cold Creek Ranger Station recalls (personal communication with the authors) the blowout that was reported to Ken McCrae of the AFS at 1820 hrs on October 17, 1982. His activities reflect the considerable involvement that AFS staff had with other land users within the forest.

"On October 18, Gordon Jaap (Chief Ranger at

View of the Lodgepole blowout after fire erupted on November 1, 1982
Ken McCrae

the same station) and I went to the site late in the afternoon to meet with Amoco representatives, although we were not allowed on the site because of the sour gas. At another meeting on October 25 we met with Amoco staff again regarding a temporary bridge across the Pembina River and Zeta Creek. Amoco wanted this bridge as an emergency escape route to the south, in case the only access from the north became cut off for any reason. Amoco had constructed a road from the 1978 blowout site down to the river. As we walked along the road I saw Amoco had placed Scott Packs (emergency breathing apparatus) for us to use in case there was a wind shift and we had to scramble out of there.

"On November 1, Amoco tried to release the pipe stem in the wellhead and the pressure of the gas shot the pipe out through the derrick. I was told a piece of the pipe hit the derrick causing a spark, which started the first fire. When Ken McCrae and I flew the area at 1615 hrs about 12 to 16 hectares had burned.

"On November 3, I was on site with the ERCB to discuss plans for cleaning up the condensate from the well that had settled on the ice on the Pembina and Zeta. Now that the rig was on fire it was safe from a poisonous gas point of view. AFS staff were required to make a number of decisions, such as giving permission for Amoco to clear trees on the site, because they were soaked in condensate and had burned. On November 4, I gave Amoco permission to clear a new wellsite to the north for a relief well. There was some question later about what authorization I had to do this, but heck, it was an emergency and I made an emergency decision - right or wrong.

"The wellhead fire was extinguished on November 14 but it re-ignited on November 25, burning until December 23. We had a joint meeting with Amoco about clean-up and reclamation in early January, 1983. A large area around the blowout was sprayed with condensate and it was felt the trees would die. We felt these trees should be removed to prevent a fire hazard and not to waste the timber.

"Later in January we laid out the area for tree removal. No one had any idea of how much condensate spray it took to kill a tree, so we went to the edges of the affected area and if there was about six millimetres of condensate on the snow and the trees felt oily, we felt they would probably die and should be removed. The salvage operation wasn't fully completed right away, and our predictions eventually proved to be correct."

Seedhouse said that as for the condensate that sprayed out from the ruptured wellhead, it filtered down through the ground and flowed along the bedrock to recovery pits dug by Amoco.

Women in Forestry

Lola Cameron started work in the office of the Director of Forestry in 1946, retiring in 1981. She became secretary to Director Eric Huestis when he was appointed in 1948 and served in that capacity through three successive directors. Well organized and remarkably capable she assumed responsibilities of a director before such positions were common and women were considered capable of assuming them. In contemporary times she would have been a Director of Administrative Services. She functioned as such without the official recognition and related salary. Cameron saw that decisions were carried out through a combination of phone calls, telegrams (later teletype messages), memoranda and directives. Not one to usurp or assume authority, she worked effectively within her responsibilities, but with a flair and understanding that inspired teamwork and action. She got to know all the field staff, forestry, wildlife and fisheries, by face and name as they made periodic visits. She committed them to her capacious memory, which helped immensely to bind the organization into a semblance of a like-minded family.

Cameron deserves singular recognition for what she achieved in communication, personnel relations, records and myriad executive services as the agency grew in size and complexity during that post-war period. She epitomized the spirit of the Forests and Wildlife family during her 35 years of service.

Lola Cameron at her retirement party, Edmonton, September, 1981
Sheryl Gogerla

Construction of road access to the Fort McKay Ranger Station from a winter crossing on the Ells River, Fort McMurray, Lac La Biche Division, mid-1960s
Corinne Huberdeau

Reg Loomis was Senior Superintendent of Forest Surveys from 1949 to 1959 and Director of Forest Management from 1959 to 1968. Eric Huestis brought Loomis to Alberta in the fall of 1949 to make certain that the forest inventory was done correctly. Loomis' vision for Alberta forestry was "to set up the whole province on a sustainable basis." When the first inventory was nearing completion in 1955 he helped to negotiate the first Forest Management Agreement with North Western Pulp and Power Ltd. at Hinton (now Hinton Wood Products, West Fraser). He was successful in ensuring a working commitment to sustained yield forest management, including prompt reforestation of harvested areas. He led the forest management planning initiative for all Alberta's provincial forest lands. In the process he developed silviculture programs; got the saw milling sector committed to sustained yield; and with the growing impact of oil and gas activities in the forested areas, set up a land-use unit to manage for values other than timber.
Alberta Government, AFHPC

Forestry Dodge power wagon in front of Fort McKay Ranger Station, Lac La Biche Division, mid-1960s
Corinne Huberdeau

Larry Huberdeau, 1st NAIT Graduating Class, Forest Technology School, Hinton, May, 1966
Alberta Government, AFHPC

Access road to the Skyline Ranger Station, Crowsnest Forest, 1960s
Alberta Government, AFHPC

Bruce MacGregor standing beside an AFS International Travelall outside Doucette Tower. MacGregor was part of cruising party on timber licence S4-L7, Slave Lake Forest, October, 1966
Bruce MacGregor

Motor Vessel Athabascan leaving Ells Junction with cat on barge, Fort McKay area, Lac La Biche Division, mid-1960s
Corinne Huberdeau

Rangers Gary Davis (L) and Wally Walton cruising along the Moosehorn River in waist high snow, Slave Lake Forest, 1960s
Lou Foley

Rangers Gary Davis (L) and Lou Foley, on timber cruise near Moosehorn River, repairing Nodwell track, Slave Lake Forest, 1960s
Lou Foley

Ranger Gary Mandrusiak checks poorly-installed culvert. 'Hanging culverts' are barriers to passage of fish and other aquatic life
Alberta Government, AFHPC

Staff at Edson Ranger Meeting, Edson Forest, 1969
(L to R): Archie Miller, Ben Shantz, Don Fregren, Hank Ryhanen (Forest Superintendent)
Alberta Government, AFHPC

Forestry float entered by the Edson Forest winning a prize, Town of Edson, 1966. Ranger Dan Jenkins left side of float, Edwin Preece holding trophy
Alberta Government, AFHPC

Basic Ranger Course, Forest Technology School, Hinton, 1967
Back Row (L to R): Gary Schneidmiller, Rocky-Clearwater; Howard Gray, Footner Lake; Eric Young, Whitecourt; Lou Foley, Slave Lake. Middle Row: Les Welsh, Slave Lake; Dennis York, Slave Lake; Ron Blauel, NWT Forest Service; Wally Manchester, Edson; Pat Wilson, Slave Lake; Bill Bereska, Athabasca (Fort McMurray). Front Row: Harold Boissy, Footner Lake; Edwin Preece, Edson; Bob Welch, Lac La Biche; Keith Franklin, Bow River; Wade Lamoureux, Parks Canada, Whitehorse
Alberta Government, AFHPC

Lou Foley, Basic Ranger Course, Forest Technology School, Hinton, 1967
Alberta Government, AFHPC

Bill Bereska, Basic Ranger Course, Forest Technology School, Hinton, 1967
Alberta Government, AFHPC

Howard Gray, Basic Ranger Course, Forest Technology School, Hinton, 1967
Alberta Government, AFHPC

Dennis York, Basic Ranger Course, Forest Technology School, Hinton, 1967
Alberta Government, AFHPC

Alberta Forest Service

Joe 'Lucky' Lieskovsky was awarded the Supreme Award, Adult Category at the International Competition of Forest Fire Prevention Posters in Ottawa, September, 1969
Revie Lieskovsky

Alberta Forest Service Bertie Beaver sign made by Joe Lieskovsky, February, 1964
Revie Lieskovsky

Bertie Beaver Fire Danger sign made by Joe Lieskovsky, February, 1964
Revie Lieskovsky

Footner Lake Headquarters staff, Footner Lake Forest, fall, 1978 or spring, 1979
(L to R): Ed Gillespie, Phil Dube, Don Fregren (Director Timber Management, Edmonton), John Best, Dennis Driscoll, Steve Luchkow (hidden), Cliff Henderson (Superintendent), Ed Ritcey, Ken Porter, Rick Manwaring, Henry Desjarlais
Rick Smith

District Staff, Footner Lake Forest, fall, 1978 or spring, 1979
(L to R): Rick Stewart (partially cut off), Bill Bereska, John Rizok, Kelly O'Shea, Randy Panko, Conrad Bellows, Bernie Gauthier, Paul Steiestol, Ian Hancock, Rick Smith, Brian Meads, Jurgen Moll, Len Wilton, Brian Wudarck.
Missing: Dale Huberdeau
Rick Smith

Forest Officers outside Fox Creek Ranger Station, Whitecourt Forest, February, 1983
(L to R): Shawn Milne, Rick Smith, Mike Lambe, Ian Hancock
Fred Paget

Bridge crew camp for construction of Firebag bridge, Fort McMurray, Athabasca Forest, winter, 1968. Ernie Ferguson's car at camp
Corinne Huberdeau

Construction of the Firebag Bridge, Fort McMurray, Athabasca Forest, 1968. Photo shows east end of 125-foot bridge
Corinne Huberdeau

Advanced Ranger Class, Forest Technology School, Hinton, 1969
(L to R): Jim Young, John Klassen (Yukon), Andy Kostiuk, Dave Brown, Gerry Campbell (Parks - Waterton), Doug Quinnell, Dennis Howells, Hylo McDonald, Ken Hennig, Vern MacRoberts (NWT), Gordon Jaap, Gary Giese, Wilfred (Jock) Kay, Ron Langeman (Parks - Jasper), Steve Zacharuk, Lawrence Johns, Fred Schroeder, Maurice Verhaeghe, Harold Ganske, Laverne Larson
Alberta Government, AFHPC

Participants in the first Fire Behaviour Course given to Fire Control Officers and Technicians, Forest Technology School, Hinton, October, 1969
Standing (L to R): Ken Janigo, Ben Shantz, Art Peter, Chuck Rattliff, Bob Diesel, Emanuel Doll, Del Hereford, Bert Varty, Lou Boulet, Jack Naylor, Colin Campbell, Bernie Simpson, Carson McDonald, Art Lambeth, Dick Mackie, Irv Allen, Norm Rodseth, Jim White (Parks), Irv Frew. Sitting: John Benson, John Booker, Joe Kirkpatrick, Len Allen, Ernie Ferguson, August Gatzke
Alberta Government, AFHPC

Forester Bill Fairless cruising in the Fox Creek area, 1970
Cliff Smith

1966 - 1984   191

Dale Huberdeau hauling 30 kegs of aviation gas (2400 lbs) for a timber management cruise to Clausen's Landing, 35 kilometres north of Fort McKay, Athabasca River, Athabasca Forest, September 26, 1970
Corrine Huberdeau

Forest protection planning meeting at the Provincial Forest Fire Centre (Depot), Edmonton, early 1970s
(L to R): Joe Smith, Lorne Goff, Carson McDonald, Hank Ryhanen, Dennis Cox, Bob Miyagawa, Dave Kiil (CFS), George Chrosciewicz (CFS), Art Peter
Alberta Government, AFHPC

Provincial Forest Fire Centre dispatch room, Edmonton, 1970s. Dispatchers Fred Schroeder and Mike Dubina
Alberta Government, AFHPC

Second Year NAIT Forestry Class, Forest Technology School, spring, 1970
Back Row (L to R): P. Michon, B. Fisher, D. Ryder, D. Ireland, K. Heemeryck, M. Posniak, B. Davidson, D. Laing, B. Chlan, R. Dixon, B. Provo, D. Krangnes, R. Woods, P. Wearmouth, R. Verhaeghe, E. Rutt, J. McQuarrie, B. Wilson. Front Row: W. Patterson, B. Allen, R. Olsson, K. McCrae, K. Sampson, P. Stoochnoff, J. Halvarson, J. Gould, G. Christensen
Alberta Government, AFHPC

Fire Control Officers and Forest Superintendents at a fire control meeting, Forest Technology School, Hinton, 1970
Back Row (L to R): Not Identified, Cliff Henderson, Art Peter, Don Fregren, John Booker, Bert Coast, Rex Winn, Hank Ryhanen. Third Row: Jack Macnab, Dick Mackie, Del Hereford, August Gatzke, Mike Lalor, Chuck Rattliff, Not Identified, Bill Kostiuk Second Row: Ben Shantz, Bernie Simpson, Lou Boulet, Jack Naylor, Bernie Brouwer, Lou Babcock, Jack MacGregor, John Benson. Front Row: Chuck Geale, Bob Thachuk?, George Deans, John Morrison, Bert Varty, Peter Murphy (Instructor)

Carson McDonald stated, "John Morrison was the Instructor i/c of Fire Control at FTS. John was a remarkable individual, who was a retired fire control person from the USFS. He was with us for a three to five year period and can be credited for being the impetus behind a major upgrade/updating to the Forest Service's fire control training program."
Alberta Government, AFHPC

Piles of pine cone sacks ready for loading and hauling to the provincial tree nursery. AFS staff would hire local people and firefighters to pick cones for future seed extraction and tree growing at the nursery. Worsley Ranger Station, Peace River Forest, 1971
Glen Gache office collection

Pine cone sacks loaded and ready to haul to the provincial tree nursery, Worsley Ranger Station, Peace River Forest, 1971
Glen Gache office collection

PINE-CONE PICKERS DO WELL

The Worsley Ranger Staff, Ken Porter and Terry Van Nest, on behalf of the Alberta Forest Service, would like to personally thank all the people who co-operated and assited in our successful Pine-Cone collection.

A total of one thousand, five hundred and twenty bushels of pine cones were collected.

These cones will be shipped to Edmonton for the extraction of their seeds. The seed at present will be stored. In future years this seed will be used to grow seedlings for planting or else seeded over an area for reforestation purposes. Once again 'Thanks for your assistance.'

Advertisement placed in the local newspaper by Rangers Ken Porter and Terry Van Nest, thanking people for their help in the collection of 1,520 bushels of pine cones, Worsley Ranger Station, Peace River Forest, 1971
Glen Gache office collection

Oil spill from a Federated Co-op pipeline, Virginia Hills area, Whitecourt Forest, May, 1972
Jamie McQuarrie

Slave Lake Forest Headquarters staff during the opening of the new Slave Lake office, Slave Lake Forest, 1971
Back Row (L to R): Dave Patterson, Dave Laing, Bruce MacGregor, Ted Cofer, Gordon Bisgrove, Dave Blackmore. Middle Row: Gerry Carlson, Gordon Brown, Dean Isles, Keith Branter, Rick Keller, Howard Townsend, Kevin Dewhirst, Keltie Wright, Bud Sloan. Front Row: Carson McDonald (Superintendent), Maurine Sprowl, Darlene Turenne, Lynn Cardinal, Marlene Gamble, Gloria Gatzke, Gemma Cere, Wendy Cooper, Norma Lund, Steve Radkevich
Rick Keller

Intermediate Fire Behaviour Course, Forest Technology School, Hinton, March, 1972
Back Row (L to R): Harold Enfield, Ken McCrae, B. Fisher, Mansel Davis, Cliff Henderson, Bob Plankenhorn, M. Rose, E. Skjonsberg. Middle Row: Gary Dakin, Bud Sloan, Glen Sloan, Frank Jones, Ted Loblaw, B. Walburger, Brian Carnell. Front Row: O.J. (Hap) Schauerte, Don Sarafinchin, Karl Altschwagger, Ed Beebe, Bill Kostiuk, Ray Hill
Alberta Government, AFHPC

University of Alberta Forestry students 2nd year Spring Camp, Forest Technology School, May, 1972
Back Row (L to R): J. Bruhn, B. Brodie, D. Phillips, V. Skalicky, R. Olson, J. Spencer, P. Gommerud, J. Hushagen. Fourth Row: A. Prelusky, H. Westgate, B. Allan, J. Hammond, E. Hotte, Keith Branter, J. Chittick, C. Newhouse. Third Row: Doug Sklar, B. Simpson, Rod Simpson, B. Olsen, D. Burr, T. Baines, B. Kwasny. Second Row: Dr. J. Schultz, M. Fulcher, A. Randall, C. Wilson, B. Schneider, G. Hoehne, D. Leblanc, J. Spencer. Front Row: Bill Gladstone, Instructor Jim Beck, J. DenHeyer, D. Pasiuk, G. Wray, C. Mellon
Alberta Government, AFHPC

Intermediate Fire Behaviour Course, Forest Technology School, Hinton, March 12-16, 1973

Fifth Row (L to R): Bob Strang, Jorn Thomsen, Lou Foley, Roger LeCerf, Darryl Rollings, Terry Turner, Dan Slaght. Fourth Row: Doug Jourdie, Ian Tempany, Rick Hirtle, Floyd Schamber. Third Row: Al Sturko, Norman Olsen, Bob Petite, Tom Oliver, Wray Adams, Tom Trott, Graham Tyson. Second Row: Luther Ferguson, George Benoit, Tony Stuart, Mark Ross, Andrew Forbes, Glen Peterson. Front Row: Del Lee, Brian Cutrell, Don Podlubny, Pete Spencer, Gordon Graham, Rick Manwaring, Fred Anderson, Ed Stashko (Weather Section, Edmonton)
Alberta Government, AFHPC

Advanced Service Course, Forest Technology School, Hinton, March 26-30, 1973

Back Row (L to R): Tony Stuart, Ken Janigo, Dale Huberdeau, Bill McPhail, Roger Olson, Bill Anderson, Not Identified. Third Row: Bud Sloan, Magne Steiestol, Dennis Cox, Herb Walker, Bob Miller, Len Sawatzka. Second Row: Dave Schenk, Jim Young, Don Law, John Beraksa, Wayne Cole, Vic Hume. Front Row: Walter Gacek, John Graham, Dan Jenkins, Bill Kostiuk, Harry Jeremy
Alberta Government, AFHPC

Fire Prevention Course, Forest Technology School, Hinton, March 26 to April 2, 1973

Fourth Row (L to R): Fred Kuipers, Bill McPhail, Dave Brown, Ted Loblaw, Hylo McDonald, Nick Galon, Ray Dubak. Third Row: Jim Skrenek, Ken Wheat, Murray Doherty, Dennis Sanregret, Pete Bifano, Ken Porter. Second Row: Stan Clarke, Dave Schenk, Harold Enfield, Bill Bereska, Maurice Lavallee, Mansel Davis, Andy Kostiuk, Dennis Howells, Len Allen, Gerry Armfelt. Front Row: Donald Harvie, John McQueen, Jim Hereford, Bill Kostiuk, Dennis York, Dick Robertson, Harry Jeremy, Chuck Rattliff (Instructor)
Missing: Dale Huberdeau, Bob Plankenhorn, Bill Wuth
Alberta Government, AFHPC

Fire Boss I Course, Forest Technology School, Hinton, April 16-20, 1973

Fifth Row (L to R): Greg Keesy, Wolfgang Richter, Jorn Thomsen, Ron Boisvert. Fourth Row: Rick Keller, Rene Cunningham, Gordon Graham. Third Row: John Morrison (Instructor), Pat Hendrigan, Ian Tempany, Jim Stewart, Ray Frey, Graham Tyson. Second Row: Brian Cutrell, Gordon McClain, Neil Christensen, Bob Whittle, Dick Talbot. Front Row: John Auger, Allen Overholt, Marty Sader, George Evert, Eugene Caldwell, Cecil Cross
Alberta Government, AFHPC

Crew Boss Course, Forest Technology School, Hinton, April 24-28, 1973
Fifth Row (L to R): Ken Mulak, James White, Robert Shaeffer, Ray Kynoch. Fourth Row: Brian Harris, Don Mickle, Michael Courtoreille, Chris Schober. Third Row: Fred Vermilion, John Tunke, Albert Lund, Philip Grey, Joe McKinnon. Second Row: George Hurrell, Rene Auger, Neil Holder, Rod Frazier. Front Row: John Morrison (Instructor), Henry Gladue, Bob Barker, Gerald Chamber, Don Clark, Gordon LaFleche, Lloyd Boucher
Alberta Government, AFHPC

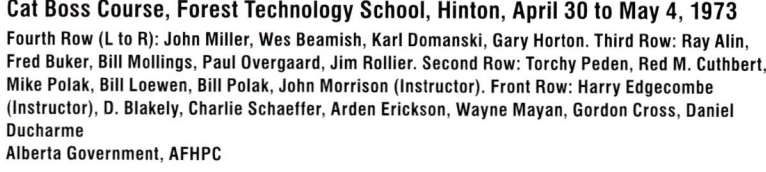

Cat Boss Course, Forest Technology School, Hinton, April 30 to May 4, 1973
Fourth Row (L to R): John Miller, Wes Beamish, Karl Domanski, Gary Horton. Third Row: Ray Alin, Fred Buker, Bill Mollings, Paul Overgaard, Jim Rollier. Second Row: Torchy Peden, Red M. Cuthbert, Mike Polak, Bill Loewen, Bill Polak, John Morrison (Instructor). Front Row: Harry Edgecombe (Instructor), D. Blakely, Charlie Schaeffer, Arden Erickson, Wayne Mayan, Gordon Cross, Daniel Ducharme
Alberta Government, AFHPC

Protection Technician Dale Huberdeau (L) and Forester Bob Grey, Fort McMurray Headquarters, Athabasca Forest, 1973
Alberta Government, AFHPC

(L to R) Lou Babcock, Forest Superintendent; Oliver Glanfield, Land Use Officer; and Bernie Brouwer, Fire Control Officer, Fort McMurray Headquarters, Athabasca Forest, 1973
Alberta Government, AFHPC

Fire Control Officers attending a Mixmaster Course for the mixing, circulation, storage, delivery and loading of fire retardants onto the airtankers. This course later became part of the Airtanker Base Managers Course. Slave Lake Airtanker Base, Slave Lake Forest, April 30 – May 4, 1973

(L to R): Ken Janigo, Fire Control Technician, Rocky Mountain House; Lou Boulet, Fire Control Officer, Calgary; Bill Kostiuk, Fire Control Technician, Edson; Lorne Goff, Fire Control Officer, Slave Lake; Len Allen, Fire Control Technician, Slave Lake; Ken South, Fire Control Technician, Peace River
Alberta Government, AFHPC

TM Crew members trying to repair Alpine Twin Track Skidoo, High Level District, Footner Lake Forest, winter, 1973-74. The Footner Lake Timber Management crew were doing reconnaissance cruising in the Peace valley, west of La Crete. Natural challenges faced were snow depth, ice pressure ridges and associated overflow on the river. Other challenges included keeping the Twin Tracks operational – they were prone to overheating and motor failure
Kevin Freehill

No luck in repairing Skidoo – on a sled and being towed, High Level District, Footner Lake Forest, winter, 1973-74
Kevin Freehill

Recreation land use assessment of the Peerless Lake – Graham Lake area, Slave Lake Forest, June, 1974. (This planning assessment was a joint exercise between AFS Forest Land Use Multiple Use Planning staff, Fish and Wildlife staff and Parks staff.)

Bottom Row (L to R): Cliff Lacey; Heather King, AFS FLU MUP; Dean Isles, Forest Officer I, Red Earth; Dell Guest, Fish and Wildlife Officer I, Slave Lake; Ray Lapitski, Forest Officer I, Red Earth; Ted Johnson, AFS (MUP) Biologist; Archie Landals, Parks Planner; Peter Eligh, AFS MUP Forester; Dennis York, Forest Officer II, Red Earth. Top Row: Ed Johnson, Forest Officer III, Slave Lake; Thomas Bellrose, Cook; Lou Foley, Land Use Officer, Slave Lake; Not Identified, Laborer Flunkey; Ken Wilson, MUP - Head MUP; Jim Skrenek, Chief Ranger Wabasca; Jim Smith, Parks; Paul Short, MUP Biologist; Gord Armitage, Head Forester, Slave Lake. Top, behind helicopter: Pilot Axel Porsild
Lou Foley

Ranger John Branderhorst's 'Speeder Ticket' for the Northern Alberta Railway, 1975
John Branderhorst

Hank Ryhanen, Director of Forest Protection, Provincial Forest Fire Centre, Edmonton, 1976
Alberta Government, AFHPC

25th Anniversary gathering of the first 1951 Forestry Training School KFES class graduates, Forest Technology School, 1976
Back Row (L to R): Bill Bloomberg, Harry Edgecombe, Jack Macnab, Buck Rogers, Sandy Brown, Jim Hereford, Johnny Doonanco. Front Row: Eric Huestis, Bob Cosmock, Jim Clark, Vic Heath, Des Crossley, Mike Reap, Not Identified, Not Identified
Alberta Government, AFHPC

Attending 25th Anniversary gathering of the first 1951 Forestry Training School KFES class graduates, Forest Technology School, 1976
(L to R): Johnny Doonanco, Dexter Champion, Harry Edgecombe
Alberta Government, AFHPC

Fire Prevention Course, Forest Technology School, Hinton, December, 1976
Back Row (L to R): Gordon Krassman, Not Identified, Darryl Rollings, Ralph Oberg, Not Identified, Kevin Dewhurst, Doug Joudrey. Middle Row: Not Identified, Wayne Cole, Rick Manwaring, Ralph Woods, Walter Radowits, Otto Losel?, Randy Rawe. Front Row: Howard Gray, Len Wilton, Frank Lewis, Rick Stewart, Ross Graham, Dave Laing, John Graham, John McLevin. (Two of the not identified could be K. Owen and R. Desroches from NWT)
Alberta Government, AFHPC

Chief Ranger Dale Huberdeau assessing new Bombardier amphibious personnel carrier, Rainbow Lake District, Footner Lake Forest, 1970s. First test-drive into the water and 'there it stayed'
Corinne Huberdeau

Rangers Dennis Howells (L) and Gary Mandrusiak at the Virginia Hills stop-over cabin, doing land use inspections, Whitecourt Forest, 1978
Rob Thorburn

Alberta Forest Service staff providing fire protection duties during Queen Elizabeth II's commemoration of Lac Cardinal Provincial Park, Peace River Forest, late 1970s
(L to R): Garth Berg, Gary Dakin, Terry Van Nest, Bob Hilbert and Don Thompson
Terry Van Nest

Alberta Forest Service Northwest Cup Hockey Tournament at Nampa, Peace River - winning team, 1978. (Other teams Playing in tourney that year: Footner, Slave Lake and Grande Prairie)
Front Row (L to R): Larry Berube, John Bradley, Paul Lannen, Tony Sikora, Derek Hanebury. Back Row: Frank Lewis, Gary Dakin, Garth Berg, Ken Baldry, Don McKeever, Lyall Gill
Frank Lewis

Transporting container seedlings by Bow Helicopters Bell 204 to a planting project in the Chinchaga River area, Peace River Forest, 1978
Dennis Driscoll

1966 - 1984

Chief Ranger Symposium, Forest Technology School, Hinton, December, 1978
Back Row (L to R): Ken Wheat, Robb; Owen Bolster, PFFC; Bud Sloan, Smith; Don Harvey, Grande Cache; Mag Steiestol, Blairmore; Wayne Robinson, Hinton; Dennis York, Red Earth. Middle Row: Wray Adams, Slave Lake; Dan Jenkins, Edson; Stan Clarke, Grovedale; Ted Loblaw, Rocky; Fred Kuipers, Cold Creek; Lou Foley, Beaver Lake; Bob Diesel, Swan Hills; Gary Giese, Fox Creek; Bob Glover, Wabasca. Front Row: Irv Allen, Turner Valley; Vic Hume, Ghost; Kelly O'Shea, High Level; Bill Bereska, Fort Vermilion; Ed Dechant, Whitecourt; Glen Sloan, Kinuso
Alberta Government, AFHPC

Firefighters hired to bag cones picked during a cone-picking project near Calling Lake Ranger Station, Lac La Biche Forest, 1979. Cones were being readied for transport to the Pine Ridge Forest Nursery
Kevin Freehill

Howard Gray (L), Forest Protection Officer Slave Lake and Dale Huberdeau, Fire Boss discussing strategies at the Saulteaux Airstrip base camp, Saulteaux Fire, Slave Lake Forest, 1979
Howard Gray

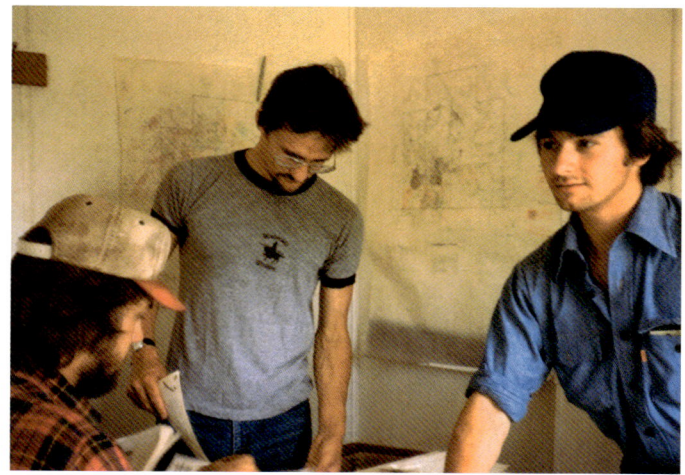

Members of the Footner Lake timber management crew review maps and plot information, Fort Vermilion, Footner Lake Forest, 1979
(L to R): Wayne Becker, Stan Kavalinas, Mike Poscente
Dennis Driscoll

Sector Boss Course, Forest Technology School, Hinton, December, 1979
Back Row (L to R): Kurt Frederick, Doug Beddome, John Brewer, Randy Panko. Front Row: Paul Steiestol, Jim Gordon, Glenn MacPherson, Steve Tuttle
Alberta Government, AFHPC

Time Officer Course, Forest Technology School, Hinton, March, 1980
Front Row (L to R): Gord Baron (Instructor), Gerald Thom, Susan Zimmerman, Ken Orich, Maxine Lightfoot, Darryl Johnson. Middle Row: Lowell Lyseng, Darryl Allsop, Peter Stoochnoff, Not Identified, Dave Laing. Back Row: Dave Schenk, Terry Smith, Bud Sloan, Doug Nichol, Wes Eror, James Metchoayeah
Alberta Government, AFHPC

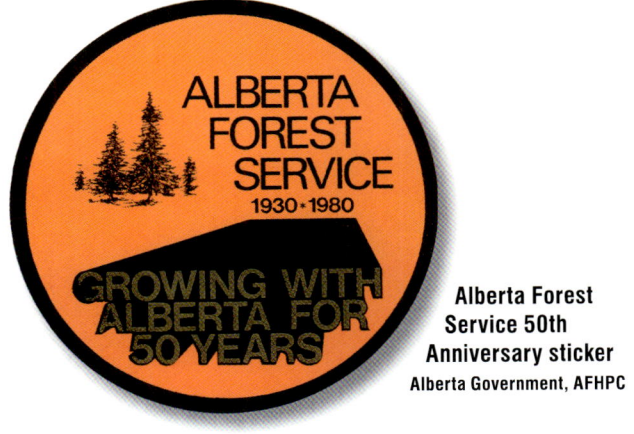

Alberta Forest Service 50th Anniversary sticker
Alberta Government, AFHPC

Ranger Rob Thorburn at his regeneration survey camp, Fort Vermilion Ranger District, Footner Lake Forest, 1980
Rob Thorburn

AFS Chief Rangers attend a Kepner Tregoe Management Seminar, Forest Technology School, Hinton, March, 1981

Back Row (L to R): Jurgen Moll, High Level; Philip Dube, Manning; Irv Allen, Turner Valley. Third Row: Ken South, FTS Hinton; Ed Dechant, Whitecourt; Gordon Japp, Cold Creek; Dennis York, High Prairie; Mansel Davis, Sundre; Gary Giese, Fox Creek; Bob Glover, Rocky Mountain House; Dan Jenkins, Elbow. Second Row: Ted Loblaw, Nordegg; Harold Dunlop, Beaver Lake; Don Harvey, Grande Cache; Kelly O'Shea, Swan Hills; Joe Burritt, Recreation Bow/Crow; Fred Kuipers, Valleyview; Bob Petite, Smith; Dave Brown, Spirit River. Front Row: Bill Bereska, Beaver Lake; Wayne Robinson, Hinton; Ken Wheat, Robb; Mag Steiestol, Blairmore; John Graham, Fort Vermilion; Dennis Cox, Slave Lake; Vic Hume, Fort McMurray; Dale Huberdeau, Rainbow Lake
Alberta Government, AFHPC

Hooking up the Water Buoy water bucket to the cargo hook and attaching the electrical cord on Bell 206 Jet Ranger JEL, Kananaskis Ranger Station, Canmore District, Bow Crow Forest, spring, 1981

(L to R): John MacAulay, Bruce Mayer, pilot Jim Lipinski
Alberta Government, AFHPC

Operational Cruise Compilation Course, Forest Technology School, Hinton, April, 1981

Back Row (L to R): Kurt Wentzell, Brian MacIntosh, Not Identified, Murray McDonald (Instructor), Dan Slaght. Front Row: Not Identified, Not Identified, Not Identified
Alberta Government, AFHPC

Forestry Caucus Committee tour, Lac La Biche Airtanker Base, early 1980s

(L to R): Larry Huberdeau, LLB Forest Superintendent; Con Dermott, Director of Timber Management; Cliff Smith, Assistant Deputy Minister; Fred McDougall, Deputy Minister; Bud Miller, Associate Minister of Public Lands and Wildlife; Jack Campbell, MLA; Shirley Cripps, MLA; Not Identified; Frank Appleby, MLA Athabasca
Alberta Government, AFHPC

Operational Cruise Compilation Course, Forest Technology School, Hinton, February, 1982
Back Row (L to R): Ed Albrecht, Rick Smee, Tim Burggraaff, Dennis Frisky, Murray McDonald (Instructor). Middle Row: Russ Braham, Tony Schlenker, Gary Davis, Steve Blanton, Rick Horne. Front Row: George Stevens, Gordon Crowder, Tracy Cove, Rick Prince, Jeff Monty
Alberta Government, AFHPC

Helicopter Management Course, Forest Technology School, Hinton, Feb-Mar, 1982
Back Row (L to R): Wayne Robinson, Darryl Rollings, Norm Olsen, Gary Schneidmiller, Dave Brown, Floyd Schamber, Mike Dubina, Ken Janigo. Third Row: Stan Olszowka, Dollard O'Connor, Ross Graham, Doug Ellison, Ted Cofer, Rob Thorburn, John Beraska, Mike Hancock, Herb DeMars. Second Row: Ken Porter, Conrad Bello, Keith Branter, Rick Bambrick, Ken McCrae, Phil Dube, Larry Warren, Jamie McQuarrie. Front Row: Russ Stashko, Wayne Bowles, Murray Doherty, Joe Smith, Don Welsh, Glen McPherson, Frank Nuspel, Ray Olsson
Alberta Government, AFHPC

Advanced Fire Behaviour Course, Forest Technology School, Hinton, April, 1982. This inaugural course shaped fire behaviour training within Alberta
Back Row (L to R): Marty Alexander (Instructor), Lou Foley, Brian Wudarck, Horst Rohde, Mike Thompson. Fourth Row: Dennis Cox, Gary Dakin, Terry Van Nest, John Branderhorst, Bob Hilbert, George Benoit. Third Row: Jurgen Moll, Don Law, Ken South, Irv Allen, Hylo McDonald, Dennis Quintilio (Instructor). Second Row: Chuck Rattliff, Bill Wuth, Ed Dechant, Don Harvey, Murray Doherty, Frank Nuspel, Bob Petite. Front Row: Gary Mandrusiak, Brian Meads, Ken Porter, Rob Thorburn, Rick Hirtle
Alberta Government, AFHPC

Transporting seedlings by sling, with an Alberta Government Bell 206 (C-GFSA), Fort McMurray District, Athabasca Forest, 1982
Alberta Government, AFHPC

Tree planters planting a cutblock in the Fort McMurray District, Athabasca Forest, 1982
Alberta Government, AFHPC

Bruce MacGregor, Land Use Officer, handing out seedlings to students during the National Forest Week celebration, Slave Lake Forest, May, 1982
Bruce MacGregor

Sector Level Fire Suppression Course, Forest Technology School, Hinton, fall, 1982
Back Row (L to R): Dennis Driscoll, Dale Asselin, Not Identified, Gerald Sambrooke. Third Row: Don Pope, Paul Bassendale, Gary Dudinsky, Jim Deitrich, Paul de Coursey, Russ DiFiore. Second Row: Greg Cunliff, Scott Hennigar, Norm Hawkes, John Hogue, George Panici, Dave Martel. Front Row: Kurt Wentzell, Gary Dolynchuk, Wes Eror, Not Identified, Glen Peterson?
Alberta Government, AFHPC

Fire Prevention Course, Forest Technology School, Hinton, December, 1982
Back Row (L to R): Rick Stewart, Joe Niederleitner (Instructor), Doug Ellison, Ted Cofer, Hugh Boyd. Third Row: Pat Hendrigan, Andy Gesner, Kurt Wentzel, Don Zwicker. Second Row: Frank Nuspel, Mansel Davis, Stew Walkinshaw, Glenn MacPherson. Front Row: Frank Lewis, Brian Stanton, Bill Black, Dennis Driscoll
Alberta Government, AFHPC

Division Level Fire Suppression Course, Forest Technology School, Hinton, February, 1983
Back Row (L to R): Dennis York, Wayne Robinson, Rick Hirtle, Dennis Quintilio (Instructor – in back) Darryl Rollings, Tony Znak, Dave Bartesko. Third Row: Jurgen Moll, Don Welsh, Gary Dakin, Stan Clarke, Rick Smith, Leonard Kennedy. Second Row: Ken Porter, Chuck Rattliff, Ralph Woods, Dick Seamen, Ed Dechant, John Branderhorst, Jamie McQuarrie. Front Row: Leon Graham, Russ Stashko, Len Wilton, Ray Olsson, Wayne Cole, Mike Dubina, Not Identified (US Instructor)
Alberta Government, AFHPC

Helicopter Management Course, Forest Technology School, Hinton, February, 1983
Back Row (L to R): B. Moerkoert (out of province), Rick Arthur, Brian MacIntosh, Russ Braham, Rod Simpson, Craig Quintilio, Bob Yates. Third Row: Brian Wudarck, Gord Graham, Don Podlubny, Terry Kennedy (NWT?), Gary Mandrusiak, Rick Horne, Jorn Thomsen. Second Row: Chris Hale, Bill Black, John Stepaniuk, Fred Schroeder, Dennis Halladay, Barry Gladders, Brian Meads. Front Row: John Hogue, Ken Orich, Dave Redgate, Brian Cutrell, Bill Kostiuk, Don Pope, Bruce Cartwright
Alberta Government, AFHPC

Fire Behaviour II Course, Forest Technology School, Hinton, April 5-8, 1983
Back Row (L to R): Rod Blades, Leo Drapeau, Lyall Gill, Bill Bereska, Dave Brown, Mike Hancock. Front Row: Tom Grant, Bob Lenton, Kelly O'Shea, John Graham, Butch Shenfield
Alberta Government, AFHPC

First Rappel Training and Certification Course, Forest Technology School, Hinton, April, 1983
Back Row (L to R): Dennis Sanregret, Dean Richards, Todd Poproski. Fourth Row: Not Identified, Phil Langford, Rick Jarvis, Al Johnson. Third Row: Greg Lynch, Sefton Smith, Jim Beare?, Ken Karpov, Todd Fedorak. Second Row: Rene Spielman, Daryl Stewart, Rob Roehler, Dave Myers, Jim Morrow, Bill Bishop. Front Row: Steve Jeremy, Don Pope, Steve Inaba, Steve Heemeryck, Rino Keskisalmi, Ray Ault
Alberta Government, AFHPC

Fire Prevention Course, Forest Technology School, Hinton, December, 1983
Back Row (L to R): Lloyd Seedhouse, Rob Thorburn, David Hughes, Rick Smith, Joe Niederleitner (Instructor). Middle Row: Bruce Coutanche (Yukon or NWT), Gary Walsh, Rick Horne, Norm Hawkes, Ralph Kermer (Yukon or NWT). Front Row: Darryl Johnson, Wayne Bowles, Bob Lenton, Tony Schlenker, Dave Scott
Alberta Government, AFHPC

Fort McMurray Ranger District staff, Athabasca Forest, 1983
Back Row (L to R): Ken South (Chief Ranger), Don Pope, Bill McDonald, Don Brewer, Philippe Robert, Dwayne Desjarlais. Front Row: Rick Hirtle, Dale Gee, Gary Dodsworth
Alberta Government, AFHPC

Sector Level Fire Suppression Course, Forest Technology School, Hinton, December, 1983
Back Row (L to R): John Eaton, Fred Paget, Pieter Broersen, John McLevin, Wayne Becker, Barry Congram, Ed Barnett, Dave Brown. Third Row: Tracey Cove, Cory Chouinard, Stewart Fairweather, Collin Janvier, J. Minoose, Vern Seib, George Panici, Gary Dudinsky. Second Row: Bill Cooper, Cliff White (Parks), Mike Berry, Brian McIntosh, Brian Stanton, Jim Lunn, Norm Begin, Bruce Coutanche. Front Row: Scott Hennigar, Don Zwicker, Len Stroebel, Vern Danes, Jeff Monty, Dave West, Merv Byron
Alberta Government, AFHPC

Helicopter Operations and Safety Course, Forest Technology School, Hinton, February, 1984
Back Row (L to R): Darrell Hemery, Brent Bochan, Dave Bartesko, Rod Blades, Dan Slaght, Dan MacPherson. Third Row: Gord Masson (Wood Buffalo), Leonard Norman, Roger Litke, Kevin Dewhirst, Don Harrison, Not Identified, Gord Taylor. Second Row: Murray Melon, Norm Hawkes, Tony Znak, Dave Hughes, Brian Stanton, Jeff Brooks. Front Row: John McLevin, Murray Heinrich, John Brewer, Dennis Driscoll, Ian Dunk, Steve Harrison
Alberta Government, AFHPC

Integrated Resource Management Planning Course, Forest Technology School, Hinton, fall, 1984
Back Row (L to R): Dave Blackmore, Ed Dechant, Rory Thompson, Larry Skinkle, Bill Bindon, Jerry Sunderland, Greg Branton, Mark Johnston. Third Row: Rod Houle, Dan Blackmore, Darryl Johnson, Keith Branter, Fred Vanderzee, Vern Danes. Second Row: Phil Dube, Al Malcolm, Rick Stewart, Larry St. Antoine, Lorne Fisher, Mark Anielski, Not Identified. Front Row: Susan Corey, Facilitator Not Identified, Nancy McMinn, Corrie Fordyce, Not Identified, Don Gelinas
Alberta Government, AFHPC

Sector Level Fire Suppression Course, Forest Technology School, Hinton, December, 1984
Front Row (L to R): Paul Hammond, Dale Darrah, Beverly Wilson, Mike Dempsey, Kevin Heartwell, Dave Cook. Middle Row: Tony Schlenker, Aaron Doepel, H. Madill, Rick Prince, Henry Grierson, Bill MacDonald. Back Row: M. Zrum, Jack Budd, Tom Grant, Wayne Bowles, Bill Black, Gerry Matthews
Alberta Government, AFHPC

Fall Forest Protection Officer Fire Management workshop, Chateau Louis Hotel, Edmonton, November, 1984
Back Row (L to R): Mag Steiestol, FPO Fort McMurray; Ed Johnson, FPO Whitecourt; Lou Boulet, Operations Supervisor PFFC; Gordon Bisgrove, FPO Slave Lake; Howard Gray, Manager Fire Control Operations PFFC; Don Law, FPO Lac La Biche; Ben Shantz, FPO Rocky; Cliff Smith, Director Forest Protection Branch; Lou Foley, FPO Peace River. Front Row: Bill Wuth, FPO Grande Prairie; Kelly O'Shea, FPO Calgary; Dale Huberdeau, FPO Footner Lake; Joe Smith, FPT Edson (sitting in for Hylo McDonald FPO)
Howard Gray

Slave Lake Forest Hockey Team, playing in the North West Cup, 1988
Back Row (L to R): Bud McKeown, Brian Tanner, Wayne Becker, Howard Gray, Rick Moore, Jamie Thompson, Barry Gladders, Tony Schlenker. Front Row: Bart Elliot, Howard Herman, Brian Panasiuk, Wayne Bowles, Mark Storie, Rick Alguire, Larry Skinkle
Howard Gray

Huestis Bonspiel

In 1965, discussions began between Chuck Geale, John Kokotilo, Doonie Donovan and Erling Winquist to organize an annual provincial curling bonspiel to promote cameraderie amongst the staff of the Alberta Forest Service. They agreed that the following year, a bonspiel would be held in the middle of February to break up the long winter. Each of the 10 Forests was invited to enter a team along with the six Branches in Edmonton. In some cases, playoffs took place to determine which team would represent their Forest or Branch in this prestigious event.

For many years the bonspiel was held at the St. Albert Curling Club with the banquet held at the Bonaventure Motor Hotel on St. Albert Trail. Mr. Eric Huestis donated the trophy for the "A" event winner, and from then on the annual event was known as the Alberta Forest Service "E.S. Heustis" Bonspiel. For most of the years, the trophy was piped into the banquet and dance on Saturday night followed by the winners of the past year.

The Bonspiel ran continuously for 31 years until 1996. In its later years, the Bonspiel was held in Whitecourt, Ardrossan and Peace River. Erling Winquist organized the Bonspiel until his retirement in 1988 and then Lowell Lyseng took over. The E.S. Huestis Curling Bonspiel was one of the "losses" during the downsizing and reorganization that took place in the 1990s.

All photos provided by Erling Winquist and Lowell Lyseng

Advertisement for the 2nd Annual Alberta Forest Service E.S. Huestis Curling Bonspiel Banquet and Dance, 1967

Eric Huestis presenting the E.S. Huestis trophy to Pat Wilson, Slave Lake skip, 1967

Edmonton team winning the inaugural E.S. Huestis Bonspiel, 1966
(L to R): Fred McDougall, John Kokotilo, Louis (Doonie) Donovan, W. (Jock) McLean

Slave Lake winning team, E.S. Huestis Bonspiel, 1967
(L to R): Don Bunbury, Laverne Larson, Pat Wilson, Lorne Goff

Bob Steele (R) presenting winning trophy to the Clearwater-Rocky team skip Dick Radke, 1968

Clearwater-Rocky winning team, E.S. Huestis Bonspiel, 1968
(L to R): Ken Janigo, Ron Lyle, Dave Clauge, Dick Radke

Mr. and Mrs. Eric Huestis at the 1968 E.S. Huestis Bonspiel

Peace River winning team, E.S. Huestis Bonspiel, 1969
(L to R): Brian Carnell, Walter Gacek, Andy Engebertson, Jack Naylor

Edmonton team, E.S. Huestis Bonspiel, late 1960s
(L to R): John Kokotilo, Fred McDougall, Larry Bunbury, Not Identified

John Kokotilo with winning rock in 13th end, E.S. Huestis Bonspiel, late 1960s

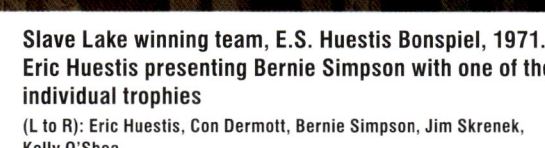

Slave Lake winning team, E.S. Huestis Bonspiel, 1971. Eric Huestis presenting Bernie Simpson with one of the individual trophies
(L to R): Eric Huestis, Con Dermott, Bernie Simpson, Jim Skrenek, Kelly O'Shea

Slave Lake winning team, E.S. Huestis Bonspiel, 1970
(L to R): Con Dermott, Bernie Simpson, Pat Wilson, Roger Olson

(L to R): Jim Skrenek, Kelly O'Shea and Con Dermott being piped into the banquet, E.S. Huestis Bonspiel, 1971

Watching rock stop in the house, E.S. Huestis Bonspiel, early 1970s

Jack Lunan piping in the E.S. Huestis trophy to the evening banquet, early 1970s

Art Peter (front left) and Alf Longworth (front right) in snowshoe race at the E.S. Huestis evening banquet, early 1970s

Slave Lake winning team, E.S. Huestis Bonspiel, 1973
(L to R): Jim Skrenek, Eric Huestis, Kelly O'Shea, Randy Rawe
Missing from the picture is Scott Mayston

Measuring to see which rock would count, E.S. Huestis Bonspiel, 1970s

Eric Huestis with curling trophies he would present for the Alberta Forest Service bonspiel named in his honour
Alberta Government, AFHPC

212 Alberta Forest Service

Edmonton team, E.S. Huestis Bonspiel, 1974
(L to R): Not Identified, Fred McDougall, Lowell Lyseng, Larry Bunbury

Grande Prairie winning team, E.S. Huestis Bonspiel, 1988
(L to R): Dave Beck, Bill McDonald, Stan Clarke, Lowell Lyseng

Edmonton team, E.S. Huestis Bonspiel, 1988
(L to R): Don Fregren, Bob Miyagawa, Fred McDougall, Chuck Geale

Whitecourt winning team, E. S. Huestis Bonspiel, 1990
(L to R): Jeff Henricks, Gordon Bisgrove, Darryl Johnson, Dennis Driscoll

Grande Prairie team, E.S. Huestis Bonspiel, 1991 or 1992
(L to R): Dan Brink, Bill McDonald, Blaine Renkas, Marian Cowan

50th Anniversary Celebrations

Special guests at the Alberta Forest Service 50th anniversary celebration
(L to R): Bob Steele (DM 1973-1978), Fred McDougall (DM 1978-1989), Henry Ruste (Minister 1965-1969), Lt. Gov. Frank Lynch-Staunton (1979-1985), Al Brennan (ADM Forestry 1979-1985), Merv Leitch (Minister 1978-1982), Robin Huth (Author, Horses to Helicopters), Allan Warrack (Minister 1971-1975)
Alberta Government, AFHPC

Jim Pearse, Program Support Branch, with Alberta Forest Service 50th Anniversary display, Convention Inn South lobby
Alberta Government, AFHPC

Lt. Governor Frank Lynch-Staunton addressing guests at 50th Anniversary celebration
Alberta Government, AFHPC

Hank Ryhanen, Director of Forest Protection, addressing guests at 50th Anniversary celebration; Fred McDougall, Deputy Minister and Mrs. Leitch, wife of Minister Merv Leitch also at head table
Alberta Government, AFHPC

Jack Lunan, back left, shaking hands with Hank Ryhanen; Lt. Governor Frank Lynch-Staunton, front row left, and Chuck Gaele
Alberta Government, AFHPC

Al Brennan cutting the Alberta Forest Service 50th Anniversary cake
Alberta Government, AFHPC

Forest Officer Tim Hale displaying container-grown seedlings at the Pine Ridge Forest Nursery, mid-1980s
Alberta Government, AFHPC

CHAPTER 7
The Use of Aircraft

The first use of aircraft for federal forest patrols in Alberta dates from 1920 in the Rocky Mountains Forest Reserve with the single-seat Avros and two-seat de Havillands. It was not, however, until the discovery of oil in Alberta in 1947, and the post-Second World War boom, that the economy in Alberta supported more funds for the protection of its forested areas and expansion of aircraft use.

Open-Cockpit Patrols

Flying was an adventure back in the early days of open-cockpit biplanes shortly after the First World War. An interview with forester and navigator Lloyd Van Camp provided the following recollections. At that time the planes were used for fire detection and prevention patrols. The planes flew from the forestry air base in High River in the 1920s. [1]

"In 1921 I was anxious to test the effectiveness of dropping some leaflets announcing our new aerial forest fire patrol on a town east of High River. The local paper showed a rodeo would be in progress that day so a good crowd would be there to pick up our triangular leaflets. We were flying at about 2,000 feet over the fairgrounds and zoomed down across the racetrack while the blizzard of leaflets blew into the stands. What we had not foreseen was the stampede of broncos from the infield when the strange roaring bird swooped down on them. The pilot decided we had better disappear before someone read our identification letters, so we flew away in a fast climb.

"On another day, the clouds were low over the forests north of the Bow River, so instead of going back to High River, the pilot George Howsam decided to fly us up the Bow River for a look at Banff. We went part way to Lake Louise, and as we came east over Mount Rundle, a "waterfall" type of downdraft dropped the plane about 2,500 feet. With the Banff scenery spread out below, my desire to get a good photo from the air had led me to unfasten my seat belt and kneel on the seat facing back for a shot over the tail. The viewfinder went from showing a great view of the mountains to suddenly showing blue sky. I realized the tail was pointing straight up and we were diving into the Bow Valley. Also, I found that the plane was a few inches lower than I was, but I caught up with a jolt when Howsam pulled back hard on the stick and the dive flattened out. We had to weave our way down the valley past Canmore and out through the Gap before we had enough height to clear the south ridges and head for High River.

"The pilots were supposed to learn the territory by flying, with me on board to help show the way and give them some training. They would later fly alone and report fires by radio. This radio idea had several faults that soon became apparent. The sending set in the plane ran from a small electric motor, powered by a small wooden

de Havilland aircraft at High River Airport, c. 1921
Dominion Forestry Branch, AFHPC

216 Alberta Forest Service

propeller under the top wing. The wave emitted was continuous, meaning that a sound at one tone was heard by the listeners at the High River receiving station. To make Morse Code signals, the operator in the plane pressed a key that changed the tone. This gave a series of higher and lower notes, much like a monotonous whistle.

"Sending Morse Code this way was an awkward chore. First of all, we were muffled in very heavy flying clothes for the sub-zero 130-mile gale at 13,000 feet in an open cockpit. Sheepskin, leather pants and jacket, lined helmet and goggles, plus silk inner and sheepskin outer gloves were necessary. To send a radio signal, it was first necessary to unwind a 100-foot copper aerial below the observer's floor, with a heavy lead weight to keep it trailing below the plane. This had to be done after takeoff, and we lost several aerials by having someone forget to reel them in before landing.

"To send you had to remove the heavy outer gloves, and your hands soon became numb with only the thin silk gloves on the icy key. I had little trouble sending Morse Code from learning it in Boy Scouts at Port Parry, and improving my sending speed in naval air service training in England. The pilots, all fighter veterans, most without any experience of Morse Code, were very slow or even resentful of learning. It soon became obvious that the idea of pilots sending back fire reports by code was not workable.

Navigator Lloyd Van Camp (L) and pilot Major George Howsam, High River Airport, 1921
Dominion Forestry Branch, AFHPC

Vickers Vancouver, 1930. These aircraft were used for forest inventory and survey work
Alberta Government, AFHPC

"It was also galling to the forest rangers that they had to sit at their cabins as long as planes were in their area, waiting for any fire reports we might send. They were all old-time cowboys, loggers or ex-ranchers, more accustomed to saddle-horse patrols. Having to wait for some college boy or "fly boy" to relay a phone call through High River was a very sore point with them.

"A.A. Leitch, called 'Ack-Ack' from the war time phonetics for anti-aircraft fire, flew the south patrol and quickly learned the territory after two flights with me. He was a non-conformist, with little regard for rules. On our last flight together we tried for an altitude record on the way home and reached over 19,000 feet in the biplane. The plane was sluggish at that height, and we were near the oxygen limit for altitude. However, A.A. had fun and we spiralled down from much greater height than usual as we came into the High River aerodrome.

"The last flight I took was almost that, literally. By now we had two new pilots trained to fly patrols. My flight with Major Croil was to establish a distance record north toward Jasper Park and back. South-bound flights started up the Highwood Valley and normally followed the south side of the river, with a left turn heading towards the U.S. border (past the Gap). The north-bound flight did the opposite on the north side of the river. Major Croil was almost over the river as we turned right to head north. Just then the south-bound patrol plane turned left and we passed head-on, one plane only a few feet below the other. Of all the miles of open air, the only two planes flying in Alberta avoided a collision by a very scant margin! The pilot of the plane which passed just below us flew until 1923, when he crashed in the Porcupine Hills. I found his plane after hearing of the crash. He had sliced through timber but was not injured, and had already walked away from the wreck before I got to it."

Vickers Vedette aircraft supplied timber reconaissance camp, 1928
Alberta Government, AFHPC

The Modern Era

Notes compiled by Bob Petite, a provincial air attack officer, show that in the early 1950s very little use was made of aircraft for either detection or fire suppression, due to the lack of funds and aircraft.

In July, 1953, the Department endorsed procedures for the cooperation of the Royal Canadian Air Force and commercial aviation companies in reporting wildfires in the absence of a government fleet for detection and fire suppression. Regular fire patrols were not recommended except under certain conditions. It was believed, however, that aircraft had a definite role in fire suppression, especially in the quick delivery of men, supplies and equipment while the fires were still small.

In 1954, however, a brief to the Alberta government by the Canadian Institute of Forestry, Rocky Mountain Section (*Forest Fire Protection in Alberta*) recommended serious consideration be given to utilizing aircraft in delivering men and equipment to wildfires in remote areas of the province.

In 1956 Norman Willmore, Minister of Lands and Forests, and Frank Platt, Supervisor of Fire Control, attended a conference in Missoula,

Lands and Forests Minister Norman Willmore, 1954-55 to 1964-65
Alberta Government, AFHPC

Montana, on aerial delivery systems.

That conference led them to develop a plan for the use of aircraft in fire suppression in Alberta. This plan included a basic provincial government fleet with seasonal contracts for other aircraft. Casual charter was to be used when the need was evident, such as suppression of wildfires without ground access. At this time, the U.S. agency aircraft were all controlled from a central dispatch centre so they could be used on agency-wide fire priorities and so their movements could be controlled for aircraft safety. This came to form the cornerstone of Alberta's policy.

The breakthrough event was the purchase in 1957 of the first in a fleet of AFS aircraft – the Helio Courier CF-IYZ *(photo page 231)*.

One of the first aerial attacks was initiated by forest superintendent Bob Steele in 1958. Lightning had started a fire in the Atlas Lumber Company Camp 15 area on a weekend. He phoned Wetaskiwin AirSpray's Dave Harrington who flew down with his 200 U.S. gallon capacity Stearman crop duster. Harrington landed on the old gravel Highway 11 near the Shunda Ranger Station. He and Steele flew over the fire and decided water bombing would be effective.

Rangers loaded the plane on the highway and soon the fire was controlled. Steele commented that he was reprimanded for not obtaining approval first from head office, but no one was available to call at the time.₂ Harrington also used his Stearman to attack the Maskuta Creek fire near Hinton that same year.

The first helicopter contract was in 1957 with Associated Helicopters of Edmonton, for a Bell 47J (CF-JRG). The first radio service trip to a lookout tower was undertaken during that contract and completed in half a day. The same service by horse had taken six days.

The fire centre in Edmonton was charged with tracking all aircraft hire and movement. All aircraft activity was coordinated under the central Aircraft Dispatch unit, established in 1958.

First Contracts

The first fixed-wing contract was in 1959 with Pacific Western Airlines. The contract was for a Beaver bush plane.

The de Havilland Otter contract with Gateway Aviation in 1971 was more successful than the Beaver in the movement of firefighters and equipment. The Beaver was phased out in favour of the Otter.

In 1968 the Alberta Forest Service (AFS) signed a contract with Mercury Flights for a Cessna 150 patrol aircraft. Although inexpensive, the Cessna 150 was too small and limited in its utility for other jobs, so this was a short-lived contract.

The 1972 contract for a Cessna 180 patrol aircraft with Lethbridge Air Services was more flexible. The plane was used to ferry personnel, haul freight and firefighting equipment.

The Fixed-wing Fleet

The AFS set up an initial contract with Associated Helicopters of Edmonton to store, maintain and pilot the government fleet of helicopters and fixed-wing aircraft. AFS management of fixed and rotary-wing aircraft ended when that duty was transferred to provincial Government Services in 1975.

The initial purchases of used aircraft came with the original registration letters. With the purchase of new aircraft, a block of registrations was reserved by Transport Canada specifically for AFS aircraft.

Loading Pacific Western Airlines Beaver at Peace River, 1965
Bob Stevenson

Preparing de Havilland Otter for a trip
Alberta Government, AFHPC

Some of the first fixed-wing aircraft owned by the province and managed by the AFS were:

CF-IYZ - A Helio Courier H-391B, short takeoff and landing, three passenger, purchased in 1957. It was later sold and replaced by CF-AFB, a Dornier.
CF-AFA - Helio Courier H-391B, short takeoff and landing, purchased in 1960. AFA was later sold in 1972 and replaced by CF-CKM, a Beechcraft Queen Air.
CF-AFB - Dornier DO 28 twin engine, five to six passenger, purchased in 1964.
CF-AFC - A second Dornier, purchased in 1965.
CF-AFD - Beechcraft King Air 100, six to nine passenger, purchased in 1972.
CF-IAE - A Douglas DC-3C twin motor, 28 passenger, purchased in 1968.
CF-CKM- A Beechcraft Queen Air M70, a twin motor six to nine passenger, purchased in 1972.
C-GFSB - A Beechcraft King Air 200 twin motor, nine passenger executive type, purchased in 1975.

Alberta's CF-IAE Douglas DC-3C, now in Reynolds Museum in Wetaskiwin. The aircraft was manufactured in 1944 and bought from Shell Canada Ltd. for $50,000 in 1968. It was used for years to carry AFS fire crews
Alberta Government, AFHPC

Some of the early fixed-wing pilots hired by the AFS were Rene Boudais, Dale Harris, Paul Slager and Leo Ulrich.

The DC-3C was originally used as a prop in the film industry in Hollywood. It was instrumental in transporting many firefighters (in 25-man crews) to fires both within Alberta and in other provinces. It was also initially used in the para cargo program. As of 2005, CF-IAE is on display at the Reynolds Museum in Wetaskiwin, still in Alberta government colours.

Airtanker Contracts

It was quickly recognized that initial aerial attack could best be done by the rapid delivery systems offered by bomber-type aircraft. Slurries of water and various thickeners and fire retardants were developed, because water alone tended to disperse and evaporate. In 1958 the AFS used Harrington Air Services in its first firefighting campaign, with a fleet of Stearman

AirSpray Stearman aircraft being loaded for a drop
Alberta Government, AFHPC

aircraft using calcium borate as a water thickener. In 1960 bentonite, commonly available as drilling mud, was used in place of calcium borate in the slurry. Aerial application of fire retardants was proving successful, so by 1962 a formal contract was drawn up with Harrington Air Services for the Stearman aircraft.

Dave Harrington was the first pioneer of air attack in Alberta. He started Harrington Air Services, which became Wetaskiwin AirSpray, to service the agriculture sector with pest control applications. Harrington experimented with tanks, dropping gates, additives and dropping techniques with his Stearman and later his Grumman TBM Avenger tanker, a converted torpedo bomber. In 1967 he incorporated as Air Spray (1967) Ltd. He and AFS staff learned together during those trial years.

Early airtanker contracts included:

1956 - first TBM Avenger airtanker used on the North Western Pulp and Power lease fire, using water and a birddog aircraft flying from Jasper Park airstrip.

1958 - AirSpray Stearman used on three fires in Edson Forest.

1961 - four Stearman aircraft and three Avengers worked three fires.

In 1966, the AFS signed the first Snow Commander aircraft contract with Wetaskiwin AirSpray Limited. The Snow Commander, manufactured by Snow Aeronautical Corp., was capable of delivering 250 gallons at a time. The Thrush Commander, comparable to the Snow but made by Rockwell's Aero Commander Division, was contracted from Mercury Flights in 1968.

The Stearman, Snows and Thrush were gradually replaced by the PBY Canso and Douglas B-26 aerial tankers.

The Canso airtanker was first used in 1966, followed by the first contract with Norcanair of Prince Albert, Saskatchewan, for the same aircraft in 1967. The Canso was an amphibious twin engine "flying boat" initially designed as a coastal patrol aircraft. When modified with an internal water tank it was well suited as a water bomber, capable of scooping up 800 gallons of water from the surface of any suitable lake within striking distance of a forest fire.

AirSpray Snow Commander, mid-1960s
Joe Smith

The B-25 aircraft was first used in 1967 but was discontinued in favour of the more reliable B-26 airtanker, capable of 760-gallon loads. The B-26 was initially contracted in 1970.

Don Hamilton, another pioneer in aviation, was also becoming established in Alberta. Raised in Moose Jaw, Hamilton was an adventurous pilot who got his licence in 1946 and invested in a new Cessna 120 after wartime service as a navigator with the RCAF. He barnstormed at fairs for $3 a ride and flew fish from northern Alberta lakes to a packing plant at Cold Lake, among other ventures. As a sales representative for Gateway

Flying Fireman Canso airtanker at Edmonton's downtown airport
Alberta Government, AFHPC

Aviation in 1956 he demonstrated the Helio-Courier to Frank Platt, which led the AFS to buy one in 1957. He later sold another Courier and two Twin Dorniers to the AFS.

Dave Harrington invited Don Hamilton to partner with him in 1970 to pool their converted B-26 airtanker and a leased Cessna 310 birddog aircraft in a one-year trial contract with AFS. It was a successful season – they got another B-26 and Cessna 310 and another contract. In 1972 Hamilton became sole owner. During those early years with the AFS, and especially at first through Frank Platt, they continued to refine airtanker operations and air drop techniques. Air Spray (1967) Ltd. grew to develop a fleet that also provided air attack services in British Columbia and Yukon. Their fleet in 2004 comprised seven Electras, 15 B-26s, five Twin Commander turbine aircraft, three Cessna 310s, three Aerostars, one Cessna 340, a Cessna Citation jet and an F86 Sabre jet for their various enterprises.

By 1975 the need for even larger and faster planes for aerial delivery of fire retardant was recognized and resulted in the first contract for the DC-6B land-based airtanker.

The Canadair CL-215, capable of carrying 1,200 gallons of water, plus retardant slurry or foam delivery, replaced the ageing Canso. The Canadair, designed and produced in Canada by Bombardier, was designed to scoop water from suitable lakes for a faster turnaround on fires.

Birddog Officer Course, May, 1983
Front Row (L to R): Rob Thorburn (AFS Instructor), John Beraska (AFS), Hugh Freeman (BC Instructor), Pete Spencer (AFS Instructor), Larry Warren (AFS Instructor), Mark Campbell (Saskatchewan). Centre Row: Paul Maczek (Saskatchewan), Wally Peters (AFS), Don McKeever (AFS), John Brewer (AFS Instructor), Jonathan Klink (AFS), Jeff Brooks (AFS). Back Row: Marlon LaBach? (Saskatchewan), Steve Semenuik (AFS), Wayne Rutter (AFS), Brent Bochan (AFS), Revie Lieskovsky (AFS Instructor), Eric Bell (Parks Canada)
Alberta Government, AFHPC

Training

Along with the increased use of aircraft, especially airtankers, came the need for specialized training courses.

The first Birddog Officer (an airborne position responsible for the coordination of aerial and ground forces) course was held at the Natural Resources Building in Edmonton in April, 1967.

The first Mixmaster course was held jointly at the Forest Technology School in Hinton and at the Entrance airstrip in August of 1968. A Mixmaster specialized in the mixing and loading of various fire retardants for aerial application on forest fires.

A Mystery Solved

On September 28, 1958, a single engine de Havilland DHC2 Beaver, registration CF-DJM, piloted by Maynard Richard Bolam departed from Eureka River airstrip west of Peace River. On board was one passenger, Daniel Roberts

Refuelling Grumman Avenger
Alberta Government, AFHPC

Helicopter and 300 gallon monsoon water bucket, with Fire Control Officer Bernie Simpson, Slave Lake, 1968
Alberta Government, AFHPC

from Manning, 1,000 pounds of flour and one barrel of diesel fuel headed for fire 28A-2-58 approximately 45 miles to the north.

Shortly after takeoff the pilot radioed his estimated time of arrival at the Doig Tower airstrip, a bush airstrip near the fire. He was never heard from again. An intensive air search was conducted but the plane was not found until 19 years later.

In August, 1977, while flying in a Bell 206B helicopter on a tower servicing trip from Hotchkiss Tower to Doig Tower, Alan Overholt, a Detection Officer in Peace River, spotted the downed Beaver aircraft. The plane had gone down in a dense stand of mature spruce trees less than a mile from the bush airstrip on the Doig fire and was only visible from a specific angle. Evidence later pointed out that the main bolt holding the wing strut to the wing had come loose and the wing had folded up. The plane went straight in without leaving much of an opening in the forest canopy other than knocking the tops off four trees.

A recovery operation was completed and a memorial plaque has since been erected.

Helicopters and the AFS

In 1954, Whitecourt Forest Superintendent Rein Krause inquired about using helicopters on fires during a Superintendents' meeting in Edmonton. It was agreed that this was a very expensive proposition. The Minister for the Department of Lands and Forests was to contact the federal government to see if it would be possible to use search and rescue helicopters stationed in Edmonton. The Alberta government-owned Stinson aircraft was being considered for use in dropping supplies and equipment on small fires. In addition, the Forest Superintendents and the Edmonton office were to make inquiries as to what assistance might be expected from local aviation companies.

In 1955, Slave Lake Forest Superintendent Bill Wood inquired at the Superintendents' meeting in Edmonton about the feasibility of using helicopters to locate new lookout tower sites. He was told that this had been considered but the cost would be $1,600 dollars to test three sites. The Department would not approve the use of helicopters due to the high cost. Lac La Biche Forest Superintendent Bert Coast asked about getting more freedom for local forest rangers to hire aircraft when needed. He was told that the hiring of aircraft required the Minister's approval or in his absence, the Deputy or Director could authorize the use of aircraft for fire control purposes. Forest Surveys Branch landform maps were used as a basis for a helicopter aerial survey of 15,000 square miles in the province during the fall. This is believed to be the first use of rotary-wing aircraft by the Department of Lands and Forests.

In 1956, Edson Forest Superintendent Donald Buck inquired at the annual meeting of Superintendents in Edmonton about using helicopters in the selection of lookout tower sites instead of travelling by foot and climbing trees. He was told that no funds were available for this purpose but pressure continued for permission to utilize helicopters in forestry work. In June of the same year, Frank Platt, Assistant Superintendent, Forest Protection Branch, discovered a wildfire (Edson # 23B-6-56) during an aerial fixed-wing patrol north of the Athabasca

River in the foothills west of Edson. Platt was a proponent of aircraft use in forest protection. Word was sent to Chief Ranger Walt Richardson at the Entrance Ranger Station to request the use of an Okanagan Bell 47 helicopter contracted out to the Triad Oil Company. The helicopter was engaged in oil and gas exploration near the community of Muskeg. The helicopter was used to transport 12 men and equipment to the fireline early the next day – leaving some of Triad's crew on a mountain overnight while the machine was commandeered for use in firefighting operations.

Newly-purchased AFS helicopter CF-KEY inspected by (left to right) Eric Huestis, Director of Forestry; Heber Jensen, Deputy Minister and Minister Norman Willmore. The aircraft was purchased on March 19, 1958, for $72,720.50. The logo on the door has 'Government of Alberta, Forest and Wildlife Division'
Alberta Government, AFHPC

Later that month, Platt sent a report to the Director of Forestry in regard to fire protection. He stated: "A definite increase in the use of aircraft this season has proved of great value. Without aircraft, the ground firefighters cannot be aware of the true picture of the fire and in some cases cannot be supplied properly with food and equipment. It is felt that there is an imperative need for both aircraft as well as helicopters. In one case a very serious fire north of Obed was spotted by a conventional aircraft and later reached by helicopter in time to bring it under control. Any other transportation at this time would have resulted in serious delay in getting men to the fire. In another case, a fire near Berland Lake was prevented from spreading by a helicopter giving the Forest Ranger a lift over the fire and later transporting him to a bad spot fire, which could not be reached due to river flood conditions. The use of conventional aircraft was found to be of great value in keeping track of the fire's progress and allowing key personnel to evaluate the situation and coordinate all efforts. The use of helicopters was almost prevented owing to the fact that all such aircraft were out under long-term contracts and in most cases were not available. This condition was also found to a lesser degree in regards to conventional aircraft."

Platt went on to write: "It is the writer's recommendation that at least three helicopters be placed under contract from March 31 to July 31 and be based in Edmonton, Slave Lake, and Peace River. A de Havilland Beaver aircraft on wheels, floats, and skis should be purchased with provisions for dropping equipment and supplies. This would be a specialized aircraft not normally available. It is evident that the use of aircraft would require a number of limited airstrips close to lookout towers as well as fuel caches for helicopter work."[3]

Responding to Platt's suggestions in July, Ted Hammer, Senior Superintendent, Forest Protection, said in a memo to the Director of Forestry: "For air transportation there can be no doubt helicopters are the answer for efficient fire

Bell 47G at Calling Lake airstrip, early 1960s
Ernie Stroebel

Bell 47J at Wabasca River east of Wadlin Lake, 1966, pilot Frank Arman
Bob Stevenson

Bell 206-B helicopter picks up fire retardant from portable tank
Alberta Government, AFHPC

suppression plus conventional aircraft for patrol and transportation over larger distances."[4]

In November, 1956, the AFS approached Associated Helicopters of Edmonton to evaluate the Bell 47J Ranger four-place helicopter for suitability in forestry use. Associated was the first commercial helicopter company in Alberta, in operation since 1950. Tellef Vaasjo, who was a pilot for Associated Helicopters, told how Frank Platt persuaded his company to buy one of the new 47J helicopters, even though the AFS could only offer a four- to-six week contract for that year. This was a tremendous act of faith on both sides. So, in 1957 Associated Helicopters purchased one of the first Bell 47J helicopters in Canada for use on a trial two-month period during May and June under contract to the AFS. The pilot of CF-JRG was Jack Lunan. The trial was very successful.

A year later, Associated Helicopters obtained a three-month contract for its Bell 47J. AFS purchased its first helicopter in the spring, another Bell 47J (CF-KEY). One of the first considerations in purchasing the Bell 47J was to have it come with an electric hoist kit to be used to lower fire personnel as close as possible to the wildfire where there were no suitable landing sites. The hoist cost $3,000 dollars, was bulky and weighed 70 pounds. Soon afterwards Associated Helicopters found a solution in the Sky Genie, used by the U.S. Army for rappelling from helicopters. Weight for this rappelling equipment was only 10 pounds and the cost just $200 dollars. Records are sparse, but it appears that operational rappelling into fires was not carried out to any great extent in the early years. Training on the use of the Sky Genie was still being carried out into the early 1970s at fire crew training courses for crew bosses, straw bosses, and some Forest Officers at the Forest Technology School in Hinton.

The AFS eventually purchased a total of four Bell 47Js for forest protection and related environmental work, before moving to the turbine-powered Bell 206 Jet Ranger. One Bell 47J, CF-KEY, was written off in 1966, 70 miles north of Fort McMurray. It was destroyed by fire after the log landing pad collapsed, causing the helicopter to shift. The nose of the helicopter dipped violently and the main rotor hit the ground. The tail section broke off and leaking fuel ignited. The pilot escaped without injury.

Following are some of the helicopters purchased by the Department after 1958:
CF-KEY - A Bell 47J piston powered three-passenger helicopter, purchased in 1958. This was the first of the

Bob Heighington, pilot with Associated Helicopters, June 6, 1971. Bob has been part of the Alberta Forest Service story ever since helicopters were introduced to the Alberta government
Jamie McQuarrie

helicopter fleet and after an accident in 1966 was written off and replaced with helicopter CF-AFI.

CF-AFI - A Bell 47J-2 purchased in 1966 to replace CF-KEY. It was upgraded to a Bell 47AJ-2 and later sold in 1972 to be replaced by CF-AFH, the first of the turbine driven fleet.

CF-AFJ - A Bell 47J-2 when purchased in 1962, upgraded to a Bell 47AJ-2 supercharged piston powered three-passenger helicopter in 1967. The "A" in the model (47AJ-2) designates that the motor was upgraded by Associated Helicopters to a turbo charged engine giving it improved horsepower over the original model. The supercharged model used 100/130 octane aircraft fuel while the 47J-2 used 80/87 octane.

CF-AFK - A Bell 47J-2 super charged piston-powered three-passenger helicopter, purchased in 1960 and upgraded to a Bell 47AJ-2 super charged piston powered three-passenger helicopter in 1967.

CF-AFH - A Bell 206-B, the first of the turbine driven helicopters owned by the AFS, purchased in 1973.

C-FAFL - A Bell 206-B purchased in 1973. It was also the first helicopter purchased that used the new registration letters required by the Federal Department of Transport (C followed by a four letter registration).

C-FAFM - A Bell 206-B, purchased in 1973.

C-GFSA - A Bell 206-B, purchased in 1973.

Some of the early helicopter pilots in this time of growth for the AFS were Lloyd Anderson, Sandy Donaldson, Bob Heighington, Paul Kristapovich, Jack Lunan, Rudy Rothermel, Clayton Thayer, Palmer Peterson, Harvey Trace, Axel Porsild, Bruce Carr, Ed Szeliga, Doug Hartnell and Anton Wakulchyk.

There was a very close call when helicopter C-GFSI went down in the Clear Hills north of Hines Creek in 1982. This was the brand new Alberta-owned Aerospatiale Twin-Star five-passenger machine (modified with two engines) that was working fires that summer. There were six people on board: pilot Ed Crahn; Cliff Smith, Director of Forest Protection; Howard Gray, Head of Fire Operations; Lou Foley, Forest Protection Officer at Peace River; Fred Schroeder, Aircraft Dispatcher and Con Dermott, Director of Reforestation and Reclamation. Cliff Smith recalls that they had flown from Peace River on Sunday June 20, to check on a fire on the B.C. border. On their way

Bell 212 dropping water on Moosehorn fire, 1981
Alberta Government, AFHPC

History of Alberta Government-Owned Helicopters

Registration	Aircraft	Year of Registration	Year Disposed	Remarks
CF-KEY	Bell 47J	1958	1966	Accident
CF-AFK	Bell 47J	1960	1979	Sold
CF-AFJ	Bell 47J	1962	1979	Sold
CF-AFI	Bell 47J	1966	1972	Sold
CF-AFH	Bell 206B	1973	1978	Accident
C-FAFL	Bell 206B	1974	1994	Sold
C-FAFM	Bell 206B	1974	1989	Accident
C-GFSA	Bell 206B	1979	1993	Accident
C-GFSD	Bell 206B	1979	1994	Sold
C-GFSE	Bell 206B	1979	1993	Accident
C-GFSI	AS 355N	1982	1982	Accident
C-GFSI*	Bell 222U	1983	1995	Sold
C-GFSO	Bell 206B	1986	1994	Sold

*C-GFSI was reused.

Aerospatiale twin turbine A-Star owned by Phoenix Helicopters of Fort McMurray operates helitorch for back-burning. Back-burning is a way of fighting fire with fire. This aircraft later became Air One for Edmonton Police Service
Rick Arthur

back they spotted an unreported smoke and decided to look at it since they were nearby and could probably control it. As Cliff Smith described it: "On the final approach the pilot made a turn, and just as he got the helicopter around to land, about 30 feet off the ground, there was a loud snap behind us [from the tail rotor drive shaft]. We were wondering, 'what was that?' whereupon the pilot said, 'Oh shit!'"[5]

Then things happened fast. They knew the helicopter was in trouble and the pilot was having problems controlling it. "All of a sudden we landed, but on an angle and as the right skid hit, the helicopter turned on its side. Lou was the first one out of the helicopter because he was in the left-hand seat and was up in the air. The pilot was sitting next to the ground and I was sitting next to the ground on top of Con in the back seat. As Lou got out he said, 'You've got to move it guys, it's starting to burn.' So we scrambled out very quickly, but when we looked back we realized the pilot was in some difficulty. So we went back and assisted him. He had broken his foot, but had stayed in there to try and get everything shut down to minimize the possibility of fire. Then we stood back on the creek and started to watch this pride of the Alberta government, the million-dollar helicopter, go up in smoke. I call it the most expensive spot-fire ever in the history of the Forest Service!

"While we were standing there, Lou remembered that despite the travesty that was taking place, we did have a mission – to control that fire. So we proceeded up the hill to the fire with no equipment whatsoever. I had a buck-knife on my belt and we cut some balsam boughs and started to beat the fire out of the trees - the fire was just starting to torch out at that point. We thought we didn't have much of a chance but we were successful in beating the flames down. Within minutes there was another helicopter overhead with a bucket. I can remember that so well because it was hot and we were standing right out in the middle of this burn, and this water was coming down - it was probably the most enjoyable shower I've ever had in my life!"

Para Cargo Program

The AFS's use of para cargo (freight delivered by parachute) began in the winter of 1945, when Gordon Fowlie, a young radio operator working at Adams Creek Lookout in the Edson Forest, was asked by AFS Radio Superintendent Tony Earnshaw to help develop a para cargo system as part of the Winter Works Program.

With the assistance of literature from the U.S. Forest Service in Missoula, Montana, Earnshaw had parachutes made locally using burlap in two sizes, one a nine-foot diameter and the other a 42-inch square. Earnshaw and Fowlie carried out test drops at Cooking Lake in the spring of 1946. Gordon Fowlie continued the project during the winter of 1946, designing the light cargo harness that was used into the mid-1970s. A project report was completed and submitted on August 26, 1947, showing favourable results for the continued use of para cargo.

From 1947 until 1954, very little was accomplished in the use of para cargo due to lack of facilities and aircraft. Frank Platt, Assistant Senior Superintendent of Forest Protection Branch and a proponent of aircraft usage, tried to rejuvenate the para cargo program in 1953 after re-examining its capabilities. Unfortunately, the aircraft earmarked for this, the government-owned Stinson, crashed and burned on a passenger flight at Cochrane setting the program back several years. In 1957, the AFS purchased a Helio Courier aircraft, and again, hopes were raised for the para cargo program. After completion of further test drops and experiments, a set of instructions on packing, rigging and drop zone procedures was written up by Platt and sent to each Forest with its complement of parachutes, 20 large and 10 small. From 1957 until 1970, para cargo use was minimal.

In August, 1970, Pete Spencer, a former RCAF parachute rigger and jump and drop master, joined the AFS as its Para Cargo Officer. He began training personnel from various Forests that winter. In 1971, para cargo began to establish itself as a low cost and speedy method of providing equipment and supplies for active fires. The program was successfully demonstrated was identified because many lightning-caused fires have difficult ground access and as a result, may grow out of control before a firefighting crew can reach them by land. Rappelling onto these fires saves time, thereby reducing the size, cost and intensity of the fire. It increases the chance of rapid control, thus achieving the provincial containment objectives.

Gordon Fowlie (L) and Jack Dempsey inspecting burlap chutes used for para-cargo trials in the mid-1940s. Note equipment pack on the ground attached to the parachute
Ron Linsdell

in May, 1971, when an outbreak of fires occurred in the Lac La Biche Forest and a total of 29,790 pounds was dropped to firefighters within 10 days by a single Otter aircraft. A training and demonstration drop was tested on July 15 at the Wabasca airstrip using the AFS-owned DC3. The aircraft was equipped with a roller conveyor and in a single pass six bundles were dropped, weighing a total of 2,300 pounds, the heaviest weighing 700 pounds and the lightest 250 pounds. [6]

Records show that in October, 1973, Paul Rizzoli, a heavy equipment operator in the Lac La Biche Forest, became one of the first successful candidates of the para cargo training course. Training was conducted by Pete Spencer, assisted by John Carson, a senior parachutist and jumpmaster. Fire Control Supervisor Frank Platt presented Paul with his drop master and parachutist qualifications. The capacity and speed of the helicopters used later meant air cargo drops were discontinued.

Rapattack Program

Rapattack crews (Rappel Attack) are wildland firefighting crews trained and equipped to rappel from a hovering helicopter. The need for rappel

Carl Larson, Edson, with weights for a para cargo drop
Alberta Government, AFHPC

Paul Rizzoli (C), flanked by para cargo course director Pete Spencer (R), and assistant instructor John Carson
Paul Rizzoli

Rappelling from helicopters in Alberta has taken place since the early 1960s, but the formal organization of the modern Alberta Rapattack Program was established in 1983 with nine crews, two of which were capable of rappelling from the helicopter. Over the years the number of crews and helicopters has evolved. In 2004, the configuration was 10 seven-person crews and eight contract Bell 205 & 212 helicopters, all of which were rappel capable. All new and returning rapellers and spotters (crew leaders) are certified each spring on the rappel tower at Hinton Training Centre. This centralized certification format has produced an excellent safety record among all 10 rapattack crews.

Each crew has one spotter and six rappellers along with enough firefighting equipment to split the crew into three two-person units during multi-fire situations. Once the assigned crew members have rappelled into the area where the wildland fire exists, the rappel rope is dropped to the ground, then the cargo bags, which carry the fire equipment, are lowered by webbing. The helicopter is then available for water bucket support. Rappel crews immediately fight the fire and if needed, cut a helispot close to the fire. The rappel helicopters can extract cargo bags out of the area with the use of a rope system attached to the helicopter cargo hook. This saves the time, effort, and cost of opening a helispot.

These crews are provincially mobile, self-contained with fire equipment and a ground transport vehicle, and are committed for the duration of the fire season. They supplement the local initial attack crews during periods of high hazard, are self-sufficient for a minimum of 24 hours and play a variety of roles on larger fires.

Rappelling is called for when walking distances are greater than 20 minutes or 1.5 kilometres, or when helicopter landing areas are too hazardous. Whether on small initial attack fires, campaign fires or project work, rappel personnel have been used to construct helispots for conventional access to worksites.

Editor's note: Our thanks to Bob Petite who helped ensure the accuracy of information in this chapter.

Rapattack firefighter on Lost Creek fire, 2003
Kathleen Thomson

Rappelling from Conair Bell 205-A1
Alberta Government, AFHPC

Equipment packs containing swede saws, axes, shovels and water pails ready for dropping to firefighters on the ground as para cargo, mid-1940s
Ron Linsdell

Alberta Government Helio Courier registration CF-IYZ, mid-1950s. This was the first fixed-wing aircraft owned by the Alberta Forest Service
Alberta Government, AFHPC

AirSpray Stearman completing a test retardant drop, late 1950s
Alberta Government, AFHPC

Thrush airtanker, registration CF-XLD, 1960s
Alberta Government, AFHPC

Alberta Forest Service Bell 47J2 (not turbo charged), registration CF-AFI, 1960s
Alberta Government, AFHPC

The Use of Aircraft

AirSpray Snow Commander, mid-1960s
Cliff Smith

AirSpray B-26, mid-1960s
Cliff Smith

AirSpray fleet, mid-1960s (L to R): Consolidated-Vultee L-13 birddog aircraft, Stearman (biplane) and Thrush
Cliff Smith

Alberta Forest Service Dornier DO-28, registration CF-AFB, sitting outside the Associated Helicopters hanger, c. 1964. This was the first Dornier purchased by the AFS replacing the Helio Courier
Alberta Government, AFHPC

Dornier DO-28 on floats, registration D-IBAV, 1964 Rex Winn, Superintendent Slave Lake Forest standing on shore
Alberta Government, AFHPC

Exchanging skids for floats on a Bell 47J, mid-1960s
Cliff Smith

The Use of Aircraft

Aerial view of Alberta Forest Service fleet of fixed-wing and rotor-wing aircraft, Edmonton, 1964. (L to R): 2 AFS Helio Couriers, 4 Bell 47J Helicopters, 3 de Havilland Beavers. CF-IYZ, the first Courier on the left, was sold in February 1965 and replaced by CF-AFC Dornier DO-28
Alberta Government, AFHPC

AirSpray Stearman airtanker, registration CF-JRK, 1965
Alberta Government, AFHPC

Sketch of cabin with airtanker working fire in background
Artist Lorna Bennett

Skyway Avenger airtanker, Slave Lake Forest, 1968
Alberta Government, AFHPC

Flying Fireman Canso Tanker 7, Edmonton, 1973
Alberta Government, AFHPC

Mitchell B-25 Tanker 38, 1970s
Alberta Government, AFHPC

Aerial view of Air Spray fleet, 1975
Alberta Government, AFHPC

Alberta Forest Service Bell 47AJ2 (turbo charged), registration CF-AFK, mid-1970s
Alberta Government, AFHPC

The Use of Aircraft 235

Air Spray (1967) Ltd. B-26 airtanker, registration CF-PGF, 1970s. This was the first B-26 in Alberta
Alberta Government, AFHPC

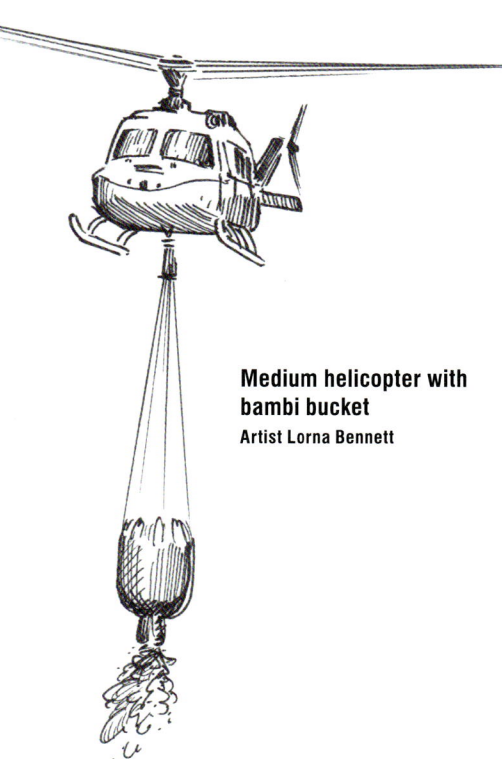

Medium helicopter with bambi bucket
Artist Lorna Bennett

Aerospatiale Gazelle registration C-FTCS, Peace River, late 1970s
Terry Van Nest

Alberta Forest Service DC3 being loaded with gear and firefighters at Panny Base, Slave Lake Forest, 1980
Dave Brown

Alberta government fleet of Bell Jet Ranger 206Bs, Edmonton Municipal Airport, 1980s
Alberta Government, AFHPC

Associated Helicopters Bell 212 C-GIRZ hauling all-terrain vehicles in an external net, Red Earth District, Slave Lake Forest, MOF reforestation program, June, 1981
Jamie McQuarrie

Alberta government helicopter Bell 222 registration C-GFSI, 1980s
Alberta Government, AFHPC

Hughes 500C registration C-GUXE, early 1980s
Terry Van Nest

The Use of Aircraft

Northern Mountain Jet Ranger registration C-GNMI, early 1980s
Terry Van Nest

Aerial view of Air Spray B-26 airtankers and Cessna 310 birddog aircraft, Penhold, fall, 1982 or spring, 1983
Air Spray (1967) Ltd.

Air Spray fixed-wing B-26 Tanker 4 in flight, 1980s
Alberta Government, AFHPC

Air Spray fixed-wing B-26 airtankers in flight, 1980s
Alberta Government, AFHPC

Air Attack Officer Strategies and Tactics Seminar, Forest Technology School, Hinton, 1986
(L to R): John Brewer, Hy Baker, Rob Thorburn, Wayne Rutter, Don Lalonde, Steve Otway, Paul Rizzoli, Wally Peters, Pete Spencer, Revie Lieskovsky (PFFC), Leo Drapeau, Don Zwicker, Steve Semeniuk, Larry Warren, Buck Dryer, George Panici
Alberta Government, AFHPC

Canadair CL-215 Tanker 49, 1980s
Alberta Government, AFHPC

Bow Helicopters Bell 206 Jet Ranger (CF-BHI) drawing water out of portable water tank with Water Buoy water bucket. This was at a joint AFS and Parks Canada training session. Pilot is Jim Davies, Banff, 1970s
Bob Stevenson

The Use of Aircraft

Bow Helicopters Bell 204, registration CF-BWR, early 1980s. This was the first Bell 204 in Canada
Alberta Government, AFHPC

Alberta government Bell 206B registration C-FAFL lighting piles with aerial torch, 1980s
Alberta Government, AFHPC

Alberta government Bell 206B registration C-FAFM flying, early 1980s
Alberta Government, AFHPC

Bow Helicopters Bell 205A, registration CF-FHA, early 1980s
Alberta Government, AFHPC

Bell 212 helicopter in flight, early 1980s
Alberta Government, AFHPC

Bow Helicopters Bell 47 G3B with AFS DC3 in background, early 1970s
Alberta Government, AFHPC

Okanagan Bell 206B lighting brush piles with aerial torch, 1980s
Terry Van Nest

Maple Leaf Helicopters Bell 204B registration C-GRGY, early 1980s
Terry Van Nest

The Use of Aircraft 241

Canadair CL-215 Tanker 203, late 1980s
Alberta Government, AFHPC

Canadair CL-215 skimming lake to pick up water load. The water is then injected with a foam concentrate before dropping on a fire. This mixture extends the effectiveness of the water and will help to 'smother' (remove the oxygen from) a surface fire if left untouched
Terry Van Nest

Conair DC-6 at the Manning airtanker base, Peace River Forest, 1991
Ken McCrae

Alpine Helicopters Bell 212, July, 2005
Sean MacNeil

Mike Dubina (L) and pilot Roger Tessier at Falls Lookout, Rocky Clearwater Forest, 1991
Alberta Government, AFHPC

Alberta government Bell 222 registration C-GFSI at Raspberry Lookout, Rocky Crowsnest Forest, 1991
Alberta Government, AFHPC

Air Attack Officer Strategies and Tactics Seminar, Forest Technology School, Hinton, early 1990s
Back Row (L to R): Brian Wesolowsky, Ken Baldry, Roger Marchand, Fred Paget, Steve Semeniuk. Middle Row: Rob Thorburn (Instructor FTS), John Brewer, Leo Drapeau, Wayne Rutter. Front Row: Chris Killoran, Revie Lieskovsky (Instructor PFFC), Buck Dryer, Phil Robert, Herb MacCauley
Alberta Government, AFHPC

Firefighters moving fuel barrels to the helicopter landing area, 1990s
Terry Van Nest

The Use of Aircraft 243

AFS helitack member rappelling from a Highland Helicopters Bell 212 registration C-GLZG, 1990s
Alberta Government, AFHPC

AFS helitack member rappelling from a Canadian Helicopters Bell 212 registration C-GOKW, 1990s
Alberta Government, AFHPC

Alberta government Bell 206B registration C-FAFL, 1990s
Alberta Government, AFHPC

Group of CL-215s and a DC-6 Group (Tanker 44) at the Slave Lake airtanker Base, Slave Lake District, Slave Lake Forest, 1991
Alberta Government, AFHPC

Canadian Air-Crane Sikorsky S-64 used during the spring Whitecourt and Slave Lake fires, 1998
Alberta Government, AFHPC

Air Spray B-26 airtanker dropping load of retardant on Cherry Hill Fire, Crowsnest Pass, Southwest Region, 2000
Alberta Government, AFHPC

Conair Air Tractor 802 tanker, c. 2000
Alberta Government, AFHPC

Flying Tankers Inc., Martin Mars water bomber on Slave Lake, 2002. This airtanker is capable of holding 6,000 gallons and was brought into Alberta to work on the late spring and early summer fires in Slave Lake and Lac La Biche
John Brewer

Flying Tankers Inc., Martin Mars water bomber on Slave Lake, 2002
John Brewer

Fixed-wing FHN checking smoke near Copton Lookout, Grande Prairie Forest, 1961
Artist Robert Guest

The Use of Aircraft

CHAPTER 8

1985 - 1992

During the later years of the 20th century there was an evolving determination to achieve sustainable forest management (SFM) for a wide range of values, prompted by greatly expanded activity in the forest.

The surge of industry development was pushed by a growing global demand for forest products and the Alberta government's policies that responded to it. Dramatic growth occurred through allocation of previously uncommitted coniferous annual allowable cut (AAC). In addition the deciduous-based industry experienced unprecedented growth as research and development showed Alberta's poplar varieties were suitable for high-value products such as oriented strand board (OSB) and pulp and paper. During the period 1985-1992 eight pulp mills were expanded or constructed, along with several new sawmills and structural board facilities.

Sustainable Development

The concept of SFM grew from the principles of sustained yield, multiple use and integrated resource management. SFM also incorporated economic, environmental, social and cultural values.

The Brundtland Commission report, *Our Common Future*, was issued by the United Nations in 1985.[1] It recognized the importance of both the economy and the environment, and urged "sustainable development" of natural resources. The Canadian Council of Forest Ministers (CCFM) issued its first *Forest Sector Strategy for Canada* in 1987 and followed with Canada-wide consultations for the *National Forest Strategy* and the *National Forest Accord* in 1992. These documents defined SFM, a concept supported by the Alberta government and forest industry. The *National Forest Strategy* stated that through SFM: "The long-term health of Canada's forests will be maintained and enhanced, for the benefit of all living things, and for the social, cultural, environmental and economic well-being of all Canadians now and in the future."[2] Alberta's support for and participation in evolving national and international events were particularly noteworthy at this time. The province was maturing as a jurisdiction through its willingness to become a global player in events that could directly affect its capabilities to manage its own affairs.

Paul Short, Senior Manager, Intergovernmental and Forest Policy for Sustainable Resource Development, explained that the signing of the 1992 Accord was only part of a sequence of federal/provincial initiatives related to national and provincial forest strategies and accords that began in 1981. Alberta's support for the Accord was significant historically for two reasons. First, it showed Alberta's commitment to a national agenda and its willingness to work alongside other provincial/territorial/federal agencies for better, more sustainably managed

Shunda Creek, Clearwater Forest – a landscape managed for multiple values
Bob Stevenson

Lac La Biche Forest staff, 1984
Back row (L to R): Dave Brown, Don Law, Rory Thompson, Barry Gladders, Garry Ehrentraut, Glen MacPherson. Middle rows: Wes Nimco, Kevin Freehill, Steve Hill, Rick Stewart, Colin Williams, Leon Graham, Gordon Brown, Vern Seib, Brock Allen, Ken Yackimec, Dave Lind, Bill Bereska, Kenton Miller, Dan MacPherson. Front row (kneeling): Gordon Taylor, Don Sarafinchin, Larry Huberdeau (Superintendent), Paul Rizzoli, Denis Hebert
Alberta Government, AFHPC

forests. Second was the evolution of Alberta's understanding of its accountability for international agreements signed by Canada.

Following the 1992 *National Forest Accord*, which was taken to the international UN Sustainable Development Conference in Rio de Janeiro, national delegates agreed that criteria and indicators for SFM should be drafted for the major global forest regions. Canada and the CCFM joined in the deliberations to develop Criteria and Indicators for the Conservation and Management of Temperate and Boreal Forests. This became known as the Montreal Process, based on the location of the first meeting. As part of this process, in 1995 the CCFM developed a set of Canadian criteria and indicators for SFM, to which Alberta subscribed.[3] The discussions also had an impact on

Minister Don Sparrow, 1985 - 1987
Bob Stevenson

Cliff Smith, Director of Forestry in 1985, became Deputy Minister in 1989
Bob Stevenson

federal agreements on protected areas that resulted in Alberta initiating the Special Places program. There was a greater awareness of the need to manage other values beyond the demands of local and regional influences.

Administration

Events during this period of 1985-1992 started with a transition from the Department of Energy and Natural Resources to the new Department of Forestry, Lands and Wildlife. Don Sparrow was appointed Minister of the new Department after serving as Associate Minister for Lands in the previous Department. Fred McDougall retained the role of Deputy Minister in this new Department, and in 1985 appointed Cliff Smith as Acting Assistant Deputy Minister for the Alberta Forest Service (AFS). The new Minister, Don Sparrow, suggested several themes in his 1985-1986 Annual Report.[4] Three points are particularly worth noting:

1) "When the renewable resources sector was separated from Alberta Energy

Minister Leroy Fjordbotten, 1987 - 1992
Bob Stevenson

and Natural Resources early in 1986, a new Department was initially called Alberta Forestry. Later, however, the name was expanded to Alberta Forestry, Lands and Wildlife, which more accurately reflects the Department's main areas of concern in carrying out its mandate: To ensure the wise use of the province's natural resources for the maximum benefit of all Albertans.

2) "Protecting environmentally sensitive areas is a high priority with the Department and with this objective in mind five new Forest Land Use Zones (FLUZ) were established to deal with conflicting recreational activities in certain areas in the eastern slopes. These zones allow outdoor enthusiasts to enjoy the scenic wilderness while protecting the areas for the future.

3) "In forestry, a major step was taken for developing one of the province's under-used resources – the hardwood forests – when a Forest Management Agreement (FMA) was awarded to Pelican Spruce Mills, Ltd. For the first time in Alberta the agreement calls for harvesting hardwoods as well as the traditional softwoods. The agreement means further diversification of the economy and more jobs for Alberta."

The FMA to which Sparrow referred supported construction of a second OSB mill, this time at Drayton Valley, along with a sawmill. Minister Sparrow also noted that while "this has been a period of budgetary restraint, the Department's dedicated staff has continued to provide a high level of service to the Alberta public, and I want to compliment them here for their fine work."

The new Department comprised six major divisions in 1985-86: Alberta Forest Service; Public Lands; Fish and Wildlife; Resource Evaluation and Planning; Bureau of Surveying and Mapping; and Foreign Ownership of Land Administration. The Forest Technology School at Hinton was administered under Personnel Services within the Deputy Minister's office.

Acting Assistant Deputy Minister Cliff Smith stated in the same Annual Report that the objective of AFS was "to manage Alberta's forest lands to ensure a perpetual supply of benefits and products while maintaining a high quality forest environment."[5]

Industry Development

In 1986 Minister Sparrow created the Forest Industry Development Division (FIDD): ". . . to assist new and existing forest companies and private investors to upgrade and utilize Alberta's vast uncommitted forest resources for a variety of forest products; and to attract investment and private sector involvement in upgrading and developing renewable resources other than the forest industry."[6] In this same year Al Brennan was appointed Executive Officer of FIDD, equivalent to Assistant Deputy Minister, and Cliff Smith was named Assistant Deputy Minister of Forestry on a full time basis.

The major objectives of this new FIDD included increasing employment and competitiveness, supporting forest products research and development, identifying equity funding, and increasing diversification and

Weyerhaeuser OSB mill (previously Pelican Spruce Mills) Drayton Valley
Weyerhaeuser

Mostowich Lumber mill near Fox Creek, late 1980s
Alberta Government, AFHPC

products and marketing. It was also to serve as a sole point of contact for potential investors to coordinate review of proposals among the Departments involved. This approach worked very well as set up under Sparrow's leadership. With his support, Brennan and his staff including economist Nick Gartaganis and research director Ted Szabo, worked effectively to meet FIDD objectives. Among the projects planned, completed or under way in 1985-86 were:

- Pelican Spruce Mills OSB plant at Drayton Valley
- Weldwood-Slave Lake's conversion from Aspenite waferboard to OSB
- Mostowich Lumber's new sawmill at Fox Creek
- Blue Ridge Lumber's new medium-density fibreboard (MDF) mill at Blue Ridge
- Jager Industries' Calgary I-beam facility
- Millar Western's chemi-thermomechanical (CTM) pulp mill in Whitecourt
- Procter and Gamble's mill modification to produce aspen pulp at Grande Prairie (did not materialize)
- Weldwood of Canada's (Hinton Division) doubling its pulp mill capacity
- BC Forest Products modernizing its sawmill at Grande Cache.

LeRoy Fjordbotten, who succeeded Don Sparrow as Minister, commented in his Annual Report for 1986-87: "A major step in forestry was taken with the establishment of the Forest Industry Development Division. Designed to promote greater development of Alberta's vast timber resource, this division is an outstanding contribution towards this government's goal of economic diversification. While the Department encourages resource development, it is understood that resource use should not take place at the expense of the environment. To this end, a new forest land use zone in the Bow-Crow Forest of southwestern Alberta was established, while five forest land use zones created in 1985 received permanent status. Such zones ensure the preservation of these areas for the future enjoyment of all Albertans."[7]

In 1987 Minister Fjordbotten commented that: "The Department's Forest Industry Development Division played a major role in contributing to the Alberta government's initiative towards economic diversification. A number of significant, new forest industry projects were announced, placing Alberta at the leading edge of forest industry development in North America . . . these outstanding achievements were long in the making. Years of hard work to establish or maintain suitable areas of marketable timber began to bring results in the past two years."[8]

Alberta-Pacific Forest Industries pulp mill
Alberta Government, AFHPC

**Ken Higginbotham,
Assistant Deputy Minister,
1990 - 1995**
Bob Stevenson

The Annual Report for 1989-90 announced that $3.5 billion in new capital had been attracted to Alberta through investments in the forest industry, including the Daishowa Canada pulp mill at Peace River and expansion of the Weldwood pulp mill at Hinton. The following year, the Alberta Energy pulp mill at Slave Lake, Alberta-Pacific Forest Industries aspen pulp mill near Athabasca and the Procter and Gamble pulp mill expansion in Grande Prairie were announced.

Fred McDougall resigned as Deputy Minister

1985 - 1992

in 1989 and moved to Weyerhaeuser Canada as Vice-President of the company's Alberta operations. Cliff Smith was promoted to Deputy Minister. K.O. (Ken) Higginbotham was appointed as Acting Assistant Deputy Minister for the AFS and was permanently assigned that position in 1990. Ken Higginbotham's previous positions included professor of forest ecology at the University of Alberta, starting in 1976 and head of the AFS Forest Research Branch as of 1988.

Forest Management and Public Interests

The report of the Phase III forest inventory, which covered the entire forested area, was issued in 1986. It was to play an important role in the evaluation and allocation of areas under new and renegotiated Forest Management Agreements (FMAs). Forest landscape management guidelines were also printed that year to form an integral part of the provincial Timber Harvest Planning and Operating Ground Rules. Together, they would provide a basis for managing the timber resource along with the visual and other values within the forested areas of Alberta.₉

The FMA and Quota agreements that had been negotiated with forest industries provided for forest regeneration on harvested lands and the basis on which SFM could be practised. During this period, however, there were legal actions and other events that questioned the policies and practices.

The first of these had its roots in 1982 when a coalition of lumber producers in the United States launched a trade countervailing action against Canada, alleging that Canadian forest companies were being unfairly subsidized through non-competitive allocation of timber. In Alberta the agreements required industry to practice forest management at its own expense, thereby incurring costs that were in addition to the timber dues they had to pay. Canada, including Alberta, its provincial colleagues and lumber trade associations strongly argued against the charges, which led to a long series of inquiries, hearings, trade talks, high-level meetings, draft bills and legal actions within and between the United States and Canada that seemed to add to the complexity. In December 1986 a pact was signed that called for Canada to impose a 15 per cent tax on exports of lumber to the United States, over the next five years. Although that was not a lasting resolution, the countervail action stimulated the AFS and the forest sector to reassess policies in Alberta and their implications, and to discuss how they might be refined.

Signing of the Weldwood Forest Management Agreement, Hinton, June, 1988
(L to R): MLA Dr. Ian Reid, Minister Leroy Fjordbotten, Tom Buell, Dick Sainsbury, Colin Warner
Rick Keller

Prescribed burn at Steen River, Footner Lake Forest tested a constructed hand line against a line constructed using prima-cord, 1973
Bruce MacGregor

The second event was triggered by the announcement in 1988 of the Daishowa Canada (now Daishowa-Marubeni Industries or DMI) FMA, signed in 1989, to support construction of a pulp mill at Peace River. The DMI agreement, which followed a succession of other forest industry developments, prompted quick and strong opposition from some Peace River area farmers and environmentalists. When public opposition failed to halt the development, farmer Peter Reese, the Alberta Wilderness Association, Peace River Environmental Society and Sierra Club of Western Canada brought court action against the Minster of Forestry Lands and Wildlife and Daishowa Canada.[10] The charges, in general terms, were that the FMA was not legal, that it did not comply with the *Forests Act* and that it could not achieve the "perpetual sustained yield" specified in the agreement. The two-week court hearings in October, 1991, brought expert witnesses for both sides, including several AFS staff who had to explain in testimony and cross examination that the forest management system could deliver SFM as proposed. In his 40-page Reasons for Judgment, dated January 23, 1992, Justice D.C. McDonald dismissed all charges after presenting his reasons for his conclusion. The experience demonstrated the importance of developing sound plans in support of SFM. Perhaps of equal importance, it highlighted the importance of public consultation to better determine public interests and to address their concerns in plans and proposals.

Public Participation

When the Alberta-Pacific Forest Industries pulp mill project was announced in 1989, there were more protests. Public meetings and open houses were held in 40 communities to hear public views and explain development and forestry policies. These events led to public hearings and additional studies. The rapid growth of the forest industry over the past several years coincided with increased worldwide interest in and concern about the state of the environment. In response to the wide range of concerns being expressed to government, Minister Fjordbotten appointed an Expert Panel on Forestry, chaired by Dr. Bruce Dancik of the University of Alberta with Lorne Brace, a forest research scientist with the Canadian Forest Service, Dr. John Stelfox, a research scientist with the Canadian Wildlife Service, and Bob Udell, a professional forester with Weldwood of Canada, Hinton Division. The panel held hearings, collected public opinions, and submitted its report and recommendations on May 16, 1990. The major subjects, as raised by the public, included: forest practices, the Forest Management Agreement process, fish and wildlife management, integrated resource management and environmental conservation. The report was supportive of some of the current policies and critical of others. It set the stage for changes and refinements over the next decade.

The report stated: "The present forest management system includes security of tenure, detailed forest management planning, and reforestation requirements unmatched elsewhere in Canada."[11] One of the major needs identified, however, was for greater public involvement in the forest management planning process. Assistant Deputy Minister Ken Higginbotham announced in the 1990-91 Annual Report: "A public involvement process was introduced for forest management planning on FMAs. This process requires FMA holders to consult with the public in the development of their detailed forest management plan, and then provide

Public access and use are among the many forest values
Alberta Government, AFHPC

Enjoying a picnic in the back country above Vimy Creek, Clearwater Forest
Bob Stevenson

opportunities for public review of the completed draft plan. In addition, companies are encouraged to establish public liaison committees, comprised of local interest groups, to provide ongoing advice on forest management issues."[12] While not a clearly stated requirement, most companies also presented lower-level detailed plans such as the Annual Operating (logging) Plans for public review and comment.

In its response to the Expert Panel report the government acknowledged there were areas where more needed to be done. Such work included greater consideration of non-timber resources in forest management decision-making and greater involvement of the public in many of the more significant policy and planning decisions.

Hinton is home base for the Foothills Model Forest
Bob Stevenson

In 1990 Department staff developed a mandate and mission statement defined in the 1991-92 publication *Creating our Future*. It was created as a strategic planning initiative for Alberta as part of a province-wide effort to strengthen natural resource management into the 1990s and beyond. It signalled another period of major change and transition for the AFS. Under its umbrella came development of a *Natural Resource Policy Framework* and a *Forest Conservation Strategy* (the latter developed between 1994 and 1997). Both of these were expected to provide overall policy direction and fulfill key recommendations made in 1990 by the Expert Panel on Forestry. The Natural Resources Conservation Board (NRCB) was established in 1991 under the *Natural Resources Conservation Board Act*. Its mandate was to provide an open, fair and impartial review process for projects that would or might affect the natural resources of Alberta and to contribute to the achievement of sustainable natural resource development reflecting the public interest. Creation of the NRCB brought a level of oversight designed to ensure a balance of environmental, societal and economic value for Albertans. Its purview extended to forest-sector project reviews, and the process of environmental impact assessments was introduced.

Research & Development

The Canadian Forest Service (CFS) continued to support advances in forestry research and development as part of its national responsibilities. In addition to its national research centres and operation of the Northern

Canadair CL-215 makes a practice drop
Bob Stevenson

Forestry Centre in Edmonton, CFS has overseen several national federal-provincial agreement programs. From 1985 - 1992 two such agreements were completed and in 1992 a new one started. From 1984 to 1989 the Canada-Alberta Forest Resource Development Agreement (CAFRDA) provided about $4.6 million per year in federal and provincial funding in support of several key objectives: to ensure security of timber supplies and industry viability, to promote the efficient utilization of forest resources, especially hardwood resources, and to increase economic diversity and employment opportunities in the forest sector. Projects included reforestation of federal and provincial lands, development and demonstration in vegetation management and studies of growth and yield.[13] With funds derived from Canada's export charges on softwood lumber sold into the United States about $29 million was reinvested in silviculture projects and research. This was known as the Public Lands Development Program (PLDP).

On April 15, 1992, Canada and Alberta signed a three-year, $30-million Canada-Alberta Partnership Agreement in Forestry (CAPAIF), a program that succeeded the CAFRDA. CAPAIF was designed to reinforce current initiatives in reforestation, intensive forest management, research, public involvement and public education. In tree planting and stand tending activities alone, under CAPAIF, over 13,000 person days of employment were generated in the first year.[14]

In 1992, Weldwood of Canada in Hinton, with support of other community partners and encouragement from the AFS, was successful in its bid for a Model Forest within the Canadian Model Forest Network. This was a national program initiated by the federal government's Green Plan. Joined by Jasper National Park in 1995, it became the largest model forest in the national system. The Foothills Model Forest (FMF) received $5 million over five years to develop pilot and demonstration activities that included community forestry, technology transfer, and large-scale demonstration projects in integrated resource management. The FMF has also been invaluable as a "show-me" centre to which delegates of foreign study or trade groups might be taken to demonstrate how SFM is being pursued on a multi-disciplinary basis.

Junior Forest Wardens celebrate Arbour Day in Edmonton, 1987. Minister Don Sparrow second from right, with federal Minister of State for Forestry, Gerald Merrithew
Bob Stevenson

Junior Forest Rangers' work experience camp, Calling Lake, Lac La Biche Forest, 1985
Bob Stevenson

It also clearly illustrates the size and diversity of Alberta's forest management areas. The FMF program at Hinton was extended to 2007.

In 1992 Alberta signed the National Forest Accord, a federal-provincial agreement designed to put into practice Canada's *National Forest Strategy* and the country's international commitments to sustainable development. During the same year the Alberta legislature unanimously endorsed its broad vision of sustainable development as defined by the

Sheep used in an experiment to reduce grass and deciduous competition in coniferous reforestation project, Calling Lake, 1987
John O'Brien

Alberta Round Table on Environment and Economy and stated in its Annual Report: "Alberta, a member of the global community, is a leader in sustainable development, ensuring a healthy environment, a healthy economy, and a high quality of life in the present and future."[15]

An Alberta Forest Research Advisory Council (AFRAC) was established in 1988 with membership from forest industry, provincial and federal governments, Alberta Research Council and the University of Alberta.[16] The AFRAC objective was to advise on research priorities and to encourage collaboration among research agencies. It was established partly in response to a recommendation of the Canadian Council of Forest Ministers. As a result of reorganization, the AFS Forest Research Branch was disbanded in 1991, and staff reassigned to other branches.

Forest Protection

In 1982 the province entered into an agreement with the federal government to enhance aerial firefighting capability. Under this agreement the province purchased two Canadair CL-215s in 1986 and received an additional two from the federal government in 1988, for the nominal fee of $1 each. The two new CL-215s saw immediate action on a total of 25 fires in 1986. These airtankers were additions to the AFS fleet of three DC-6s and 11 B-26s. Alberta's membership in the multi-agency Canadian Interagency Forest Fire Centre (CIFFC) agreement enabled movement of aircraft and personnel to areas of Canada which required them most. AFS has been an active participant, sending crews, aircraft and equipment across Canada and the United States, and receiving assistance when needed. In the early 1980s, the AFS was also equipped to receive data from remote automatic weather stations to enhance the fire weather network.

Highlights of the Department's 1985-86 Annual Report included recognition of the growing importance of fire as a resource management tool.

Planting the next generation of trees
Alberta Government, AFHPC

Taking seedlings to planting site. Seedlings are packed in boxes designed to keep them fresh until planted
Alberta Government, AFHPC

This was reflected in five prescribed burns, covering about 260 hectares, conducted in 1985 and almost 3,000 hectares in six separate projects in 1986. The report also noted that in 1985 the mountain pine beetle outbreak in the Crowsnest area was in decline. This serious infestation had first been reported in 1977 and vigorous control efforts were undertaken in cooperation with the Canadian Forest Service. By 1990 monitoring showed no evidence of the beetle, suggesting that the control efforts had been effective.

In 1986 the AFS took over the forest insect and disease survey from the Canadian Forest Service. A significant increase in spruce budworm defoliation in three northern forest areas had been identified in 1987. By 1989 it had expanded to 85,000 hectares, with the largest outbreak located in the Footner Lake area. For the first time in Alberta, the biological insecticide Btk (Bacillus thuringiensis var. kurstaki) was used to spray 1,000 hectares in the Grande Prairie Forest.

The fire season of 1988 started with hazardous spring weather conditions and winds of up to 100 kilometres per hour. The abnormally high temperatures and winds gusting to 100 kilometres per hour occurred again in December, which caused a wildfire south of Hinton to spread to 1,200 hectares. On August 6, 1990, a one-day record was set with 114 fire starts. During the 1990 fire season Alberta was also able to respond to 10 out-of-province requests for assistance, and received help from outside agencies on nine occasions.

As part of a comprehensive forestry and youth awareness program, the Junior Forest Wardens (JFW) program continued to grow in popularity, reaching 3,500 members and 750 volunteers. There were 125 JFW clubs in 92 communities across Alberta in 1990.

Forest Management

During this time of industry expansion and increasing impacts of forest land uses, the forest management group ensured that the forest inventory data were kept up to date. They also ensured that calculations of allowable cuts were based on sound information and realistic projections. As more and more FMAs were signed, the primary responsibility for writing detailed forest management plans was assumed by the individual companies. AFS staff still had overall responsibility for all provincial forests so they worked closely with industries to review their plans and resolve concerns. These plans describe all aspects of the forest, such as tree species, their ages, growth and health, as well as the forest ecosystem itself. The plans state the objectives for management, show how the forest will be harvested and regenerated and how the other forest values and uses will be protected. Calculations also project how the forest will be sustained over the next 100 or more years. The plans are very detailed and are hundreds of pages in length.

Determining reliable estimates of rates of growth of forest stands and yields of usable wood at harvest times is fundamental to sound planning. Volume and growth estimates are based on detailed measurements of diameters and heights of selected individual trees at various ages. Since trees grow differently, depending on the sites on which they are growing and other conditions, AFS established a series of Permanent Sample Plots (PSPs) starting as early as the 1960s. Trees on the plots, usually 0.10 hectares in area, are tagged for long-term identification and plots are re-measured every five to 10 years. The data show the growth

Adult mountain pine beetle
Alberta Government, AFHPC

and mortality of trees, and illustrate the dynamics of different stands. AFS and forest industries combine and share their PSP data through a Growth and Yield Cooperative agreement.

Reforestation

Silvicultural treatments and reforestation programs continued to expand as forest harvesting increased. Additional site preparation, seeding, planting, spacing and similar projects continued under the Maintaining Our Forests program that expired at the end of its seven-year duration in the spring of 1987. Enhanced silviculture-related projects continued with funding provided by CAFRDA, PLDP and CAPAF.

The forest site classification project was continued in cooperation with Resource Evaluation and Planning Branch (REAP) and the Canadian Forest Service (CFS). An important field guide to aid in the identification and management of forest ecosystems in west-central Alberta was developed in 1986 by forest scientists Ian Corns and Richard Annas.[17]

Brian Rodriguez measuring stump diameter while on timber cruise in the Watt Mountain area, Footner Lake Forest, December, 1988
Craig Brown

The Annual Report for 1985-86 mentioned that in addition to the ongoing regeneration surveys, a Juvenile Stand Survey (JSS) was initiated on 255 cutblocks totalling 2,851 hectares. The stated objective was to: "describe the composition of our regenerated forest 10 years after establishment."[18] Although regeneration surveys had been indicating satisfactory restocking on 80 to 90 per cent of the cutover areas, it was becoming increasingly evident that the stocking levels declined on some sites after the surveys. An expanded JSS on all sites showed that after 10 years about one-third of the sites were still satisfactorily restocked (SR), a third needed remedial work such as weeding, release or fill-in planting, and the other third was unsatisfactory, largely as a result of competition from grasses and weeds, and needed re-treatment.

Silviculture forester Keith Branter initiated a "Free to Grow" (FTG) study in 1986 to "develop and validate new regeneration standards for recently harvested cutovers."[19] Existing standards based on tree age and time-on-site alone were clearly not adequate for measuring reforestation success.

As well, a new silviculture planning section was set up in 1987 to undertake cost-benefit studies for silvicultural decision-making and to serve as a liaison with planning functions in other units of the AFS. The intent was to more effectively integrate silvicultural planning in advance of harvesting. Forester Jim Kitz moved from a field position in Whitecourt to work with Branter in the Reforestation Section to develop the FTG standards cooperatively with forest industry. The new standards were officially introduced on March 1, 1991. At that time Minister Fjordbotten stated that: "The Department took a giant step towards achieving (management for) sustainability with the implementation of some of the toughest reforestation standards in the world. The Free to Grow standards set new benchmarks for tree establishment and growth and will be significant in ensuring and maintaining sustainability of new forests. As under past standards, industry will be responsible for meeting these reforestation standards."[20] As noted in the Department's next Annual Report, the standards: "have caused forest managers to take a critical look at reforestation prescriptions, resulting in an emphasis on ecologically-based reforestation."[21] These requirements applied to government and industry.

The new Free to Grow standards measured number, distribution, vigor and height growth of new trees through two successive surveys, and company reforestation responsibility was extended from 10 years to a minimum 14 years. The first, or establishment, survey was conducted between years four and eight on coniferous cutovers and between years three and five on deciduous cutovers. The survey determined per cent stocking levels, species composition and growth performance of seedlings against specified height requirements. The second, or performance, survey was conducted between years eight and 14 on coniferous cutovers that met the establishment survey standards. In addition to the stocking and growth measurement, this survey also measured whether crop trees were free to grow, and were free of repressive competition. If the standards were not met, the site would be treated again and resurveyed until criteria were met. If growth was impeded by competition, stand tending would be done.

In recognizing the need for change in the provincial reforestation program, Cliff Smith reflected that: "Alberta had always prided itself on its regeneration success and its cutblock tracking system. We also became aware during the 1980s that as successful as we had been, there were some serious problems developing on a lot of those areas that had been restocked. It was mainly because the regeneration had been established, but had not been nurtured or maintained enough. Some of these areas threatened to become, or in some cases had become, NSR (not satisfactorily stocked) and there needed to be improved standards."[22]

To meet the increasing demand for seedlings, due to expanded harvesting activity and new regeneration standards, container seedling capacity at Pine Ridge Forest Nursery was expanded between 1991 and 1993. This included the addition of a new greenhouse of 6,000 square meters in 1991. Seedlings shipped that year totaled 50.8 million. The Forest Genetics and Tree Improvement Section had also been established at the Pine Ridge Forest Nursery under Narinder Dhir. Its activities included setting up seed orchards, development of a cooperative seed orchard with forest industry, identification and propagation of superior trees and maintaining a seed bank.[23]

Watershed Assessment course participants along main access road into Weldwood OSB Slave Lake FMA, Westblock operating area, north of Slave Lake, 1991
(L to R): Doug Ellison, Patti Campsall, Wayne Becker, Darby Paver, Wally Born, Al Hovan, Barry Northey, Glen Bergstrom (Section Head, Watershed Forest Land Use Branch) and Bob Anderson
Al Hovan

Forester Con Dermott explained[24] there was another challenge with respect to the 1986 quota renewal. There was a need to tighten up the utilization standards to make sure timber harvesting actually reflected the forest profile on which the AAC was calculated. In 1986, utilization of individual trees was changed from a 19 cm butt, 13 cm top to a 15 cm butt, 11 cm top. This change emphasized the important role of planning to ensure that if an area was harvested, it could subsequently be reforested.

Forest Land Use

Numerous land use activities were underway during the period covered in this chapter.
- The Range Management Program in the Green Area supported 208,000 animal-unit months (AUMs) in 1985. One AUM is the equivalent of one adult animal or one cow and one calf together, grazing for one month.
- The AFS managed and maintained 150 recreation areas and 22 group camps. Most were auto-accessible campgrounds.
- Procedures for pre-disturbance assessment of watershed conditions were developed.
- The timber harvest had increased to 6.96 million cubic metres, of which almost 10 per cent was deciduous. Pine Ridge Forest Nursery shipped 14 million container seedlings and

Alberta Government, AFHPC

Motorized recreation can leave its own unique footprint on the landscape
Butch Shenfield office collection

continued with its forest genetics and tree improvement research.
- A bio-monitoring program was established in the Lodgepole sour gas well blowout area to assess soil and vegetation damage and to document recovery.
- Resource planning was managed by REAP, which noted: "Public involvement is an important component of integrated resource plans. The public involvement program provides the opportunity for public input and comment on plans throughout the planning process."[25]
- The Integrated Resource Planning (IRP) program continued to play a key role in achieving the Alberta government's philosophy of integrated resource management.

The Allison/Chinook forest land use zone was established in 1986 in the Bow-Crow Forest to resolve conflicts between motorized and non-motorized recreational use by separating those activities as much as possible. In addition, the Bighorn Wildland Recreation Area was established west of Nordegg in April of 1986, comprising almost 4,000 square kilometres. As stated in the Annual Report for 1986-87, this area: "offers more opportunities to the outdoor recreationist than any other comparable recreation area across North America. It is a vast unique area, managed to provide a wide range of wildland recreation activities."[26]

Land Use Challenges

Land use issues demanded increased planning and zoning with public consultation, inspections and concurrent reclamation of disturbances as needed. The Department declared two new Natural Areas, containing 1,480 hectares. This was the start of a new Natural Areas protection initiative. By 1992 a total of 118 Natural Areas were designated, comprising a total of 35,545 hectares. An additional 181 sites totalling 75,300 hectares were under protective notation. As well, AFS maintained 114 kilometres of cross-country ski trails and 707 kilometres of snowmobile trails. An estimated 600,000 people used the 206 AFS recreation areas. There was no charge for use of trails and campgrounds.

Women in Forestry

By this time, women had become an integral and important component of the Forest Service, in positions ranging from firefighter through to ranger, forester, scientist and senior executive.

Two of the first women foresters hired in senior roles in Edmonton were Evelynne Wrangler and Susan Corey. Ms. Wrangler became the Section Head Management Planning, Timber Management Branch while Ms. Corey became the Section Head Integrated Resource Planning, Forest Land Use Branch. A number of significant changes occurred in the field in the late 1970s and early 1980s when women began to hold the traditionally male-dominated Forest Officer or District Ranger positions. In June 1981 Beverly Wilson (Fort Assiniboine), Marian Cowan (Smith) and Joyce Kendrick (Turner Valley) were among the first women posted as Rangers to field offices. Joanna Bush became a Forest Officer III in Fort Chipewyan that year as well. Tracey Cove went to LaCorey as a ranger in November 1981. Marian Cowan recalls 'heads turning' in Slave Lake when three vehicles, all driven by female rangers, drove down the main drag in the early 1980s. Marian also recalled that at that time when she was considering forestry as a career, the write-up on the Forest Technology Program at NAIT said something like "women should seriously consider the limited employment opportunities for them in this field before applying to this program." Marian remembered applying to private industry for summer employment and getting a couple letters back saying that "due to the fact that they had no camp facilities for women, my application would not be considered. It never occurred to me to be upset about it, it's just the way it was." Beverly Wilson is a graduate of both the NAIT and U of A programs and is now a Senior Resource Analyst working in Edmonton. She recalled the challenges of being a woman forest officer. For example, when company representatives would call to talk to a ranger, the call was transferred to her and they would still ask for a ranger, she would have to explain that she was the ranger. Wilson was one of the women who initiated some shifts in thinking.

Evelyne Calliou was a graduate of both the

Alberta Vocational College Forestry Crew Worker Program and the Forest Technology program at NAIT before being posted to Fort Vermilion. Women also played an instrumental role in the development of the Alberta Genetics and Tree Improvement Program and the Pine Ridge Forest Nursery.

Gail Tucker, Area Manager in High Level, recalls a discussion with a local farmer when stationed at the East Peace Ranger District in the early 1990s. "When I responded to his request for information, he looked directly at me (in my uniform) and asked to speak to a ranger. When I told him I was a ranger, he asked to speak to the ranger responsible for the Weberville area. I told him that I was responsible for that area. He then asked to speak to one of the men. Unfortunately for him, I was the only ranger in the office that day and he had to deal with me." Tucker also says that " … for every member of the 'old boy's club' who might have been wary of a woman in forestry, there were a couple who gave me a break or I wouldn't be an area manager now."

Margarete Hee graduated in forestry from the University of Alberta in 1977. A Canadian Institute of Forestry Gold Medal winner, Hee was one of the first three women in the forestry program. Her first summer forestry job was in 1974 in High Level with the Alberta Forest Service on one of the first "mixed" timber management (TM) crews. Hee and Deb Tomchuk (now Wortley) were the first two women hired as field foresters upon graduation. In 2004, editor David Holehouse asked Margarete Hee about women in forestry. Her response focused on the " … wealth of women and how we have moved, particularly over the last 25 years, into every facet of forestry, and at nearly all levels – although not in droves." Others making significant professional contributions include Deanna McCullough, Terry Zitnak, Jan Schilf and Anne McInerney. The names go on … [27]

Note: Many women have played a very significant role in building the Alberta Forest Service, and its successor organizations, into today's first-class organization. Those listed here are but a few of the many deserving of recognition for bringing excellent managerial, technical and administrative capabilities to roles in Edmonton and the various field locations.

Timber Management Cruising Party on the Waskahigan River, 1978
Back Row (L to R): Pieter Broersen, John Brewer, Glen Crammer, Otto Angst, Jean Lussier, Ken McCrae, Palmer Peterson (Pilot). Front Row: Margarete Hee, Barb Forbes, Patches the dog
Ken McCrae

Some of the many policy documents generated by the Department
David Holehouse

A number of manuals were created by the Department during this time period
David Holehouse

"The Departure of Fred McD"

There are strange things seen in the White and Green
By men of the forest crew;
By the old farmhands of the Public Lands and by Fish and Wildlife, too.
There are strange tales told that would leave you cold
About regs and policy,
But they all would smile at the "straight up" style
Of the Deputy, Fred McD.

Now, we've come a ways since the "homestead days"
And the horseback forestry,
With the gas and oil came a sudden boil in our economy.
There was much to do; the Department grew to a never-heard-of-size,
And at its head, the Minister said, "We must reorganize!"

"What I need," said he, "is a Deputy with sense of purpose keen,
And who'll take a stand, for there's big demand
On these lands both White and Green.
Now McD's the man, or else no one can get this outfit well in hand.
For my little dream of a complex scheme called an in-te-gra-ted plan!"

Well, the rest you see is just history and the stories spread and grew,
How this Fred McD was exemplary to the civil service crew.
For the Deputy, just to save a tree, never used a memo form,
But he had his say in a "marginal way" and in language often warm.

Throughout all those days, in the government's ways,
There were many memos wrote,
But now's the time we must write the line of our final "briefing note."
For he's moving on, and he'll soon be gone to the forest industry (!)
We all wish him well as we say "farewell" to our Deputy, Fred McD.

There are strange things seen in the White and Green
By the men of the forest crew;
By the old farmhands of the Public Lands, and by Fish and Wildlife, too.
There are strange tales told that would leave you cold
About regs and policy;
But we'll always smile at the "straight up" style
Of the Deputy, Fred McD.

By Helen Newsham (with thanks to Robert Service).

Written at the retirement from government of Deputy Minister Fred McDougall, who in departmental lore gave his name to the AFS — "Anything Fred Says"

Provincial Ranger Meeting of Directors, Superintendents, Section Heads and Chief Rangers, Forest Technology School, March 13 and 14, 1985
First Row (L to R): Al Brennan, Jim Skrenek, John Drew, Steve Ferdinand, Don Fregren, Doug Lyons, Roger LaFleche, Bill Bresnahan, Chuck Geale, Joe Soos, Cliff Smith, Con Dermott, Howard Gray, Stan Navratil, Gordon Armitage, Craig Quintilio. Second Row: John Benson, Don Harvie, Rick Bambrick, Barry Court, Roger Hamilton, Larry LaFleur, Mansel Davis, Vic Hume, Ted Loblaw, Art Evans, Ed Ritcey, Cliff Henderson, Brydon Ward, Norm Rodseth. Third Row: Don Harrison, Mag Steiestol, Hylo McDonald, Kelly O'Shea, Dennis York, Gordon Bisgrove, Art Peter, Maurice Lavallee, Dennis Cox, Ed Johnson, Gary Schneidmiller, Ken Porter, Dan Jenkins, Jamie McQuarrie. Fourth Row: Tony Sikora, Ray Olsson, Ted Flanders, Ed Dechant, Frank Lewis, Dale Huberdeau, Bill Bereska, Larry Huberdeau, Rory Thompson, Don Law, Gary Dakin, Bill Fairless, Bob Petite, Dale Darrah, Harold Dunlop. Fifth Row: Bill Kostiuk, Jim Molnar, Dave Brown, Irv Allen, Glenn MacPherson, Ralph Woods, Larry Skinkle, Dave Blackmore, Carson McDonald, Carl Leary, John Graham, Lou Foley, Vern Danes, Herb Walker. Back Row: Bill McPhail, Fred Sutherland, Jim Young, Don Dawson, Ben Shantz, Bob Lenton, Jurgen Moll, Rick Stewart, Gary Giese, Stan Clarke, Keith Branter, Mort Timanson, Bob Glover, Bob Hilbert, Phil Dube, Lorne Goff, Darryl Rollings

Alberta Government, AFHPC

Alberta Forest Service Superintendents, spring, 1985
Standing (L to R): Art Peter, Bow Crow; Bill Fairless, Athabasca; Carson McDonald, Slave Lake; Lorne Goff, Peace River; Mort Timanson, Grande Prairie; Rex Winn, Pine Ridge Forest Nursery Sitting: Larry Huberdeau, Lac La Biche; Cliff Henderson, Whitecourt; Fred Sutherland, Rocky Clearwater; Carl Leary, Footner Lake.
Missing: Norm Rodseth, Edson
Alberta Government, AFHPC

Alberta Forest Service Executive and Directors, spring, 1985
Standing (L to R): Cliff Smith, Assistant Deputy Minister; John Drew, Reforestation and Reclamation; Con Dermott, Timber Management; Don Fregren, Land Use; John Benson, Forest Protection; Chuck Geale, Program Support Sitting: Al Brennan, Forest Industry Development; Minister John Zaozirny; Fred McDougall, Deputy Minister; Joe Soos, Forest Research
Alberta Government, AFHPC

Alberta Forest Service Executive, Directors and Superintendents, spring, 1985
Standing (L to R): Art Peter, Ed Gillespie, Barry Rogers, Cliff Smith, John Drew, Carson McDonald, Lorne Goff, Con Dermott, Don Fregren, Mort Timanson, John Benson, Rex Winn, Bill Fairless. Sitting: Norm Rodseth, Larry Huberdeau, Fred Sutherland, Al Brennan, Minister John Zaozirny, Fred McDougall (Deputy Minister), Joe Soos, Carl Leary, Cliff Henderson, Chuck Geale
Cliff Smith

Pickup load of cone sacks ready to transport to the Pine Ridge Forest Nursery, Fort McMurray, Athabasca Forest, 1980s
Alberta Government, AFHPC

Dispatcher Training Course, Forest Technology School, March, 1985
Back Row (L to R): Ken Janigo, John Stepaniuk, Mike Thompson, Gerald Sambrooke, Floyd Schamber. Third Row: Rod Blades, Ted Cofer, Leo Drapeau, Wolfgang Richter, Brian Wudarck. Second Row: Paul Campbell, Dennis Halladay, Bill Kostiuk, Bob Yates, George Benoit. Front Row: Len Allen, Brian Meads, Dennis Quintilio (Instructor), Don Sarafinchin, Frank Nuspel, Joe Smith
Alberta Government, AFHPC

Alberta Forest Service staff pose in front of the first Canadair CL-215 purchased by the Alberta government, 1986
(L to R): John Benson, Director of Forest Protection; Owen Bolster, Supervisor of Wildfire Operations; Fred McDougall, Deputy Minister; Cliff Smith, Assistant Deputy Minister; Gordon Bisgrove, Manager Wildfire and Aviation
Alberta Government, AFHPC

Fort McMurray District parade float in the Blueberry Festival, 1985. Dave Pollard, Forestry Mechanic driving float vehicle
Alberta Government, AFHPC

Fort McMurray District parade float in the Blueberry Festival, 1985
Alberta Government, AFHPC

Spring Ranger Meeting Edson Forest, Forest Technology School, Hinton, April-May, 1986

Back Row (L to R): Gord Costie, Clayton Deardon, Tim Burggraaff, Jim McCammon. Fourth Row: Joe Smith, Dave Beck, Frank Staubitz, Hylo McDonald, Don Podlubny, Hugh Boyd. Third Row: Dave Schenk, Don Harvie, Tony Schlenker, Tony Sikora, Edwin Preece, Dave Brown, Brian Wallach. Second Row: Ray Olsson, Ross Graham, Brian Stanton, Stew Walkinshaw, Rick Bambrick, Murray Heinrich, Malcolm Pugh. Front Row: Walter Radowits, George Robertson, Frank Lewis, Fred Windjack, Bill Fairless (Superintendent), Ken Wheat, Bill Gilmour, Wally Manchester

Alberta Government, AFHPC

Advanced Fire Behaviour Course, Forest Technology School, Hinton, April, 1986

Back Row (L to R): Jim Cochrane, Tony Znak, Not Identified, Russ Braham, Bruce Cartwright, Bob Mazurik, Darryl Johnson, Not Identified. Middle Row: Darrell Hemery, Not Identified (Manitoba), Dave Redgate, Howard Herman, Brian (Buck) Dryer, Gord Taylor, Wally Peters, Not Identified. Front Row: Marty Alexander (Instructor), Leon Graham, Brian MacIntosh, Wayne Bowles, Gord Graham, Nick Galon, Steve Otway, Dennis Quintilio (Instructor)

Alberta Government, AFHPC

Fall Ranger Meeting, Athabasca Forest, Fort McMurray, 1986

Back Row (L to R): Dave West, Gordon Armitage, Jorn Thomsen, Gary Dakin, Tim Burggraaff, Bruce MacGregor, Kent MacDonald, Ralph Woods, Carson McDonald (Supervisor, Pine Ridge Nursery). Second Row: Bill Breshnahan, Greg Cariou, Val O'Donnell, Greg Cunliffe, George Benoit, Gary Dodsworth, Philippe Robert, Rodger Hamilton, Dave Blackmore. Front Row: Jerry Kress, Dave Redgate, Ken McCrae, Gordon Brown, Buck Dryer, Kevin Ledieu, Rick Hirtle

Alberta Government, AFHPC

Rainbow Lake District Staff, Footner Lake Forest, spring, 1986

(L to R): George Panici, Janet Scott, Paul Ronellenfitch, Mel Cadrain, Chris Hale, Conrad Bello, Jamie McQuarrie (Chief Ranger), Tom Archibald

Tom Archibald

Nordegg Ranger Station staff, Rocky Clearwater Forest, 1986
(L to R): Mike Poscente, Dave Kmet, Doreen Laing, Ted Loblaw, Jim Lunn, Ken Orich, Roger Litke and Pete the Dog
Jim Lunn

Advanced Fire Behaviour Course, Forest Technology School, Hinton, February, 1987
Back Row (L to R): Not Identified, Not Identified, Don Harrison, Not Identified, Not Identified, Kurt Wentzel, Not Identified, Not Identified. Middle Row: Not Identified, Jim Cochrane, Rick Prince, Steve Otway, Henry Grierson, Rory Thompson, John Brewer. Front Row: Hugh Boyd, Tony Schlenker, Jim Maitland, Not Identified, Not Identified, Gordon Crowder, Rick Arthur, Not Identified
Alberta Government, AFHPC

Skidder hauling Jet B helicopter fuel barrels to Margaret Lake, Footner Lake Forest, February, 1987
Corinne Huberdeau

Stuck! South of Eva Lake, 40 inches of snow, Footner Lake Forest, February, 1987
Corinne Huberdeau

Huestis Bonspiel winners, 1987
(L to R): Terry Van Nest, Bob Miyagawa, Jim Skrenek, Gordon Bisgrove
Terry Van Nest

Fire Prevention I Course, Forest Technology School, Hinton, 1987
Back Row (L to R): William Balmer, Vern Tetz, Don Carr, Darren Tapp, Gordon Baron (Instructor). Second Row: Garry Thompson, Ken Hennig, Morgan Kehr, Bob Held, Mike Berry. Front Row: Shane Armitage, Kevin Ledieu, Wally Born, Darnell McCurdy
Wally Born

Chief Ranger Ralph Woods, Fort McMurray District, Athabasca Forest, 1987. Ralph Woods writing up memorandum to office staff, "Thou shalt not take the Chief Ranger's name in vain or thou shalt get stuck with regen surveys, cruising or any number of unpleasant activities." So let it be written, so let it be done
Gary Dodsworth photo and story

Lac La Biche Forest Fall Ranger Meeting, 1987
Back Row (L to R): Len Stroebel, Eugene Baranski, Darnell McCurdy, Mark Froehler, Wally Born, Hugh Boyd, Rory Thompson. Middle Row: Kurt Frederick, Wally Peters, Gary Schneidmiller, Dave Brown, Harold Dunlop, Neil Barker, John McLevin, Dave Lind. Front Row: Kevin Freehill, Ken Yackimec, Collin Williams, Sharon Turner, Brydon Ward (Superintendent), Gord Taylor, Geoff Becker, Bill Bereska
Kevin Freehill

Ken McCrae, Land Use Officer, working the 1987 National Forest Week trade fair booth, Fort McMurray, Athabasca Forest
Alberta Government, AFHPC

Fort McMurray District staff, Athabasca Forest, 1987
(L to R): Eva Lue Reeve, Gary Dodsworth, Heather Groves, Rick Hirtle, Ralph Woods, Tim Burggraaff, Dane McCoy, Jim Witiw, Kevin Ledieu
Alberta Government, AFHPC

Slave Lake Ranger Meeting, Slave Lake Forest, 1987
Back Rows (L to R): Larry Kaytor, Barry Gladders, Tony Schlenker, Wayne Johnson, Bill Tinge, Wayne Becker, Mike Lambe, Wayne Bowles, Pat Guidera, Mark Storie, Brian Wudarck, Roy Nichols (hidden), Brian Tanner, Bart Elliot, Mark Canton. Third Row: Gerald Carlson, Gail Tucker, Rick Moore, Mark Coolen, Ralph McKeown, Pat Hendrigan, Lyall Gill. Second Row: Maurice Lavallee, Howard Herman, Bob Petite, Paul Steiestol, Dennis Cox, Don Zwicker, Russ Stashko, Howard Gray (Superintendent), Brian Anderson. Front Row: Larry Warren, Larry Skinkle, Darby Paver, Dennis York, Dean Isles, Ed Jones, Mag Steiestol
Wes Nimco

Spirit River District staff, Grande Prairie Forest, winter 1987-88
(L to R): Chris Killoran, Kurt Wentzell, Bill McDonald, Judy Gushlak, Bob Glover (Chief Ranger), Tom Archibald
Tom Archibald

Sector Level Fire Suppression Course, Forest Technology School, Hinton, 1987

Back Row (L to R): Marcel Harpe, G. Janes, Jeff Dixon, Shawn Zwerzinski, Bart McAnally, Rick Horne, Ambrose Jacobs, D. Giannotti, Doug Smith. Third Row: D. Heron, Ray Luchkow, Steve Cornelsen, Conrad Baetz, Gail Tucker, Darren Tapp, Roger Tetreault, Bruce Mayer. Second Row: Clarence Budal, Mike Pozniak, Gary Dodsworth, Ken Grover, Howard Townsend, N. Fontaine, Ian Pengelly, Doug Nichol. Front Row: N. Wentzell, Paul Rizzoli, S. Lewis, D. Quann, Evelyn Calliou, Chris McGuinty, Mitch Pence, N. Wortley

Alberta Government, AFHPC

Retirement party for Joe Niederleitner, Detection Officer, Provincial Forest Fire Centre, 1987. Nick Nimchuk presenting Joe with Bullwinkle the Moose. Nick said that Joe 'was always after this Holy Grail . . . a moose . . . but could never seem to bag one . . . but he loved the hunt regardless'

Alberta Government, AFHPC

Opening of the Huestis Demonstration Forest, Whitecourt Forest, 1988

(L to R): Cliff Henderson, Ernie Ferguson, Eric Huestis. Dale Darrah at back

Alberta Government, AFHPC

Advanced Fire Behaviour Course, Forest Technology School, Hinton, April, 1989

Back Row (L to R): Ralph Kermer (Fort Nelson), Not Identified (NWT), Paul Johnson (NWT), Peter Deering (Parks), Bud McKuen, Don Podlubny, Rick Smee, Wes Pinsent, Not Identified. Middle Row: Barry Gladders, Wayne Becker, Dean Isles, Brian Cuttrel, Dave Brown, Len Stroebel, Don Brewer, Rick Alquire, Not Identified (BC?). Front Row: Steve Cornelsen (Parks), Marty Alexander (Instructor), Don Cousins, Bill Lesiuk, Ken Orich, Mike Lambe, Rus DiFiore, Don Pope, Karl Peck, Don Mortimer

Alberta Government, AFHPC

Sector Level Fire Suppression Course, Forest Technology School, Hinton, 1988

Back Row (L to R): D. Williams, Jeff Henricks, Michelle Shesterniak, Blaine Renkas, John Belanger, Marvin Pearce, R. Greer, Greg Cariou. Middle Row: Geoff Cole, Bill Van Dyk, Sharon Mill, D. Bell, Wally Born, Ian Johnston, Mark Missal, Gordon Baron (Instructor). Front Row: Rick Moore, J. Potts, E. Auger, Tim Bean, Keith Beraska, Bill Allen, E. Hornby, Brent Schleppe

Wally Born

Erling Winquist with Bertie Beaver at his retirement party, 1988
Bob Stevenson

Cat Boss Course, Forest Technology School, Hinton, 1989

Back Row (L to R): Richard Speakman, Lloyd Helmer, Wayne Richards, Kelly Kluin, Ray Houle, Walter Gaucher, Duane LaValley. Fourth Row: Harvey Weegar, Richard Belanko, Glenn Mitchell, Ronald L'Heureux, Cyril Thacker, Ernie Wasylynvik, John McLevin (Instructor). Third Row: Paul Rizzoli, Reg Cook, Garry Mitchell, Doug Forrest, Terry Kuzma, David Finn, Brent Schleppe, Bruce Mayer. Second Row: Leonard Storoschuk, Dennis Roman, Phil Herrod, David Swindlehurst, Brian Orum, Roger Tetreault, Barry Congram (Instructor). Front Row: Mervin Bellerose, Gordon Klassen, Ron Lukan, Kennith Smith, Alfred Schmutz, Floyd Cook, Colin Johnston, Denis LaBonte

Alberta Government, AFHPC

Fort Assiniboine Ranger District staff with Cliff Henderson, Superintendent Whitecourt Forest, Fort Assiniboine, winter, 1988-1989
(L to R): Rob Gibb, Brent Schleppe, Ian Whitby, Brian Cutrell, Don Carr, Cliff Henderson, Ken Porter (Chief Ranger), Dale Asselin
Brent Schleppe

Gordon Bisgrove holding the wooden Provincial Forest Fire Centre logo received upon his transfer from Edmonton to the Superintendent, Whitecourt Forest, 1989
Alberta Government, AFHPC

High Level District staff picture, High Level, Footner Lake Forest, February, 1990
(L to R): Bob Petite (Chief Ranger), Dave Scott, Jane Brown, Dave Weir, Jim Lunn, Darren Tapp, Tom Archibald
Jim Lunn

Chief Ranger Bruce MacGregor, Fort McKay Ranger Station, Athabacsa Forest, 1990
Bruce MacGregor

AFS Fall Fire Conference, Forestry Training School, Hinton, November 21 and 22, 1990
Back Row (L to R): Gary Dakin, FPO Athabasca; Jamie McQuarrie, FPO Slave Lake; Darryl Johnson, FPT Whitecourt; Darryl Rollings, FPO Rocky Mountain House; Gary Mandrusiak, FPT Rocky Mountain House; Art Peter, Superintendent Bow/Crow; Bob Yates, FPT Grande Prairie; Jurgen Moll, FPO Whitecourt; Hylo McDonald, FPO Edson; Brian Wesolowsky, Contact Birddog Officer; Herb MacAuley, Contract Birddog Officer. Fifth Row: Rob Thorburn, Instructor FTS; Chris Killoran, Air Attack Officer Slave Lake; Bob Young, Detection PFFC; Bill Griggs, Head Communications PFFC; Buck Dryer, Air Attack Officer PFFC; Gordon Armitage, Superintendent Athabasca; Lou Foley, Manager Operations PFFC; Ken Scullion, AVC Lac La Biche; Steve Semeniuk, Contract Birddog Officer. Fourth Row: Larry Warren, FPT Edson; Mike Dubina, Training Coordinator PFFC; Terry Van Nest, Fire Behaviour Officer PFFC; John Branderhorst, FPT Peace River; John McLevin, FPT Lac La Biche; Brent Bochan, Grande Prairie; Dahl Harvey, Weather Section PFFC; Tom Archibald, FPT Footner Lake; Ken Baldry, Contract Birddog Officer; Wayne Rutter, Contract Birddog Officer; Fred Paget, Air Attack Officer Whitecourt. Third Row: Revie Lieskovsky, Air Attack Coordinator PFFC; Don Podlubny, Chief Ranger Grande Cache; Brian MacIntosh, FPT Edson; Bill de Groot, Canadian Forest Service; Cordy Tymstra, Fire Science PFFC; Ted Cofer, FPT Slave Lake; Howard Herman, FPT Athabasca; John Graham, Chief Ranger Robb; Dale Huberdeau, FPO Footner Lake; Roger Marchand, Contract Birddog Officer. Second Row: Phil Robert, Air Attack Officer, Footner Lake; Owen Bolster, Operations PFFC, Mag Steiestol, Prevention Coordinator PFFC; Sunil Ranasinghe, Insects and Disease PFFC; Ken McCrae, FPO Peace River; Brian Meads, FPT Bow Crow; Rick Stewart, FPT Rocky Mountain House; John Brewer, FPT Peace River; Kurt Frederick, FPT Peace River; Chris Hale, FPT Athabasca; Jim Kitz, Reforestation and Reclamation PHQ. Front Row: Leo Drapeau, Initial Attack Coordinator PFFC; Pat Smith, Weather Section PFFC; Nick Nimchuk, Weather Section PFFC; Ken Orich, FPT Footner Lake; John Benson, Director Forest Protection PFFC; John Hogue, FPT Whitecourt; Lorne Goff, Superintendent Rocky Mountain House; Bill Bereska, FPO Lac La Biche; Don Law, FPO Grande Prairie; Kelly O'Shea, FPO Bow Crow
Alberta Government, AFHPC

Spring Ranger Meeting, Swan Hills Fire Base, Whitecourt Forest, March 22, 1990
(L to R): Darryl Johnson, Don Gelinas, Bruce MacMillan, Lloyd Seedhouse, Geoff Smith, Jim Mahan, Barry Congram, Jeff Henricks, Don Carr, Conrad Gray, Andy Borle, Dave Finn, Kevin Topolnicki, Gerry Erdmann, Dennis York, Dennis Howells, Alfred Schmutz, Brian Wallach, Dale Darrah, Roger Light, Gordon Graham, Joe Burnstick, Ed Dechant, Gordon Bisgrove (Superintendent), Ken Porter, Dale Asselin, Brent Schleppe, Mark Storie, Dave Kmet, Bruce Mayer, Jurgen Moll, Brian Cutrell, Gerald Carlson, Sean Harris, Darren Britton, Myles Gochee, Russ Stashko, Margarete Hee, Ken Snyder, Audrey Piesinger, Dennis Driscoll, Sandy Mah
Darryl Johnson

Fire Prevention II Course, Forest Technology School, Hinton, 1990
Back Row (L to R): Harry Bodewitcz, Chris Killoran, Wayne Bowles, Dave Brown, Paul Steiestol, Jim Lunn, Tom Archibald, Ed Barnett. Middle Row: Rod Houle, Conrad Gray, Doug Campbell, Kurt Wentzell, Mark Froehler, Howard Herman, Barry Congram, Bruce Mayer. Front Row: Larry Warren, Rick Horne, Stew Walkinshaw, Rick Bambrick, Mag Steiestol (Instructor), Barry Gladders, Henry Grierson, Chris McGuinty
Alberta Government, AFHPC

Spruce budworm spray crew, Footner Lake Forest, early 1990s. Alberta Forest Service made use of aerial tankers to spray biological insecticide Btk (Bacillus thuringiensis var kurstaki) to control spruce budworm defoliation. The first contracts were awarded to Wetaskiwin Aerial Applicators who used Air Tractor AT-302 and Ayres Thrush fixed-wing aircraft. The pilots and support crew in the picture are from Wetaskiwin Aerial Applicators and Yorkton Flying Services
Left side of picture are Brian Wesolowsky, Birddog Officer; Tom Archibald, Forest Protection Technician Footner Lake; Bob Petite, Chief Ranger High Level; Steve Semeniuk, Birddog Officer
Alberta Government, AFHPC

Nova pipeline corridor above the Little Smoky River, west of Fox Creek, Fox Creek District, Whitecourt Forest, 1991. Aerial view of working space and road to river on pipeline right-of-way
Alberta Government, AFHPC

View of tracked hoe preparing creek crossing north of Berland River for installation of gas pipeline. Pipe has concrete muskeg weights installed to keep the pipe from moving too much once installed in the pipeline trench
Alberta Government, AFHPC

Footner Lake Forest Fall Ranger Meeting, High Level, Footner Lake Forest, March 17-19, 1992

Back Row (L to R): Brian Sabatier, Brenda Van Rootselaar, Joanne Pottage, Larry Knutsen, Mike Templeton, Dave Brown, Kevin Freehill, Doug Bishop. Third Row: Anita Pederson, Sheryl Moody, Barrie Onysty, Marc Gamache, Wayne Thiessen, Marianne Devries, Dave Beck, Steve Blanton. Second Row: Craig Brown, Dennis Halladay, Stuart Carter, Ian Wallis, Al Benson, Aaron Doepel, Mike Benedictson, Tracy Pearson. Front Row: Vern Neal, George Robertson, Jorden Johnston (Superintendent), Ken Yackimec, Ken Orich, Jim Lunn, Tom Archibald, Bill Lesiuk. Missing: Bob Petite, Phil Robert, Dale Huberdeau, Lance King, Len Smith, Les Croy

Ken Yackimec

Overhead Team Meeting, Swan Hills Fire Base, Whitecourt Forest, spring, 1991

Standing (L to R): Ed Johnson, Wally Kreisler, Brian Cuttrel, Russ Stashko, Butch Shenfield, Dave Bartesko, George Benoit, Bob Lenton, Leonard Kennedy, Dennis Cox, Bruce MacGregor, Andy Gesner, Brian MacIntosh, Norm Olsen, Rory Thompson, Dave Beale, Bob Petite, Tom Grant, Dave Brown, Len Wilton, Brian Wudarck, Stan Clarke, Frank Lewis, Bob Glover, Gary Mandrusiak, Ralph Woods, Bruce Cartwright, Henry Grierson, Bob Mazurik, Jamie McQuarrie, Bill Bereska, Don Law
Kneeling: Dale Huberdeau, Lou Foley, Ken Porter, Ed Dechant, Howard Herman, Rick Arthur, Dennis York, Maurice Lavallee, Darryl Johnson, Rick Stewart, Mike Dubina

Alberta Government, AFHPC

(L to R) Darryl Rollings, Gary Mandrusiak and Mike Dubina at Falls Lookout, Rocky Clearwater Forest, 1991
Alberta Government, AFHPC

Alberta Forest Service Executive, May, 1993

Standing (L to R): Con Dermott, Director Timber Management; Rod Simpson, Director Program Support; Mort Timanson, Superintendent Grande Prairie Forest; Don Fregren, Director Forest Land Use; Gordon Armitage, Superintendent Athabasca Forest; Brydon Ward, Superintendent Lac La Biche Forest; Jorden Johnston, Superintendent Footner Lake Forest; Howard Gray, Superintendent Slave Lake Forest; Carson MacDonald, Superintendent Pine Ridge Forest Nursery; Cliff Henderson, Director Reforestation and Reclamation; Gordon Bisgrove, Superintendent Whitecourt Forest; Carl Leary, Superintendent Peace River Forest; Bill Fairless, Superintendent Edson Forest; John Benson, Director Forest Protection. Sitting (L to R): Lorne Goff, Superintendent Rocky Clearwater Forest; Cliff Smith, Deputy Minister; Ken Higginbotham, Assistant Deputy Minister; Art Peter, Superintendent Bow Crow Forest

Slave Lake Level II Overhead Team, Fire DS6-25-91
Back Row (L to R): Darby Paver, Doug Smith, Ian Whitby, Wally Born. Front Row: Mike Lambe
Wally Born

Retirement party for Provincial Forest Fire Centre staff, September, 1991
(L to R): Claude Lefebvre, Senior Communications Officer; Wayne Barker, Senior Mechanical Supervisor; Owen Bolster, Supervisor of Wildfire Operations; Bill Griggs, Supervisor of Communications Section
Alberta Government, AFHPC

Bernie Simpson, 1992. Bernie started as a Ranger with the Alberta Forest Service in 1957 and retired as the Director Forest Technology School in 1990. Bernie was the Forest Protection Officer at Slave Lake during the catastrophic spring Vega fire in 1968
Bob Stevenson

Alberta Forest Service Assistant Deputy Minister Ken Higginbotham giving speech during the Wayne Barker, Claude Lefebvre, Owen Bolster and Bill Griggs retirement party, September, 1991
Alberta Government AFHPC

Dennis Cox, Bob Newstead, John Benson at Norm Rodseth's retirement party, 1992
Bob Stevenson

Con Dermott, Don Fregren, Norm Rodseth at Norm Rodseth's retirement party, 1992
Bob Stevenson

Assistant Deputy Minister Ken Higginbotham (L) presenting Norm Rodseth with an Alberta Government plaque upon his retirement in 1992
Bob Stevenson

Northern Alberta Institute of Technology Forest Technology curriculum validation workshop, Edmonton, summer, 1991
Back Row (L to R): Al Walker, Howard Anderson, Gerry Wilde, Brian Adams, Ross Risvold, Andy Neigel, Dave Fournier, Alan Tovey, Alan Pollock, Murray McDonald, Gordon Baron, Rob Thorburn, Dennis Quintilio.
Front Row: Tony Wispinski, Jim Friesen, Patty Lemke, Keith Hutton, Thor Knapp, Mike Watson
Rob Thorburn

Alberta Forest Service Superintendents meeting, Goldeye Centre, west of Nordegg, August 6, 1993
Back Row (L to R): Carson McDonald, Pine Ridge Forest Nursery; Gordon Bisgrove, Whitecourt; Carl Leary, Peace River; Jorden Johnston, Footner Lake.
Front Row: Mort Timanson, Grande Prairie; Howard Gray, Slave Lake; Lorne Goff, Rocky Clearwater; Brydon Ward, Lac La Biche; Art Peter, Bow Crow; Gordon Armitage, Athabasca (Fort McMurray); Bill Fairless, Edson. This was the last meeting of Superintendents prior to Regionalization and the retirement of Art Peter.
Howard Gray

Fort Vermilion District staff, Footner Lake Forest, September, 1993
(L to R): Bill Lesiuk (Chief Ranger), Marc Gamache, Aaron Doepel, Jim Lunn (FO III), Mike Benedictson, Juanita Rosenberger, Doug Bishop
Jim Lunn

Alberta Forest Service Jack Lunan Golf Tournament

In the early 1980s planning began on an annual Alberta Forest Service fun golf tournament in honour of Jack Lunan. Lunan had been a long time pilot with Associated Helicopters before his untimely death while flying with the Canadian Snowbirds. The golf tournament was organized by Chuck Geale, Fred Windjack, Bill Kostiuk, Ed Picota, Walter Radowits and Norm Olson with the first AFS Lunan tournament held in 1983. The original tournament format had a male and female winner – Fred Windjack and Pat Golec were the two 1983 winners. In 1987 the tournament was reformatted as a four-ball best ball tournament. Edson served as the tournament location until the mid-1990s when the tournament was moved to Lac La Biche.

Trophy presentations AFS Jack Lunan Fun Golf Tournament, 1985
(L to R): Ben Shantz, Not Identified, Dane McCoy, Bob Heighington, Rick Strickland
Fred Windjack

Trophy presentations AFS Jack Lunan Fun Golf Tournament, 1985
(L to R): Mrs. Florence Lunan, Mark Froehler, Fred Windjack
Fred Windjack

Trophy presentations AFS Jack Lunan Fun Golf Tournament, 1985
(L to R): Vern Seib?, Not Identified, Bob Lenton, Mrs. Florence Lunan
Fred Windjack

Trophy presentations AFS Jack Lunan Fun Golf Tournament, 1985
(L to R): Gary Kostiuk, Travis Heighington, Mrs. Florence Lunan, Mark Froehler, Fred Windjack
Fred Windjack

Scarification and site preparation equipment

The Martinni Plow was used on sites where deep organic material existed, thus requiring a deeper 'plow' to expose mineral soil for planting
Alberta Government, AFHPC

Bracke scarifier, one of the many mechanical tools used to expose mineral soil during site preparation, readying a site for planting
Alberta Government, AFHPC

Brush rake on dozer used to pile logging slash debris and expose mineral soil to prepare a site for planting
Alberta Government, AFHPC

Ripper plow mounted on
Caterpillar D8K preparing
cutblock for planting, early 1990s
Alberta Government, AFHPC

Cutblock being prepared for planting by ripper
scarification, Tony Creek area, Fox Creek
District, Whitecourt Forest, winter, 1991-1992
Alberta Government, AFHPC

Ripper plow mounted on International HD31
preparing cutblock for planting, early 1990s
Alberta Government, AFHPC

Pine Ridge Forest Nursery

Minister Merv Leitch officially opening the new Pine Ridge Forest Nursery, Smoky Lake, September, 1979
Darrell Hemery

10th Anniversary ceremony at the Pine Ridge Forest Nursery, October 11, 1979
(L to R): Ken Higginbotham, Acting Assistant Deputy Minister; Carson McDonald, Superintendent Pine Ridge Forest Nursery; Dr. B. Elliot, MLA Grande Prairie and Chair of the Forestry Caucus Committee
Donna Palamarek

10th Anniversary ceremony at the Pine Ridge Forest Nursery, October 11, 1979
(L to R): Darrell Hemery, Barry Court, Kathy Yakimchuk, Russell Braham
Donna Palamarek

Pine Ridge Forest Nursery and the Alberta Tree Improvement and Seed Centre staff picture, May, 1991
Donna Palamarek

Donna Palamarek (R) giving tour of Pine Ridge Forest Nursery to the Princess of Thailand, March 13, 1990
Minister Leroy Fjordbotten looking on behind the Princess; Cliff Henderson and Ken Higginbotham to the right of Donna Palamarek
Donna Palamarek

Darrell Hemery and Jerry Murphy in greenhouse at the Pine Ridge Forest Nursery, early 1990s
Darrell Hemery

Outdoor seedling beds, Pine Ridge Forest Nursery, early 1990s
Darrell Hemery

Nursery staff in transplanter machine planting seedling plugs for one to two years of additional growth in the outdoor fields, Pine Ridge Forest Nursery, early 1990s
Darrell Hemery

Nursery staff in transplanter machine planting seedling plugs for one to two years of additional growth in the outdoor fields, Pine Ridge Forest Nursery, early 1990s
Darrell Hemery

Ken Snyder (L) and Brian Wallach participating in a planting contract supervisor course held at the Pine Ridge Forest Nursery, 1996
Darrell Hemery

View of seedling harvest operation of container and bareroot seedlings at the Pine Ridge Forest Nursery, 1996
Darrell Hemery

1985 - 1992　281

CHAPTER 9
1993 - 2005

The final decade covered by this book was characterized by major changes in how business was done by what was, until 1993, the Alberta Forest Service (AFS) – and indeed by all offices of Alberta's provincial government. The focus in 1993 was on the elimination of the provincial debt. As a result, the public service was called upon to totally re-think the way services were provided and funded, and the way in which provincial resources were managed.

Government institutions at every level mirrored a private-sector trend to downsize and rationalize operations. There was a focus on regionalization and community-level service combined with a centralization of the most senior administrative staff and offices. Government declared it was getting out of the business of business, and those services that fit better with commercial enterprise than public administration were privatized. The general public and concerned stakeholders were engaged more fully in policy making through a growing number of multi-disciplinary task groups and steering committees. This approach was consistent with international conventions promoting sustainable development. Alberta's forest management ethic continued to evolve from the traditional sustained-yield objective towards an integrated system. The new system accounted for a wider range of values (such as wood, water, wildlife) across broad ecological and landscape units.

At the same time, three forces were creating tensions about the extent and kinds of activities in the forest and implications for their sustainability. First were the new and restructured forest industries that would utilize most of the unallocated allowable cut on public forests. Second was the increasing impact on the forest by the energy sector in pursuit of conventional oil and natural gas, heavy oils and oil sands. Third comprised the increasing concerns about the sustainability of the forest ecosystem and its many values and uses, including biodiversity, water quality, recreation, aboriginal traditional uses, wilderness and solitude, hunting, fishing, trapping and a host of others. During the 1980s and early 1990s social values around protection of the environment and sustainability of the province's natural resources became greater priorities. As Cliff Smith recalled, [1] the challenge became one of serving a number of different interest groups and client groups that frequently did not share the same objectives. The Department sought new ways to increase public involvement in natural resource management decision-making processes.

Minister Brian Evans, 1992 - 1994
Bob Stevenson

Fred McDougall retirement party, 1993
(L to R): Cliff Smith, Assistant Deputy Minster, AFS; Al Brennan, Executive Director, Forest Industry Development; Fred McDougall, Deputy Minister; Les Cooke, Assistant Deputy Minister, Fish and Wildlife; Murray Turnbull, Assistant Deputy Minister, Public Lands; Mike Toomey, Executive Director, Resource Data Division
Bob Stevenson

Added to this mix was the impact of several years of seemingly larger and more intense forest fires.

Some highlights of these influences are described in this chapter, along with many of the innovative programs designed to enhance achievement of forest sustainability. What is particularly significant is that while these many influences and changes were under way, staff of the AFS, later renamed Land and Forest Service and further reorganized to Public Lands and Forest Division, Forest Protection Division and Strategic Forestry Initiatives, were still going the extra mile to protect and manage the forest.

Changing Departmental Structures

The Alberta government, under the leadership of incoming Premier Ralph Klein, instituted a major reorganization of government Departments in 1993. The AFS was placed in the new Department of Environmental Protection (later named Alberta Environment), led by Minister Brian Evans. The Department's Annual Report for 1992-93 declared: "Alberta Environmental Protection is a new Department with a new mandate and new environmental legislation. Creation of this Department begins a new era for environmental protection in Alberta."$_2$

The Department merged three separate agencies. The former Departments of Environment and Forestry, Lands & Wildlife were combined with the Parks Division of the former Department of Tourism and Recreation. The Annual Report declared: "Together these agencies meet the needs of Albertans for a quality environment by providing for clean air, water and soil; protecting wildlife, forests, parks and other natural resources; and ensuring that the development of these resources is truly sustainable." The Department was "founded on an integrated resource management philosophy, which will keep Alberta on the leading edge of environmental protection."$_3$

The AFS was enlarged by the addition of the former Lands Division and given the new name of Land and Forest Services and in 1994, Land and Forest Service (LFS). Responsibilities included forest management, forest inventory; cattle grazing; watershed protection; reforestation and reclamation; fire prevention and suppression; insect and disease control; recreation; and education through Junior Forest Warden and Ranger programs. Other responsibilities included regulation of petroleum and natural gas exploration activities; oil sands, coal, geophysical and land dispositions in the Green Area and on crown lands in the White Area. The Forest Industry Development group was moved to Economic Development.

AFS headquarters in Peace River, late 1950s – before new offices were built
Alberta Government, AFHPC

By 1995 LFS had become one of four divisions in the Department; the others being Natural Resources Service, Environmental Regulatory Service and Corporate Management Service. LFS itself was organized into four main branches: Forest Management, Forest Protection, Land Administration and Program Support. Peter Melnychuk, Deputy Minister of the former Department of the Environment, was named Deputy Minister of the new Department of Environmental Protection. Former Deputy Minister Cliff Smith took early retirement and launched a successful consulting business. Ken Higginbotham continued as Assistant Deputy Minister for LFS.

The government introduced the three-year "business plan" as a basis for meeting its goals and fostering more inter-departmental coordination. This planning process helped refine objectives and assess results. The LFS vision, mission and

Land and Forest Service Managers' meeting, spring, 1998
Back Row (L to R): Kelly O'Shea, Butch Shenfield, Neil Barker, Jim Maitland, Don Harrison, Craig Quintilio, Ray Luchkow, Rory Thompson. Middle Row: Tom Archibald, Tony Sikora, Howard Gray, Darryl Johnson, Jerry Sunderland, Dennis Cox, Mike Poscente, Ed Ritcey, Russ Stashko. Front Row Seated: John Brewer, Bruce Mayer, Dale Huberdeau, Al Hovan, Jim Cochrane, Margaret Franklin, Pat Guidera
Alberta Government, AFHPC

principle statements were by this time condensed to: "Land and Forest Service is responsible for the proper use, management, allocation and protection of Alberta's public lands."[4]

The *Environmental Protection and Enhancement Act* (EPEA) and regulations, approved on September 1, 1993, achieved a consolidation of 13 separate Acts in an effort to integrate the approach to environmental protection. This development resulted in a reorganization of Alberta Environmental Protection midway through the fiscal year.

Continued restructuring through 1994 and 1995 dramatically changed the LFS field organization. Many Ranger Stations were closed upon centralization of field offices, reducing from 40 districts to 18 and later 17. The 10 Forests were reorganized into six new administrative regions, four of which were in the Green Zone.

Staff positions were further reduced by 107, working towards a staff reduction target of 856 within the Department. The six new administrative regions were designed to reflect Alberta's natural regions: Northwest Boreal, Northeast Boreal, Northern East Slopes, Parkland, Bow and Prairie. The four forested regions had headquarters at Peace River, Lac La Biche, Whitecourt and Rocky Mountain House. Ranger districts were reassigned among the Regions. The objective was to "enhance community level service … to Albertans… and allocate more resources and more decision-making responsibility to staff in the regions."[5]

In the spring of 1995 LFS Assistant Deputy Minister Ken Higginbotham resigned to become Vice-President of Woodlands with Canadian Forest Products (CANFOR) in Vancouver. Cliff Henderson was selected as the new Assistant Deputy Minister. Henderson had started full-time with AFS in 1966 as forester at Fort McMurray. He worked in Rocky Mountain House and Edmonton, later becoming Forest Superintendent at Footner Lake and then Whitecourt. In 1988 he returned to Edmonton as Director of the Reforestation Branch for five years and Executive Director of Forest Management for two years before his appointment as ADM.

Ty Lund was named Minister of Environmental Protection on October 21, 1994, replacing Brian Evans who became Minister of Justice. Minister Lund was from the Rocky Mountain House area and was knowledgeable about forest-related issues. With the new Minister came a new mission statement: "As proud stewards of Alberta's renewable natural resources, we will protect, enhance and ensure the wise use of our environment. We are a dedicated and committed team, responsible for managing these resources with Albertans. We are guided by a shared commitment to the environment and are accountable to our partners, the people of Alberta."[6]

The Minister also noted that the Departmental business plan was designed to build a prosperous

C.J. Henderson, ADM LFS, 1995
Bob Stevenson

Minister Ty Lund, 1994 - 1999
Bob Stevenson

province. It would do this by ensuring sustained benefits from renewable resources and environmental quality, contribute to the government's deficit reduction goal by reducing overall spending by 30 per cent from 1992-93, rationalize regulations and streamline processes, restructure the Department, integrate services and enhance community level service by regionalizing and consolidating facilities. The budget reduction target was met in 1997 through further reductions of staff combined with privatization. In 1997 Minister Lund said his main goal was to: "Contribute to building a strong and prosperous province by ensuring that Alberta's renewable natural resources (air, land, water, forests, fish, wildlife, parks and natural reserves) are sustained; and high environmental quality (air, water and land) is maintained."[7]

The strategies included the following:
- Ensure the public forest resource is managed in a sustainable manner, providing Albertans with long-term economic and recreational opportunities;
- Ensure forest resources are protected from fire, insects and disease;
- Ensure wise use of Alberta's public land resource, while maintaining environmental quality;
- Ensure that fish and wildlife populations and their habitats are sustained;
- Provide recreational and commercial hunting and fishing and outdoor recreational opportunities for Albertans.

The last gathering of Land and Forest Service managers, June, 2000, prior to the realignment of Fire Centres and Forest Areas, and the creation of the new Sustainable Resource Development department, spring, 2001

Back Row (L to R): Craig Quintilio (Director, Forest Protection), Bruce Cartwright (Fort McMurray), Neil Barker (Regional Director - NEB), Rick Blackwood (Calgary), Robert Stokes (Drayton Valley), Glen Gache (Manning), Hugh Boyd (Wildfire Prevention), Brent Schleppe (Hinton). Middle Row: Jim Cochrane (Lac La Biche), Butch Shenfield (Rocky Mountain House), Ken McCrae (Operations Coordinator - NWB), Margaret Franklin (Human Resources), Cliff Henderson (ADM - LFS), Dennis Cox (Edson), Mike Poscente (Regional Director - NES), Al Hovan (Slave Lake), Don Harrison (Wildfire Operations), John Brewer (Air Operations). Front Row: Craig Barnes (Operations Coordinator - NES), Jim Maitland (Valleyview), Bruce Mayer (Athabasca), Gail Tucker (High Level), Jamie McQuarrie (Warehouse), Darryl Rollings (Operations Coordinator - PBP). Lying down on the job: Dan Wilkinson (Regional Director - NWB)
Missing: Terry Zitnak (Operations Coordinator - NEB), Pat Guidera (Regional Director - PBP), Ed Ritcey (Grande Prairie), Tony Sikora (High Prairie), Tom Archibald (Peace River), Darryl Johnson (Blairmore), Dennis Quintilio (Ecological Landscape), Glenn Selland (Land Administration), Doug Sklar (Forest Management)
Alberta Government, AFHPC

Significant elements of the Department's philosophies were identified by Minister Lund in the Annual Report for 1997-98: "Alberta's prosperity and the health and welfare of Albertans are founded on a healthy environment. Albertans rely on their environment for renewable natural resources to fuel economic growth; for clean air, water and land with which to live; and for the spiritual renewal that the natural world can offer." He also stated that: "We will continue to hold sustainable development and integrated resource management as fundamental principles."[8]

The Department's name was changed to Alberta Environment in the summer of 1999. Within this Department a new Integrated Resource Management Division (IRM) was

Mariana Lake fire – Algar Tower, 1995
Rick Arthur

1993 - 2005   285

established, headed by Dennis Quintilio, previously Director of Forest Management. The main objective of the IRM Division was to lead inter-departmental implementation of the new paradigm of land and resource management. It was to act as the "accountability centre" for Alberta's commitment to sustainable resource and environmental management, and develop an IRM process model for resource and environmental decision-making.

Gary Mar became Minister of Alberta Environment in 1999, and then a cabinet shuffle brought Halvar Jonson to the position of Minister in 2000. The new Minister immediately became involved in promoting the efforts of LFS with one initiative being the decentralizing of the authority to issue commercial timber permits down to field offices. The aim was to provide a more efficient service to clients.

A provincial election in March, 2001, brought another shuffle of cabinet Ministers and a name change for portfolios and Departments. Alberta Sustainable Resource Development (SRD) was created March 15, 2001, pulling together programs from the departments of Alberta Environment, Resource Development and Agriculture, Food and Rural Development. These elements were then incorporated into four divisions: Forest Protection (FPD), Lands and Forests (LFD), Fish and Wildlife (FWD), and Public Lands (PLD).

Sustainable Resource Development was headed by Minister Mike Cardinal with Deputy Minister Bob Fessenden. The LFS (and AFS) ceased to exist as an identifiable entity, with its component parts moved to the new divisions. SRD also oversaw the Natural Resources Conservation Board, the Surface Rights Board and the Land Compensation Board. As Assistant Deputy Minister Cliff Henderson observed, the reorganization of 2001 was significant in that there were now three Assistant Deputy Minister positions, one each for Lands and Forests, Forest Protection and Public Lands.

Fire and Forest Health

The fire season in 1995 was considered the

East of Highway 40, near Gregg River – extreme winds in December, 1997 caused fire to "dance" across the young forest
Bob Udell

Minister Mike Cardinal, 2001 - 2004
Bob Stevenson

Early spring, 1998 – fire prohibition notice
Bob Stevenson

worst since the mid-1930s with respect to fire hazard levels. A total of 803 fires burned 336,148 hectares. During a 22-day period from mid-May to early June, 316 fires started, with eight becoming major wildfires. The Mariana Lake and Melvin Creek fires were responsible for 78 per cent of the burned area.

The Fire Information Resource Environment System (FIRES) was developed in 1995 as a data-entry system to encompass and track all facets of suppression efforts and associated disciplines. At the same time, a Forest Insect and Disease Operations (FIDO) single-source entry module was implemented in FIRES to capture data on insect and disease control elements.

A team led by Instructor Rob Thorburn broke

new ground with its wildfire instruction CD-ROM training program *Principles of Fire Behaviour*, which achieved international attention and recognition. As well, the fire training simulator program was upgraded to laser disc technology. These were developed cooperatively by staff in Forest Protection and the Environmental Training Centre at Hinton (formerly the Forest Technology School), and proved to be award-winning technology. The Wildland-Urban Interface Program was also enhanced in partnership with municipal governments, federal agencies and the general public. This initiative encouraged reduction of fire threats in communities through planning, design and fuels management both within the community and adjacent forests.

In 1996 a Forest Protection Advisory Committee was established to provide an ongoing forum for key stakeholders and the Alberta government on forest protection issues, policies and projects. Membership included municipalities, oil and gas industry, forest industry, power generation and distribution industry and government.

On December 14, 1997 a serious prairie fire started near the community of Granum damaging private and public holdings in the southern Porcupine Hills and near Pincher Creek. Exceedingly dry conditions and strong chinook winds whipped flames across large areas at alarming rates, a precursor to more forest fire problems the next year. This fire left a scar of burnt grass and stubble more than 35 kilometres long and 15 kilometres wide. The fire front was finally stopped by firefighters who responded from more than 20 municipalities (using a four-lane-wide stretch of highway as a fire break) about four hours after it started. [9]

Smoke column from the Granum fire, December, 1997
Local homeowner

Replanting in burned-over areas
Vanderwell Contractors

The LFS created a wildland firefighter business strategy in 1997 to improve the efficiency of firefighting capability. The strategy offered incentives for firefighters to become more proficient through advanced training, and development of a career stream with the prospect of seasonal employment. This strategy was the impetus for the Type II Wildland Firefighter program and also promoted the development of contracting capacity and knowledge within aboriginal communities.

Despite refinements to the forest protection system, the spring and summer fire season of 1998 was another record-setter in terms of area burned. In total, 1998 saw 1,600 fires consume an estimated 760,000 hectares, destroying upwards of 14.5 million cubic metres of commercial timber. This compared to 17.7 million cubic metres harvested by the timber industry in 1997, when less than 5,000 hectares were burned. Growth of vegetation during the previous two wet years created an abundance of fine fuels. Then a combination of low snow-pack, an early dry spring, unseasonably early lightning and strong winds resulted in fires that spread rapidly and escaped initial attack. Drought later in the season also exacerbated fires in coniferous stands.

Forest fires ravaged a considerable area in the drought-stricken region around Whitecourt, Slave Lake, Swan Hills and Chip Lake. Of primary concern was the immediate loss of merchantable timber to companies that would have to adjust their plans and arrange salvage operations for 9.5 million cubic metres of burned wood. Another

Aerial Recon R-44 C-GBOQ sitting on Highway 44 during the Mitsue fire, Northeast Boreal Region, Slave Lake District, spring, 1998
Bruce MacGregor

major expense of the fires was the loss of recently regenerated stands established on previously logged areas. The most noteworthy fires in 1998 were the Virginia Hills, Mitsue Lake, Agnes Lake and Roche Lake fires.

In general burned wood must be salvaged within two years. If the wood is not promptly harvested, the fibre becomes too brittle for manufacture of pulp and paper. Also, wood-boring beetle larvae may make the wood unsuitable for sawing into structural lumber.

The LFS faced criticism from stakeholders for its wildfire suppression efforts. This prompted Minister Ty Lund to state that the Department was well prepared for the early 1998 wildfire season. There had been no reduction in fire suppression budgets in the 1990s, and in 1997 approximately half a million dollars were invested in fire training programs for Department staff. The concerns raised resulted in the Department initiating an extensive review of the 1998 fire season by the consulting firm KPMG. It also encouraged greater industry participation in the analysis of fire management.

The year 1999 also saw another difficult spring with very hot, windy and dry conditions. Severe damage occurred from large burns across northern Alberta. About $100 million was spent on fire suppression, much of it in areas containing significant numbers of oil and gas wells, pipelines, compressor stations, power lines, railway and road networks and towns and private residences.

The seismic lines and connecting pipeline corridors contained extensive grasses and other "light fuels" that acted as "wicks" or fire corridors when ignited. Distinctive fire patterns developed which, when pushed by high winds, raced unhindered into and throughout adjacent forest stands. Despite this fire behaviour and the volatile nature of the petroleum infrastructure within the fire areas, the fires were brought under control and contained without loss of human life or any major shutdown of petroleum energy production.

However there was damage to mills, log decks and other property as well as considerable loss of timber.

In mid-December, 1999, KPMG presented its *Alberta Fire Review 1998* report, which included 56 recommendations. The Department organized implementation teams and by 2001 had reviewed and implemented all of the recommendations. One of the main recommendations for the Department was to focus on a more robust, proactive and independent Forest Protection Division (FPD).

In the aftermath of the 1998 fires that had damaged an extraordinary amount of immature forests, a major re-planting program was launched in 2000. The government provided some financial assistance to industry for replanting regenerated areas burned in 1998. Upwards of 24 million seedlings were designated for planting. Primary attention was focused on the Virginia Hills near Whitecourt, where more than 130,000 hectares had burned.

In the year 2000 the total firefighting expenditure, including presuppression and suppression costs, exceeded $116 million. Part of this money covered ongoing efforts to upgrade northern airport facilities for the new generation of airtankers and their birddog aircraft.

In 2001 FPD expanded cooperation among municipalities, regional fire chiefs and First Nation reserves for "mutual aid" responses to wildfires that threaten communities. Fire control agreements and annual fire control plans were developed for use in the forest communities. Fuel management studies to reduce forest fuels and the implementation of guidelines to reduce the effect of wildfire on communities were launched under the Wildland-Urban Interface program. FPD also increased its emphasis on this work as human settlement and energy-sector activity increased within the forest.

Department fire experts issued alerts of pending high fire hazard situations early in 2001. This assessment was based on analysis of conditions that resulted from the extremely low snowpack and unusually mild winter. Firefighting resources were put in place earlier than usual. By early May, extreme hazard ratings were common in the central and northern areas of the province. A major public awareness campaign, including television and radio advertisements on fire prevention, was launched throughout the province. Despite all the presuppression efforts and initial attack actions, the first of many fires soon spread to become major events. The prolonged dry weather resulted in extremely volatile fuels which, when ignited, were aided by relentless high winds to the frustration of firefighters and Forest Protection Officers. These fires moved very quickly and erratically.

Many of the fires were in very dry and flammable forest environments. Oil and gas infrastructure, towns and individual homes were widely scattered throughout these areas. Despite the severe conditions, no human lives were lost, even though private property was damaged and extensive losses occurred to the forest's fibre production, regeneration investments and logging equipment. An unprecedented fire event, known as the Chisholm fire near Slave Lake, resulted in residential losses and prompted the first-ever public fire review. The Department established an independent panel, led by Gerry DeSorcy, to analyze the fire's growth and suppression management, and to make recommendations. The panel provided five recommendations that have since been implemented and have resulted in enhancements to the Department's wildfire management, communications and education capability. The Chisholm Fire Review Committee made the following recommendations, that:

- SRD take the lead in ensuring communication is a top priority before, during and after fire events by developing and implementing a comprehensive communications plan;
- SRD implement means of improving command and resource coordination with municipal districts, the RCMP, local industries and property owners;
- SRD recognize the need for wildland-urban strategy and tactics separate from those of wildfire suppression;
- During existing and anticipated extreme fire behaviour conditions, SRD should use other strategies in addition to resource build-up to reduce the occurrence, or impact of large fires; and
- SRD place a high priority on implementing any outstanding recommendations of the KPMG report Alberta Fire Review, 1998 and review the success of the recommendations implemented before the Chisholm fire incident, in light of and in the context of the Chisholm fire.

Additional losses occurred in 2001 on productive forestlands that had already been harvested and successfully regenerated. Major adjustments in annual operating plans were

Wildfire at Chisholm and other communities brought problems of wildland-urban interface hazards into sharp focus, 2001. Sprinklers were set up to protect private property
Alberta Government, AFHPC

required to accommodate new priorities for harvest areas and to allow salvage of the fire-killed merchantable timber within the shortest time possible.

Within the Forest Protection Area a total of 10 homes and 49 outbuildings burned and 84 people were evacuated from their homes. Outside the Forest Protection Area, 12 homes and 58 outbuildings burned and 524 people were evacuated from their homes.

August and September, 2001, also proved to be a major challenge to the Department with an extreme fire hazard in the southern part of the province. For the first time in recent history the Department closed a large part of the Forest Protection Area to public access, for safety purposes. This included the area from the Trans-Canada Highway 1 south to the Canada/United States border. FPD implemented a number of initiatives developed from the Chisholm fire review, namely the use of Wildland-Urban Interface teams and Wildfire Prevention and Information teams.

At the end of September and for most of October the challenge continued with the Dog Rib fire west of Sundre. This fire burned a total of 9,898 hectares and challenged firefighters because of the remoteness, mountainous terrain and gusty winds, estimated at times to be 100 to 130 kilometres per hour.

For the year 2001 the cost of fire suppression by the provincial government was approximately $174 million. That paled in comparison to the following year, however, when the cost totaled $297 million.

The 2002 fire season was one of the worst seasons on record, with Alberta's second-largest fire in 40 years occurring at House River, north of Lac La Biche. This man-caused fire burned 238,867 hectares, almost half the size of Prince Edward Island. The House River fire was the second largest fire in Alberta since 1961, and had over 1,000 firefighters, support personnel,

Gas plant, Coleman, 1993 - energy facilities are common across the forest landscape
Bob Stevenson

hundreds of pieces of heavy equipment, and as many as 60 helicopters during the peak of firefighting.

Alberta had adopted the first generation real-time electronic lightning detection network display system in North America in 1986. On July 7, 2002, 82,000 lightning strikes in a 24-hour period were detected by the provincial lightning detection network.

The Department, as part of its forest protection strategic direction, entered into a Wildfire Reinsurance Program, the first of its kind in North America. This reinsurance program was designed to provide partial coverage for extraordinary fire costs. The program had four triggers or circumstances that had to be met in order to initiate a payout. The four triggers were: exceed 1,350 wildfires; exceed 150,000 hectares burned; Head Fire Intensity equal to or greater than Class Four more than 25 per cent of the time; and expenditures in excess of $175 million. Due to the extremely severe fire season, the Department met all the criteria and made a claim against this policy in the first year of the program.

The Department was again challenged in the summer of 2003 with the Lost Creek fire near Blairmore, which burned over 21,000 hectares. This man-caused fire started on July 23, 2003 and burned out of control for nearly a month, before being contained by hard work and determination. Very little rain fell during this fire. The Lost Creek fire pointed out the challenges of wildland-urban interface situations, with the communities of Coleman, Blairmore, Bellevue, Hillcrest, Burmis and Beaver Mines as well as acreages, ranches and the West Castle Ski Resort being threatened for most of the duration of the fire. Approximately 3,000 people were evacuated in 2003 as a result of fires, with just over 2,000 of those people evacuated during the Lost Creek fire.

On July 15, 2004, the Forest Protection Division was challenged when 120 fires started, the highest single day fire starts on record – most caused by lightning storms. Three of the fires were man-caused. More than 80 per cent of the fires were contained within the first burning period, with only four becoming 'E' class fires (fires larger than 200 hectares in size).

In the mid to late 1990s FPD staff shared their operational and fire science skills as participants in the International Crown Fire Modeling Experiment (ICFME). This was a bold series of tests initiated to calibrate a new model for predicting the spread rate and flame front intensity of crown fires. The study area, northeast of Fort Providence, Northwest Territories, also provided an opportunity to study firefighter safety and personal protective equipment, wildland-urban interface issues as well as environmental effects of wildfires. Further research on community protection continues at the Fort Providence site. FPD participates through its partnership with the Forest Engineering Research Institute of Canada (FERIC), Wildland Fire Operations Research Group, located at Hinton.

Partners in Protection, FireSmart and Wildland-Urban Interface

Residential, industrial, or agricultural developments located within or near wildland settings with natural vegetation, are at risk from wildfire. Forested and wildland areas are highly desirable places to live. What makes them so attractive, however, also makes them hazardous. While vegetation is an amenity for residents, it is a source of fuel for a fire. Prevention and control of interface fires present many unique challenges. These challenges demand that communities take collective responsibility for the problem, and that new attitudes are developed for fire. In the 1980s Kelly O'Shea, Forest Protection Officer, Bow-Crow Forest (stationed in Calgary) began working with Ken Saulit, Fire Chief Parkland County and Murray Heinrich, Fire Chief Yellowhead County to develop a program that would reduce the risk of interface fires. O'Shea saw a large urban population trying to find their piece of solitude, all wanting the convenience of city living but the ability to step back into 'nature.'

The result of years of work culminated in the formation of Partners in Protection, an Alberta-based non-profit organization formed in 1990 to address common issues in the wildland-urban interface. Partners in Protection developed the *FireSmart: Protecting Your Community from Wildfire* manual to give communities and individuals across Canada the information and tools they need to confront interface fire protection issues. The goal was that individuals, industry, communities and governments would work together to reduce the risk from interface fires, and to increase the effectiveness of fire response when it is required. A second updated version of the manual was produced in July 2003. Member agencies include the Canadian Institute of Planners (Alberta Association), Alberta Association of Municipal Districts and Counties, Alberta Fire Chiefs Association, Parks Canada, Canadian

FireSmart Home Owners manual
Alberta Government, AFHPC

Forest Service, British Columbia Ministry of Forests, Ontario Ministry of Natural Resources, Saskatchewan Environment and Resource Management, ATCO Electric, Alberta Forest Products Association and the Alberta Government through Sustainable Resource Development and Municipal Affairs.

As a follow-up to the 2001 Chisholm Fire Review, SRD created the *FireSmart Home Owners Manual* to provide individual home owners with prevention tips to reduce the risks associated with wildland-urban fires. Included with the manual is a Home and Site Hazard Assessment form that allows home owners to do a quick assessment to determine if their homes are at risk from wildfire. Home owners, working together as a community, can then use the broader *FireSmart: Protecting Your Community from Wildfire* to assist themselves in making their 'solitude' safer.

Following on the heels of a successful Partners in Protection program came the implementation of Wildland-Urban Interface (WUI) plans on communities at risk from wildfire. The initial program began in the late 1990s and was enhanced in 2002 with the identification of three WUI zones and the collaboration with communities and municipal governments. The three WUI zones are the Interface Zone, an area where various structures and other human developments meet or are intermingled with the forest and other vegetative fuel types; the FireSmart Community Protection Zone, a variable 10 kilometre radius around the community extending from the FireSmart Wildland-Urban Interface Zone; and the FireSmart Landscape Zone, the zone extending beyond the FireSmart Community Protection Zone overlapping multiple jurisdictions at a broad landscape level.[10] Work on the FireSmart Landscape Zone focuses on mitigating the likelihood of large, high intensity, high severity fires. The Department continues to work with local communities, industries and municipalities on community protection plans designed to reduce the threat of catastrophic wildfires from overtaking a community, and to design large fire breaks by manipulating cutting patterns and/or establishing control lines from which to backfire.

In 2004, 32 FireSmart plans were in various stages of implementation.

Forest Pests

Some forest insects became threats to forest health, two of them in particular. Spruce budworm and mountain pine beetle have both had a long history in Alberta and have been a perennial concern to foresters. The incidence of these insects can be greater in less-vigorous older stands where fires have been excluded.

Spruce budworm continued to be a serious pest concern in 1993, and the biological control agent Btk (*Bacillus thuringiensis var kurstaki*) was applied to more than 35,100 hectares in the Footner Lake, Athabasca and Lac La Biche Forests. In northern Alberta in 2000, more than 190,000 hectares of white spruce and balsam fir were defoliated by the spruce budworm. Aerial spraying applied Btk on 82,500 hectares to reduce budworm populations and lessen their impact in the future. In 2003 spruce budworm defoliation still remained a concern in the northern part of the province, around Fort McMurray, Wood Buffalo National Park and High Level.

Infestations of mountain pine beetle in old-growth lodgepole pine in Banff and Jasper National Parks were monitored to track any movement of these pests eastward into Alberta. This insect had been particularly voracious in

Spruce budworm
Alberta Government, AFHPC

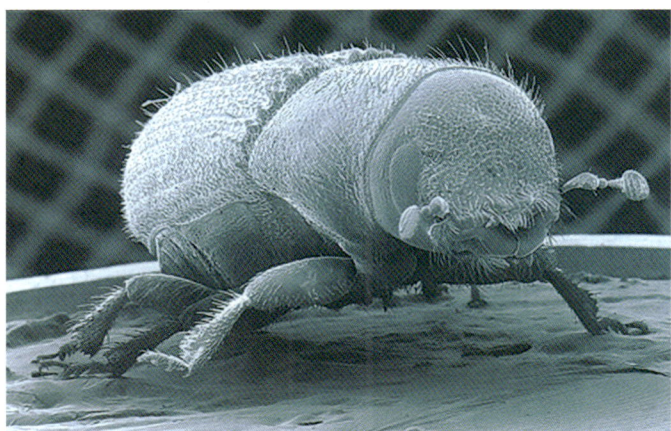

The mountain pine beetle had become a serious threat to Alberta forests in 2005. Aggressive action in Alberta and cooperative programs in British Columbia were designed to minimize the chance of a major infestation east of the Rockies. (Enlarged view)
Leslie Manning, Canadian Forest Service

British Columbia, recently spreading rapidly northward and impacting the forest industry. Colder winters in the early and mid-1990s in Alberta constrained populations. Active detection and control of outbreaks continued to be a requirement. In the early 2000s the Department continued to support the forest health strategy by successfully preventing the spread of a mountain pine beetle outbreak in the Bow Valley Corridor, where the risk of outbreak was high. In the summer of 2005, British Columbia had a major outbreak that could have long-term impacts on their forest industry. In 2005, Alberta and British Columbia with partnership from Weyerhaeuser Company and Canfor Corporation signed and invested in a cooperative agreement to fight the spread of mountain pine beetle in Alberta. During the summer of 2005, Alberta began tackling an outbreak of mountain pine beetle in the Willmore Wilderness near Grande Cache.

Industry Development

The Annual Report for 1991-92 stated that the increased pace of forestry development had led to an unprecedented level of timber management activity.[11] The new Forest Management Agreement for the Alberta-Pacific Forest Industries project had just been signed and negotiations were under way to draft or amend four other agreements. Following on the heels of this evidence of success in its efforts, the Forest Industry Development Division was moved to the Economic Development Department in 1993 during the major government reorganization. Executive Officer Al Brennan retired at that time. He died in 1994 after a battle with cancer.

The Northern Alberta Development Council released a report recommending expansion of a system of smaller or community timber permits divided amongst established forestry operators. Input was obtained from local advisory groups on how timber rights were to be awarded. Timber rights were given to registered sawmill and logging companies. Timber contractors could have timber rights if they had received a permit in the last five years.

A proposal by Ainsworth Alberta Inc. to construct an oriented strandboard (OSB) mill in Grande Prairie was accepted. The bid won out over two other applicants because proposed investment amounts were higher and the government felt Ainsworth's proposal represented the best utilization of the available wood.

The planned revision of the timber revenue system was finalized. The first step in reviewing provincial timber revenues had been completed by LFS late in 1993, and a new system to set the

Official signing of the Alberta Pacific Forest Management Agreement, August 30, 1991
Back Row (L to R): Kerry Day (Milner Fenerty), Ken Krohman (MacKenzie Fujisawa), Doug Sklar, Rick Keller, Bob Ruault (Alberta Pacific), Cliff Smith, Al Brennan, Ken Higginbotham, Vonn Bricker, Kirk Andries, Denny Thomas (Milner Fenerty), Tim Freedman. Front Row: Mack Kubo (Kanzaki Paper), Keith Fujieta (Mitsubichi), Mike Cardinal (MLA), Leroy Fjordbotten (Minister), Stuart Laing, Karl Gustafson (Crestbrook), George Jackson
Rick Keller

Typical Whaleback landscape, west side of Porcupine Hills
Bob Stevenson

timber dues for coniferous timber used in lumber production came into effect January 13, 1994. The result was a quadrupling of stumpage dues. These more market-sensitive stumpage fees doubled the base rate charged for the cutting of timber. Chris Andersen, president of the Alberta Forest Products Association (AFPA) said his members could live with this new formula "because this system fairly addresses the cyclical nature of our industry with its substantive price shifts."

The second step in the review of the timber revenue system was the introduction of a new dues system for timber used in the manufacture of OSB and pulp, the full effect of which would be seen by early 1996. The new systems for evaluating timber dues were designed to be more responsive to changes in the marketplace.

The Timber Production and Revenue System (TPRS) was implemented in May, 1997. This software program was developed to store information on timber-related dispositions, and provided the means by which the Crown could track timber production and assess fees and royalty charges for primary wood products.

By 1998 the timber harvest was 16.9 million cubic metres out of an allowable cut of 22.1 million. Timber harvesting generated $103.2 million from royalties and dues, and another $41 million was generated by land, grazing and other fees.

In 1995 there continued to be a steady barrage of questions from critics and groups about allocation of the province's timber resources. Concerns were that Alberta was running out of accessible merchantable stands, and that more Crown land should be diverted from industrial use for other purposes.

Minister Ty Lund stated in 1995: "We believe that successful environmental management must be based on sound science and sound economic practices. As a Department, we work to create a balance between development and preservation through approaches based on sustainable development."[12]

While still settling into his new office, Minister Ty Lund was receiving requests to allow logging in the Whaleback area of southwestern Alberta. The Premier was on record as stating "there will not be any logging" in this protected area, but considerable criticism was directed at the Minister even though no industrial intrusions had been permitted. The Whaleback area was later placed under Special Places protection, covered by the Black Creek Heritage Rangeland and the Bob's Creek Wildland.

In 1996 the *Status of Alberta's Timber Supply* report provided information on the forest management system to clarify how allowable annual cuts were derived on public land and show that a substantial proportion of forest lands remained unavailable for forest harvesting.

An ongoing debate continued with the proposal by Grande Alberta Paper Ltd. to establish a pulp mill near Grande Prairie. A major concern expressed was availability of adequate timber resources to sustain the operation. The proposal did not proceed.

Three proposals to harvest and process timber in the Footner Timber Development Area were received by the LFS in 1996-97. This was the last area of uncommitted timber resources in Alberta. The winning proposal was submitted by Footner Forest Products, a joint venture between Ainsworth Lumber Company Ltd. and Grant Forest Products Corp. The agreement called for Footner Forest Products to construct a $110-million oriented strandboard plant near High Level. Connected with this announcement was a timber supply agreement to permit the upgrading of LaCrete Sawmill's operations. The Forest Industry Development portfolio was moved from Economic Development to Resource Development, under the Department of Energy, in 1998. Northwest Boreal Regional Director Howard Gray was appointed to lead the new structure.

ForestCare's certified auditors check all aspects of operations
Alberta Forest Products Association

Non-forestry activities continued to have a great influence on forested lands during the period. By April 1, 1994, there were nearly 100,000 dispositions (e.g. wellsites, roadways, pipelines, grazing leases, land sales) in effect on Alberta's public lands, covering 2.8 million hectares in both the Green and White Areas. Applications for use of public land increased by 50 per cent to 8,300 that year, mostly due to greater activity in the oil and natural gas sector. For the year ending March 31, 2004 there were over 187,000 dispositions in effect on Alberta's public lands covering over three million hectares in both the Green and White Areas. There were over 18,200 new and renewed dispositions during 2003-04.

Dave Bartesko, Access Program Manager with Public Lands and Forests Division, SRD, outlined in an interview in the *Western Woodlot Conservationist* (Winter, 2004) that "improvement has been seen in the scale of disturbance by non-forestry uses on public land." The average size of dispositions approved under the *Public Lands Act* for access roads and pipelines has been reduced by 50 per cent over the past seven years, Bartesko says. Land clearing for individual seismic lines is down by approximately 65 per cent, and more than 70 per cent of the 75,000 kilometres of seismic line approved on public land each year are now low-impact. 'Seismic lines are three metres wide on average, a significant reduction from the standard eight-metre swath of 10 years ago,' Bartesko said. The geophysical sector has been innovative and has set standards that many other jurisdictions have implemented or are considering to implement.

Area Operating Agreements (AOA) continue to be a key Department tool to ensure integrated development on public land and to streamline the disposition process. An AOA is an agreement between an oil/gas company and the Department

Forest Officer David Moseley checking utilization in a cut-to-length timber harvesting operation near Round Hill Tower, Lac La Biche Area, August, 2005
David Moseley

First graduating class of the Alberta Advanced Forestry Management Institute (AAFMI) program, Forest Technology School, September, 1997
Back Row (L to R): Jim Friesen, Doug Ellison, Andre Savaria, Dr. John Lehmkuhl, Dr. Rich Everett, Dr. Mark Jensen. Middle Row: Gail Tucker, Daryl Gilday, Tim Vinge, Gordon Brown, Daryl D'Amico, Dave Patterson, Bill Gladstone. Front Row: Don Gelinas, Brian Carnell, Clark Shipka, Lindsay Kerkhoff, Diane Renaud
Alberta Government, AFHPC

outlining standards and conditions a company must follow while conducting operations within a certain area.

The Department continues to explore ways to better manage the volume of industrial dispositions, both in the approval process and in the reviewing and storing of electronic applications and land surveys. Two software programs, Application Disposition Process and Tracking (ADEPT) system and Geographic Land Information Management and Planning System (GLIMPS) continue to be modified and implemented.

Classroom display session during an Alberta Advanced Forestry Management Institute (AAFMI) module, Forest Technology School, Hinton, February, 1999
Alberta Government, AFHPC

Forest Management and Standards

The *Interim Forest Management Planning Manual - Guidelines to Plan Development* was published in April 1998 by the LFS. This was a manual designed to guide Forest Management Agreement (FMA) holders in preparing Detailed Forest Management Plans.

Throughout 1999, forest companies continued their drive to achieve forest certification for their operations. Many found their progress in this endeavor was eased by their earlier adoption of standards and codes created by the Alberta Forest Products Association's FOREST*CARE* program. The move to national and international certification was seen as another step in assuring worldwide public support for and understanding of how Alberta's forest companies manage the resource for sustainability. Certification under programs such as Canadian Standards Association, Forest Stewardship Council or the Forest Stewardship Initiative gave consumers more knowledge about sustainable forest management and confidence in how their purchases were produced.

In 2000 LFS initiated a strategy for the development of ground rules throughout the province that would, as explained on the SRD website home page, "create innovative, efficient, adaptable and pragmatic ground rules throughout the province by May 1, 2006."[13] Ground rules are the practices used in planning and conducting timber harvesting operations on FMA areas or zones agreed to by different companies. Ground rules generally include extensive standards on stream crossing, road construction, maintenance, reclamation and erosion control.

In 2003, SRD began revising the *Interim Forest Management Planning Manual* to respond to increasing complexities in forest management, and to reflect provincial stewardship expectations and illustrate best practices for forest management.

The revisions were a 'made in Alberta' approach that incorporated the Canadian Council of Forest Ministers criteria for Sustainable Forest Management performance and indicators from the Canadian Standards Association certification program. Alberta's *Forest Management Planning Standard* (2005) identifies the standards to be achieved in forest management plans.

Forest Renewal

The expansion at Pine Ridge Forest Nursery was completed in 1993. The two-year $8.6 million project, with its new 6,000 square metre greenhouse, raised the capacity of the facility to 31 million seedlings per year.

Revised Free to Grow (FTG) standards were introduced in 1994, setting additional expectations for the performance of regenerated forest stands. Further refinements better reflect differences among forest ecoregions and incorporate a better understanding of natural processes of boreal mixedwood forest regeneration as a result of ongoing forest research. Jim Kitz, who passed away suddenly in March 1994, was a 1981 University of Alberta graduate well respected for his dedication to silviculture. In addition to his work on the FTG standards, he worked on the national laser disc training package for stand tending in Canada, developed a certification course for Alberta and worked to initiate the Alberta Advanced Forest Management Institute module courses. The Canadian Institute of Forestry established the James M. Kitz award in his memory; the award recognizes young foresters who make significant, unique, and outstanding contributions in the field of forestry early in their career.

Reforestation on public land continued apace, with LFS and industry planting a combined 75 million seedlings in 1996. Establishment surveys indicated that harvested areas were 96 per cent satisfactorily restocked in conifer, and 93 per cent in hardwood.

Two concurrent events contributed to change the status of the Pine Ridge Forest Nursery. First was the forest industry's acceptance of responsibility to effectively reforest its own areas after logging, and Pine Ridge could not meet the increased demand for seedlings. Secondly, in keeping with current government philosophy, the Pine Ridge Forest Nursery was recommended for privatization. In 1998 it was leased to a consortium of forest tree nursery operators to form part of a network of province-wide private forest nurseries. The LFS continued to operate the forest genetics and tree improvement operations on the Pine Ridge site.

In the spring of 1999 a community reforestation effort by 285 volunteers occurred on the Cooking Lake-Blackfoot Recreation Area, where 34,283 white spruce seedlings were planted in an eight-hour period. This tree-planting effort is believed to be a record for Alberta. Local Junior Forest Wardens, Scouts and youth volunteers provided planting assistance with Tree Canada Foundation supplying the seedlings. Extensive white spruce forests originally dominated this area at the turn of the 20th century. In 1899, this area was the Cooking Lake Forest Reserve, managed by the Dominion Forest Service.

Policies and Strategies: Alberta Forest Legacy

In addressing improvements to resource management in 1999, Cliff Smith said: "The whole approach to resource management was evolving to reflect a broader ecosystem perspective, a move that was just beginning at that time. The concept of ecosystem management was something that most of us ascribed to but didn't fully understand in those days – a lot of the buzzwords that we use today probably weren't even in place at that time. Nevertheless, it was recognized that there was a very strong need for some different approaches." [14] In terms of harvest planning, Smith said: "We'd completely gone away from this idea of blocks and rectangles and squares which at an earlier time were thought to be good … but certainly we didn't address a lot of the other needs of the landscape. We were just moving into the early stages of ecosystem-based management in the 1990s."

Alberta launched the Alberta Forest Conservation Strategy (AFCS) consultations in 1994. As Cliff Henderson explained, "Through ministerial leadership federal-provincial accords were established emphasizing sustainable development and bio-diversity in current and upcoming forest management plans. From this initiative came the *Alberta Forest Legacy* policy and appointment of the Forest Management Science Council. Essentially, the three key policy directions comprised economic opportunities, today and tomorrow; ecological integrity of the forest; and incorporating society's options." [15]

The AFCS consultation was unique at the time. It was not a government-led process but one co-chaired by an environmental non-government organization (ENGO) representative and government member. The strategy was developed through discussions intended to lead to consensus. It published the *Alberta Forest Conservation Strategy* in 1997. In response, that same year, the government released *The Alberta Forest Legacy: Implementation Framework for Sustainable Forest Management.*

The AFCS was initiated as one of Alberta's commitments to the *National Forest Strategy* and as an element of the *Creating Our Future* initiative. The strategy was to "develop policy for the sustainable use of the province's forests and address the long-term social, economic and environmental values of the forest resources." It was developed by a multi-stakeholder working committee and involved extensive public consultation. The introduction to the *Alberta Forest Legacy* said the document reflected Albertans' desire to maintain their access to the diverse economic, cultural and recreational benefits provided by, and dependent on, sustainable forest ecosystems on provincially-owned land. The document supported the previous commitment by government and industry to achieve sustainable forest management, which represented an ecological approach based on sustaining forest ecosystems as well as economic and social values. Department managers Keith Branter and Howard Gray later visited research and management organizations in the United States to broaden their understanding of new approaches to planning on an ecological landscape basis.

Decision Making and Public Participation

By the mid-1990s the concept of multi-stakeholder and interdisciplinary committees had become well established. Designed to encourage and support public participation, the committees became a requirement for Forest Management Agreement (FMA) holders with respect to their operations. They were also established to address specific issues or proposals. The government approved a document titled *Alberta's Commitment to Sustainable Resource and Environmental Management* in March 1999. This document was developed to redefine the framework for decisions and policy-making in sustainable resource and environmental management.

In 1996 the Alberta government's *Jacques*

Backcountry Guardian Dave Ferster talking to horse riders on the trail to Lost Guide Lake, Bighorn Backcountry, 1997
(L to R): LeAnne Quintilio, Dennis Quintilio, Cliff Henderson and Dave Ferster
Bob Stevenson

Report also emphasized the importance of sustainability in the forestry sector as well as contributions to the public interest. The report was prepared by a sub-committee of elected representatives of the Legislature, chaired by MLA Wayne Jacques, and was developed after consultations with FMA holders. In concept the report recommended that as a principle in negotiating renewal of an FMA, the company should propose further investment in manufacturing capacity for value-added production and greater economic activity, enhanced forest management to ensure sustainability and increased social benefits, or some combination of these.

Field tours and demonstrations were organized for woodlot operators
Woodlot Association of Alberta

The Department responded to the government's *Aboriginal Policy Framework - Strengthening Relationships* document by partnering with other ministries from 2003 to 2005 to develop a *First Nations Consultation Policy on Land Management and Resource Development*.

Monitoring the Forests

The forest industry continued to take on more of the responsibility for forest management and SFM through their FMAs. In 2002, the government also restructured its professional legislation by establishing "Colleges" for various disciplines. Registered Professional Foresters (RPFs), previously registered under their 1985 *Forestry Profession Act* were encouraged to form a new College of Alberta Professional Foresters (CAPF) and to share the practice with members of the College of Alberta Professional Forest Technologists (CAPFT). This was a visionary approach that was then unique to Alberta but has since been emulated. Alberta government forestry staff became members of their respective Colleges along with their forestry sector colleagues, thus adding the strength of professional codes to achievement of SFM. Mindful of its stewardship responsibilities, SRD began refining its field inspections in favour of stewardship audits. Audits are backed by more clearly defined criteria for enforcement to ensure quality assurance and control.

As ADM Cliff Henderson reflected: "Although there have been significant administrative and organizational changes over the years, the traditional Alberta Forest Service spirit continues. Alberta Sustainable Resource Development has maintained the role of stewardship and is responsible for forest and public land management, fish and wildlife and forest protection. At the same time, however, the *Forestry Profession Act* of 1998 means the practitioner or industrial forester/forest technologist has been made more accountable to his/her peers and the public. In the spring of 1999 the government created the *Regulated Forestry Profession Act*. As Alberta's Registered Professional Foresters (RPFs) became more accountable for their actions the role of government changed and greater accountability from practitioners was expected."[16]

Softwood Lumber and Countervail

From 1993 to 2005, the softwood lumber countervail actions continued with no long-term resolution evident. In 1991, at the end of the initial five-year pact, the United States government imposed a 15 per cent duty on lumber from Alberta and most other provinces.

Cliff Henderson noted "in 1994 the Alberta timber dues system was changed from a flat-rate charge of about 70 cents a cubic metre to a per centage of the selling price of lumber. The change increased Alberta's timber revenues from approximately $20 million to over $100 million per year. The government created the Environmental Enhancement Fund to administer the increased revenues. The fund was to ensure sufficient resources for forest protection and reclamation and is still in existence today."

A new five year countervail duty agreement was negotiated with the United States for the period 1996 to 2001. Henderson says: "Helmut Mach of Alberta International Trade and I were the principal negotiators for Alberta, along with Garry Leithead, Executive Director of the Alberta Forest Products Association. Alberta accepted a volume agreement with the United States, and this allowed us to maintain industry operations."[17]

Unfortunately, at the end of the volume agreement in 2001, the export tax was re-imposed and, in addition, the United States imposed anti-dumping levies, both of which remain in place as of 2005.

Private Woodlots

High timber demand in British Columbia in the early 1990s led to several years of heavy logging on private and First Nation reserve lands in Alberta. The logs were exported to British Columbia to be used in their mills.

As the visual impact of this exploitation progressed across the province there were a number of reactions. Alberta forest industry sent complaints to the government and called for prohibitions on timber exports because of the loss of jobs in this province. Environmentalists lobbied for an outright ban on logging on private land or at least regulatory restrictions limiting the rights of individual timber owners.

In 1995, a variety of information services was provided in cooperation with Alberta Agriculture and Rural Development and the Canadian Forest Service (Canada-Alberta Forest Resource Development Agreement) to assist private owners to manage forest resources on their land. The next year a new system of haul permits was established under Alberta's private woodlot program to regulate transportation of coniferous logs harvested on private land. The permit system was developed to mitigate the risk of illegal harvest and export of Crown-owned timber. During 1996, 14,500 private land timber haul permits were issued and more than 1,200 logging trucks inspected.

Concurrently a group of landowners formed the Woodlot Association of Alberta on April 24, 1995. The first general meeting held in November of the same year drew 76 participants. In 2005, the membership exceeded 400. The mission of the association is to promote education and encourage sustainable forest management on private land for all values – economic, ecological and social. Many of the privately-owned woodlots are located on former Yellow Area lands that were subsequently made available for agriculture and are now growing forests again.

Special Places

The Special Places initiative originated from the World Wildlife Fund (Canada) Endangered Spaces program, which was endorsed by the federal and provincial governments. In addition to preserving the environmental diversity of Alberta's six natural regions (20 sub-regions) the Alberta program had three other goals namely, heritage appreciation, outdoor recreation and

David Coutts, MLA Livingstone-Macleod and Chair of the Coordinating Committee on Special Places 2000 talking with rancher Art Davidson during a Special Places Public Advisory Committee tour, Milk River, summer, 1997
Bob Stevenson

tourism/economic development.

At the provincial level, the Special Places program was guided by a Provincial Coordinating Committee made up entirely of representatives of widely-based environmental non-governmental organizations (ENGOs), other non-government organizations (NGOs) and industry. The Department provided logistical and technical support and resources for the committee's work. Any resident of Alberta could nominate public lands for protective designation. The Provincial Coordinating Committee evaluated all nominations and forwarded those with merit to a local committee for a further, more detailed evaluation. Again, the local committees consisted of representatives of ENGO and NGO groups as well as commercial and recreational interests. They were usually chaired by the appropriate municipal government.

The program was launched by the Department in 1994 to complete a network of areas representing the environmental diversity of the province's six natural regions. It was designed to be part of Alberta's contribution to the Endangered Spaces Program, a national and international strategy to conserve natural heritage. In 1997 the Special Places program continued to arouse concern among critics who pressed for more and bigger areas to be designated for protection. Nonetheless, substantial areas of unique landscapes meeting the goals and objectives of the program continued to accumulate. Input from local committees aided the process.

By 1998, Special Places totaling 4,900 square kilometres had been designated. Concurrent with this was a program to designate parks and nature reserves, the area of which totaled 67,855 square kilometres by 1998.

By 1999 the initiative was gathering support from many in the public and generating ongoing discussions between government and industrial interests. While the initiative was led by personnel of the Parks and Protected Areas Division, there was very active and influential

Caribou recovery and conservation, an important initiative for the 21st century
Alberta Government, AFHPC

involvement by the LFS and the forest industry. Numerous public stakeholders offered valued input from a variety of viewpoints. Substantial areas were proposed to include distinctive ecological habitats in a special protected land status. One example was in the Chinchaga River boreal foothills region of northwestern Alberta, where some members of the public opposed government proposals to allocate additional forestlands for industrial purposes. The Alberta Wilderness Association (AWA) and the Canadian Parks and Wilderness Association (CPAWS) deemed portions of these lands to be critical for protected status. Considerable debate led to designation of 80,270 hectares of the Chinchaga watershed for inclusion in the Special Places initiative. This area, known as Chinchaga Wildland Provincial Park was augmented in November, 2003, when SRD ruled out permanent timber allocation within the P8 forest management unit, just north of the park. The industrial footprint within this additional 350,000 hectare area will be greatly reduced, with a focus on watershed management, caribou recovery and grizzly bear management.[18]

Environmental special interest groups continued lobbying for more land designations for special protected areas within the managed forests. Negotiations continued with forest companies, many of whom had

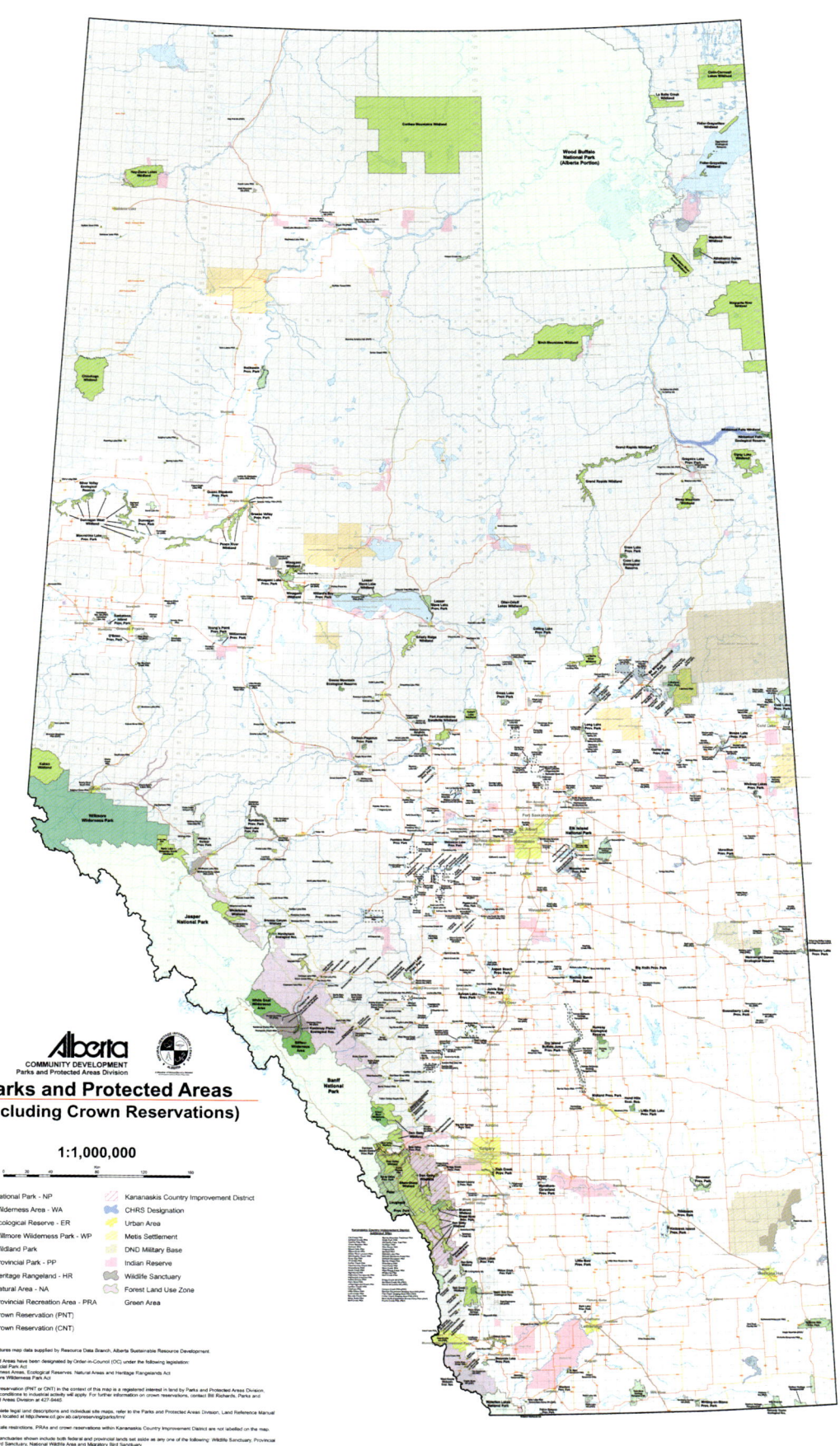

already recognized and established Special Places pursuant to the provincial guidelines and program. Some members of the public felt that the land areas should be larger than those suggested in the companies' initiatives. Nonetheless, many companies, through their FMAs, offer protection to known unique vegetation and ecological habitats. Similarly, these FMAs provide numerous opportunities for year-round outdoor recreation. Of special note is the attention given to recreation by Weldwood, Hinton Division (now West Fraser), with its user-friendly program complete with a detailed map and guide for the public.

By 1999, 50 sites within the Special Places program encompassed a total of 1,400,000 hectares of previously "unprotected" land. The Special Places program concluded in July, 2001, with a total of 81 new and 13 expanded sites adding two million hectares to Alberta's protected land base.

Woodland Caribou Conservation

The Woodland Caribou Conservation Strategy, developed by a committee of representatives from Natural Resources Service, LFS, industry, aboriginal groups, conservation groups, academics, foresters and biologists, was launched in 1995. To improve technical support for these stakeholders, landscape level

Area Manager Al Hovan (L) and Forest Officer Mike Gabourie conducting regeneration survey on Alberta Plywood Ltd. cutover, Lesser Slave Forest Area, spring, 2004
Al Hovan

planning systems were developed along with computer-based inventory data for planning and a geographic information system that enabled data to be displayed on maps. The committee identified and assessed the various factors (biological, social and economic) that might affect the vitality of caribou populations in Alberta, developed solutions to deal with those factors, and recommended specific actions to make the strategy effective.

The *Caribou Conservation Strategy* was provincial in scope. From it came three regional caribou management committees established to develop management plans for caribou in the northeastern, northwestern, and west-central regions of Alberta. They were set up to assess the effectiveness of the guidelines and habitat supply analysis, determine what further research or inventory is required, and develop a cost-sharing agreement for the management of caribou and their habitat. They continue to develop guidelines for how industrial activity will be conducted on caribou range and how adequate amounts of caribou habitat will be maintained in the short and long term.

The Boreal Caribou Committee was formed in the late 1990s, combining the northeastern and northwestern committees, to provide for a more provincially consistent organization.

Cutblocks in the Ya Ha Tinda area designed with irregular or 'feathered' edges for elk habitat, early 2000
Bob Stevenson

The main objective is to ensure caribou and their habitats are conserved while responsible resource development opportunities are maintained. This committee has membership from forest industry, oil and gas industry, government and research associations.

FRIAA

The fiscal year 1997-1998 saw establishment of a number of Delegated Administrative Organizations, with the intent of moving the administration of some activities mandated by legislation and regulation to third-party organizations. These organizations are accountable to various Ministers. They are managed by boards representing government, industry and the public. They give a major voice to those community and industry interests that have the most at stake in how funds and issues are handled.

One of these new organizations was the Forest Resource Improvement Association of Alberta (FRIAA). Its main purpose at inception was to administer the Forest Resource Improvement Program (FRIP), which was funded by a portion of the dues and levies paid by forest companies on the wood products they produced. The goal of FRIP was to enhance both the management activities and the level of understanding of all forest resources. Although activities focused on public lands, projects related to timber or non-timber forest resources were eligible if they improved the quality of the forest resource and were not considered to be the legislated responsibility of industry. Specific programs that qualified were enhanced silviculture, inventory, research, genetics and infrastructure. Training directly related to these activities was also eligible for FRIP funding. In 2002 the FRIAA mandate was extended to 2007.

Science and Research

The Forest Development Research Trust Fund provided $164,000 to support research initiatives in 1993-94. Studies addressed the forest tent caterpillar, forest growth projection methods, gall rust, aspen regeneration, and drought resistance of white spruce. In 2000, the Trust Fund was incorporated under the *Alberta Science and Research Authority Act*.

The Forest Technology School in Hinton continued to function as the location of the second-year NAIT forest technology program until 2002. The Hinton facility – renamed successively the Environmental Training Centre and Hinton Training Centre – remained as a partner in the Foothills Model Forest, and as a centre for training in wildfire management and other programs. More than 8,000 trainees attended the school and 1,000 toured the on-site museum, interpretive trail and academic building in 2002.

The Alberta Environmental Centre (later managed by the Alberta Research Council) at Vegreville became part of the new Alberta Environmental Protection Department in 1993, adding a research capability related to studies of biodiversity and other environmental and economic sustainability issues.

In an innovative move, Manning Diversified Forest Products (MDFP) established an integrated forest management research trust fund in 1993 through an agreement with the Department of Environmental Protection.

40th Anniversary celebration of the training centre in Hinton, fall, 2000. This facility was first opened in 1960 as the Forestry Training School, later becoming the Forest Technology School, the Environmental Training Centre and now the Hinton Training Centre
Directors of the facility (L to R): Peter Murphy, Bernie Simpson, Dennis Quintilio, Ross Risvold and Don Podlubny
Bob Stevenson

Alberta Forestry Youth (Junior Forest Rangers, Aboriginal Junior Forest Rangers and Junior Forest Wardens) attend World Forestry Congress, Quebec City, Quebec, October, 2003
(L to R): Bob Fessenden, Deputy Minister Sustainable Resource Development; Kathy Hendren, JFW Regional Coordinator; Gail Greenwood, JFR Coordinator; Nancy Erb, JFW; Larissa Duma, JFW; Alex Wu, JFR; Matt Engleman, JFR; Howard Gray, Assistant Deputy Minister Strategic Forestry Initiatives; Andrea Van Egmond, JFR; Chantel Quintal, Aboriginal JFR; Nolan Nicholls, Aboriginal JFR; James Atkinson, Aboriginal JFR Coordinator
Alberta Government, AFHPC

The company contributed to the fund with the understanding that it could participate with the Department in setting objectives for the research. During the period between June 1993 and April 2001, for example, $3.1 million was spent on 74 different research projects located across Alberta. The company presently administers the fund with input from local community representatives, the Municipal District of Northern Lights and the Departments of Sustainable Resource Development and Innovation and Science. MDFP also became one of the three founding partners of the Boreal Forest Research Centre in Fairview (later moved to Peace River), supporting a mandate to identify and advocate regional research, value-added products and educational priorities of importance to the northwest boreal forest region.

In 1995, Alberta Environmental Protection led the country by becoming the first provincial government Department to contribute to the Network of Centres of Excellence (NCE) in Sustainable Forest Management. The network links university, government and industry partners across the country in a coordinated program of studies into sustainable forest management issues applicable to Canada's boreal forest. The network is dedicated to providing integrated, multi-disciplinary research to ensure the sustainability of Canada's boreal forests. It was initiated by Drs. Ellie Prepas and Daryl Hebert through the University of Alberta, with support from 22 universities and research institutions across Canada. The network has supported a broad range of research projects.

Another important recognition of the role of science was the creation of the Alberta Forest Management Science Council (FMSC) in 1996. The FMSC was initiated by Dennis Quintilio shortly after his appointment as Director of Forest Management and managed by staff member Cam McGregor. It brought together eight prominent western Canadian scientists to advise the Department on forest management issues. It was chaired by Dr. Bob James, former Vice-President (Research) at the University of Alberta. The Council evaluated policies and practices, including the Canadian Standards Association SFM standard, the *Alberta Forest Conservation Strategy*, and policies in several other provinces and among forest industries. Its report in early 1999 described a set of timber supply protocols to help

Alberta Junior Forest Wardens, with Mount Fuji in the background, during an exchange program to Japan in 2002. Group included adult leaders, chaperones and wardens:
Brenda Stelmack, Monique Gregoire, Erin Ward, Naomi Martini, Bob Young (back row right), Alden Dixon, Jonathan Eeuwes, Matthew Kristoff, Phillip Poscente, Amber Garrett, Terry Garrett, Brianne Fahlman, Nicole McFarlane, Alysha Wetter, Nic Martini, Lucas Dixon, Randy Dube, Terry Cooper-Smith, Bradley Jeffery, Shelleen Lakusta (right side), Cheryl McFarlane, Kelsie Burns-Bain, Kaia Meyers-Stewart, Lida Hellqvist, Cheryl Dash, Ernst Klaszus (center back), Ryan Stelmack, Matthew Posey, Micah Laboucane, Lucas McMahon, Rob Ellen, Willy Fahlman
Alberta Government, AFHPC

1993 - 2005

achieve sustainable forest management through an applied science approach incorporating public participation and adaptive management.

In 1996-97 the LFS worked with timber companies to design new harvesting programs that did not involve traditional square-shaped clearcuts. For example, harvesting could feature "feathered cutting" on it's edges, and would leave the forest and its clearings in a quasi natural state. This technique could be used in areas where wind damage is not likely to be a problem to emulate in some respects the effects of natural disturbance. Residual patches would also be left within the block.

In the spring of 2000, LFS staff completed a computer model titled *A Desired Future Forest For Alberta* in cooperation with the FMSC. In addition, the LFS engaged in development of protocols for a comprehensive, rigorous and efficient biodiversity monitoring program for the province's forests. To make this initiative more effective a partnership was created with the University of Alberta, the Foothills Model Forest, the federal government and the Alberta forest industry.

Another priority involved an LFS concern with landscape management and the appearance of past and present logging areas. Public pressure and the LFS's desire to work with the forest industry resulted in the development of a landscape classification framework to meet effective ecological landscape and environmental requirements. This program addressed the concerns of the public about cutblocks and their impact and appearance on the landscape, while providing for the prescribed removal of wood fibre specified within Annual Operating Plans.

Educational Programs

The Junior Forest Ranger (JFR) program operated eight camps in 1993-94. The JFR program had been considered for cancellation as part of reorganization efforts, however, the Department reaffirmed its commitment to the program. With the financial assistance and support of forest industries, the program continues to offer summer employment and tours and presentations to teach about the forest and the complex challenges around responsible use of the forest. The year 2005 marked the 40th anniversary of the JFR program that has introduced thousands of students to the forest. An average of eight 12-person JFR crews have been established each year over the last two decades. Working with aboriginal communities, SRD successfully developed an Aboriginal JFR program, which in 2004 included three crews.

Community Junior Forest Warden (JFW) clubs are run by volunteers using program materials that cover the four foundations: forestry, ecology, outdoor skills and leadership. A JFW camp was established in 1981 at Narrow (Long) Lake southwest of the Town of Athabasca to provide summer programs and training workshops throughout the fall and winter. In 2004 there were 52 clubs with total membership of about 1,000

Recreational access highly valued by many
Alberta Government, AFHPC

'Respect the Land' garbage bag handed out to backcountry users in the Bighorn Backcountry, early 2000
Bob Stevenson

JFWs and over 700 dedicated adult volunteers.

In 2002, 33 JFW participants enjoyed a one-week cultural exchange to Tokyo, Japan and the Mount Fuji area. The participants studied Japanese forestry practices and utilization and were introduced to Japanese culture. In October, 2003, seven youth were selected to attend the World Forestry Congress in Quebec City. They gained valuable knowledge about forestry and the environment, and were able to see the sights of Quebec.

The Alberta Forestry Association (AFA) supported the educational curriculum and the release of the book *Focus on Forests.* This provided valuable and important insights for students and teachers. The comprehensive, well-illustrated text aids the understanding of the practice of forestry. Public education and awareness has since been provided by the Inside Education Society of Alberta, previously called FEESA, an Environmental Education Society. The society, supported by public and private contributions, offers a variety of programs in forestry and other natural resources issues such as water, energy, climate, waste and pollution. Up to 7,500 students each year visit one of three historic forest areas through cooperative educational programs run by Inside Education; Jumpingpound Demonstration Forest west of Bragg Creek, Des Crossley Educational Forest and former research block near Strachan, south of Rocky Mountain House and the Cooking Lake/Blackfoot Forest (former Forest Reserve) east of Sherwood Park.

Public Access

The use and impact of recreational all-terrain vehicles (ATVs) continued to increase throughout the province. Of particular concern to the Department in 2001 was the opposing views among various users in and around the Bighorn Wildland Recreation Area west of Rocky Mountain House. Despite usage restrictions outlined in the *Eastern Slopes Policy* of 1977 and 1984, and the subsequent *Integrated Resource Plan* of 1986, some unsanctioned ATV activity occurred in certain areas within the Zone 1 (Prime Protection) and Zone 2 (Critical Wildlife) areas. From minor intrusions in the mid-1980s evolved a major ATV usage pattern and subsequent conflict with equestrians and hikers. In addition, significant environmental degradation was occurring on some sensitive sites. The problem escalated to the point at which the Minister directed a stakeholder committee to make recommendations respecting the access issues in this popular area.

The committee conducted public meetings in the Rocky Mountain House and Sundre areas and also received numerous written submissions from recreationists and the public in the Calgary and Edmonton metro centres and from all over Alberta.

The consultation process aided development of guidelines that will help recreationists' use of backcountry areas while protecting the land from the most serious forms of degradation. A program called *Shifting Gears – Give Nature A Break* brought increased enforcement and educational efforts, as well as helpful guidelines, for those interested in off-highway vehicle (OHV) use and random camping in Alberta's Rocky Mountains and foothills. The program's website explains that Alberta's spectacular eastern slopes area is a 90,000-square-kilometre region used by outdoor enthusiasts for hiking, horseback riding, snowmobiling, mountain biking, fishing, hunting, skiing, golfing and a range of other activities.

On the enforcement side, the Department works closely with regional RCMP, especially during times of peak use such as summer long weekends. On the May long weekend, forest officers, conservation officers and RCMP conduct vehicle checks (for licence registration, open liquor) along the eastern slopes and hand out educational material.

As part of *Shifting Gears*, an educational campaign designed to help random campers reduce their impact on the landscape was launched. Campers were urged to limit themselves to temporary and portable camps, to leave no litter, and to respect the land on which they camped.

New access management plans were developed for the popular, but environmentally sensitive Bighorn Backcountry, Castle Special Management Area and Ghost-Waiparous areas of the eastern slopes. These plans included strategies for education and enforcement of legislation and land management principles, new regulations (including seasonal or permanent restrictions on trails and random camping areas) and designated use of trails, strategies for trail and infrastructure maintenance, and guidelines for environmental monitoring. [19]

The Respect the Land educational program was launched in the spring of 2003. This program was designed to deliver stewardship messages increasing the awareness of the impacts of outdoor recreation, random camping and off-highway vehicle use on public land. The program uses the motto *"Be Responsible. Do Your Part. Do Your Best. Encourage others to do the same. Respect the Land"*.

Forest Officer Ken Cox and Forest Guardian Tammy Proulx talking with horse riders while on patrol along the Onion Lake Trail, Bighorn Backcountry, September 5, 2004
Andrew Bibo

Aerial view of road being constructed north of Lac La Biche as a single access corridor for Alberta-Pacific Forest Industries and Nexen Canada Ltd., 2003. Alberta-Pacific will use the road for harvesting and log hauling access, while Nexen will use the access for development of oil and gas resources in the area. A reduced industrial footprint is the objective and result that can occur with joint planning and operational coordination between various industrial users – forestry and oil and gas being the main two
Don Pope, Alberta-Pacific Forest Industries

Land Use Strategy and Integrated Land Management

In Alberta, public lands and resources are managed under an integrated resource management philosophy that considers environmental, economic and social values in decision-making processes. Mechanisms for integrated land and resource management have had limited success in responding to the rapidly evolving economy and demographics of Alberta. The result has been an increased level of development on public land, competing land and resource demands, and competing expectations surrounding land and resource use. The Government of Alberta's 2004-07 Strategic Business Plan committed to developing a comprehensive provincial Land Use Framework for the 21st Century. SRD continues to play a significant leadership role in developing strategies for effective use of the province's land base. In 2004 the Department began developing an innovative approach to managing access to public lands to minimize industrial, commercial, and recreational footprints. This Integrated Land Management program approach reaffirms and updates the philosophy of integrated resource management and is expected to respond to current and future land and resource-use challenges. Sub-components under this strategy include Industry, Recreation and Reclamation streams.

Dave Bartesko, Access Program Manager outlined in an interview in the *Western Woodlot Conservationist* (Winter, 2004) that the Department is "raising the awareness about how we are moving toward landscape-level planning." The desire is to see resource managers, public and private, move more to the concept of integrated management at the landscape level where the cumulative effects are the main consideration as opposed to any single resource or value. This approach links nicely with other provincial strategies for biodiversity, fire protection and water.

Conclusion

Although the working environment for Department staff frequently changed over the years and additional demands created more complexities and workloads, their continued dedication ensured a creditable job of resource protection and management.

Assistant Deputy Minister Craig Quintilio reflected in 2004, "while the Department lost some good people with valuable experience during the downsizing, it also managed to keep a good many of them. Those who remained learned to live with the uncertainties inherent in the current era and managed to achieve 'some amazing things.' At the same time, more cooperative agreements with forest industry resulted in a greater sharing of the responsibilities and workload in forest management. More effort is being made within Alberta Sustainable Resource Development and other Departments to collaborate in the planning and management of Crown-owned lands, with a goal of working toward the sustainability of the forest and its many values. Integration of resource management on the landscape is mirrored by increasing integration of disciplines and mandates within the government".[20]

The only constant in the forests of Alberta has been change. If it isn't an increase in social expectations and trade irritants, it's wildfire, insect and disease or the inherent growth of people and industry within the forest itself. It is change that will most certainly continue, but with wise and innovative management, Alberta's forests will be able to provide our future generations with the same values they do today.

Crescent Falls (top) and Ya Ha Tinda Ranch west of Sundre are just two examples of the stunning and diverse beauty of the forests of Alberta
Bob Stevensen

Two founding members of the Alberta Partners in Protection program, Ken Saulit and Kelly O'Shea shake hands in front of fire prevention mascots Smokey the Bear, Sparky the Dog and Bertie Beaver, Jasper, 1992
Kelly O'Shea

Overhead Team meeting, Swan Hills Fire Base, Whitecourt Forest, spring, 1994. Provincial Forest Fire Centre staff
(L to R): Bill Bereska, Lou Foley, Mike Dubina, Con Dermott (Director Forest Protection)
Alberta Government, AFHPC

Forest Protection Officer meeting, Edmonton, spring, 1993
Back Row (L to R): Jamie McQuarrie, Slave Lake; Bruce MacGregor, Lac La Biche; Lou Foley, PFFC; Gary Dakin, Fort McMurray; Darryl Rollings, Rocky Mountain House; Revie Lieskovsky, PFFC; Con Dermott, Director Forest Protection PFFC; Don Law, Grande Prairie. Front Row: Don Harrison, Bow Crow (acting for Kelly O'Shea); Mike Dubina, PFFC; John Hogue, Whitecourt (acting for Jurgen Moll); Mag Steiestol, PFFC; Ken McCrae, Peace River; Bill Bereska, PFFC. Missing: Dale Huberdeau (Footner Lake)
Alberta Government, AFHPC

Overhead Team meeting, Swan Hills Fire Base, Whitecourt Forest, spring, 1994
Back Row (L to R): Rick Alguire, Jamie McQuarrie, Dave Brown, Gordon Japp, Rob Thorburn, Russ Stashko, Andy Gesner, Barry Gladders, Gary Schneidmiller, Tom Grant, Bruce MacGregor, Don Podlubny, Brian Wudarck, Dale Huberdeau, Con Dermott, Butch Shenfield, Lou Foley, Hugh Boyd, Bob Glover, Dave Lind, Mike Dubina, Henry Grierson, Dennis Driscoll, Darryl Johnson, Gary Mandrusiak, Don Law. Front Row: Howard Herman, Ray Olsson, Bob Mazurik, Collin Williams, Leon Graham, Bob Lenton, Kurt Frederick, Ken Porter, Len Wilton, Bill Bereska, Ed Dechant, Rick Arthur
Alberta Government, AFHPC

Overhead Team meeting, Swan Hills Fire Base, Whitecourt Forest, spring, 1994
Line Bosses (L to R): Rob Thorburn, Hugh Boyd, Bruce MacGregor, Barry Gladders, Rick Alguire, Dennis Driscoll, Bob Mazurik
Alberta Government, AFHPC

Overhead Team meeting, Swan Hills Fire Base, Whitecourt Forest, spring, 1994
Fire Bosses (L to R): Ken Porter, Russ Stashko, Andy Gesner, Bob Lenton, Brian Wudarck, Tom Grant, Dale Huberdeau, Butch Shenfield
Alberta Government, AFHPC

Overhead Team meeting, Swan Hills Fire Base, Whitecourt Forest, spring, 1994
Service Chiefs (L to R): Jamie McQuarrie, Leon Graham, Bob Glover, Collin Williams, Don Podlubny, Dave Lind, Dave Brown, Henry Grierson
Alberta Government, AFHPC

Overhead Team meeting, Swan Hills Fire Base, Whitecourt Forest, spring, 1994
Plans Chiefs (L to R): Howard Herman, Ray Olsson, Len Wilton, Kurt Frederick, Gary Schneidmiller, Ed Dechant, Don Law, Rick Arthur, Darryl Johnson, Gary Mandrusiak
Alberta Government, AFHPC

John Berry (L), CFRN Television weatherman, at the provincial fire map with Lou Foley looking on. John Berry was at the Provincial Forest Fire Centre for Secretary's Week, mid-1990s
Alberta Government, AFHPC

Aerial photo of cutblocks southwest of Calling Lake designed to meet the study needs for the *Adaptive Management, Adaptive Science and the Effects of Forest Fragmentation on Boreal Birds in Northern Alberta* study. Research was done by Fiona K.A. Schmiegelow and Susan J. Hannon with support from Alberta-Pacific Forest Industries, Alberta Forestry, Lands and Wildlife and Forestry Canada. Research was the starting point to explore effects of timber harvesting and harvest and reserve stands in relation to boreal birds
Alberta Government, AFHPC

25th Anniversary of the Provincial Forest Fire Centre (PFFC), Edmonton, 1995
Left Side, Back Row (L to R): Cordy Tymstra, Trear Matijon, Bruce Shaw, Hideji Ono, Con Dermott, Lou Foley. Third Row: Albert Sproule, Bill Bereska, Zigmunt Misztal, Patrick Smith, Lisa Jackson, Bob Young. Second Row: Rick Strickland, Miranda Ching, Ed Johnson, Nick Nimchuk, Norm Olsen, Dahl Harvey, Cinda Lau, Judy Laviolette, Mike Dubina. Sitting: Myrna Delaney, Steve Simser, Larry Brehmer, Dian Reddekopp, Revie Lieskovsky, Carla Brehmer, Debbie Perrault
Right Side, Back Row (L to R): Alvin Scott, Danny Pearn, Ron Ponipal, Rick Schauer, Marshall Yaremcio, Patrick Loewen. Third Row: Walter Kreisler, Bob Yaroslowsky, Carl Esposito, Tom Robertson, George Filan, Christine Spice (Kominek), Gerald Van Wass. Second Row: Rick Wilson, Bernie Sales, Chris Weiss, Iain Shenfield, Brenda Coughlan, Joy Stephenson, Fred Windjack, Mag Steiestol. Sitting: Wally Hare, Dwayne Lindstrom, Wendy de Groot, Debbie Thurlow, Rosalia Molinari (Balogh), Gail Elkins, Barrie Fenby, Sharon Ruta. Missing: Gordon Avery, Holly Bennett, Leo Drapeau, Mike Maximchuk, Tim Parks, Sunil Ranasinghe, Shannon Rowbottom, Nancy Shulhan, Bob Solomon, Gary Still
Alberta Government, AFHPC

Forest Protection Officer meeting and workshop, Elbow Ranger Station, March 14-15, 1995

Back Row (L to R): Jamie McQuarrie, Slave Lake; Con Dermott, Director of Forest Protection; Patrick Loewen, Information Coordinator PFFC; Gary Dakin, Edson; Bruce MacGregor, Lac La Biche; Revie Lieskovsky, Air Operations PFFC; Darryl Rollings, Rocky Mountain House; Mag Steiestol, Prevention PFFC; Bill Bereska, Fire Behaviour Officer PFFC; Don Law, Grande Prairie. Front Row: Ed Johnson, Whitecourt; Lou Foley, Wildfire Operations PFFC; Ken McCrae, Peace River; Don Harrison, (acting for Kelly O'Shea) Calgary; Mike Dubina, PFFC Missing: Dale Huberdeau, Footner Lake (moving to Fort McMurray as the Area Manager). This was the last meeting of the FPOs before the 1995 regionalization took effect - with 4 Regional Forest Protection Officers (Gary Dakin – Northeast Slopes; Ken McCrae – Northwest Boreal; Bruce MacGregor – Northeast Boreal; Andy Gesner – Prairie Bow Parkland [Southwest Slopes])
Ken McCrae

Executive Retreat, Goldeye Centre, west of Nordegg, 1994-95

Back Row (L to R): Dennis Quintilio, Director Forest Management; Lorne Goff, Regional Director Rocky Mountain House; Rick McDonald, Director Land Administration; Howard Gray, Regional Director Peace River; Cliff Henderson, Assistant Deputy Minister; Jerry Sunderland, Regional Director Whitecourt; Kelly O'Shea, Director Forest Protection; Margaret Franklin, Human Resources; Craig Quintilio, Director Program Support. Front Row: Stephania Duffee (Facilitator); Brydon Ward, Regional Director Lac La Biche; Glenn Selland, Executive Assistant
Glenn Selland

Provincial Forest Fire Centre staff dressed up for annual PFFC Klondike Breakfast, 1995

(L to R): Con Dermott, Bill Bereska, Wendy de Groot, Dian Reddekopp, Christine Kominek, Margaret Molinari, Lisa Avis, Rosalia Balogh, Mag Steiestol
Alberta Government, AFHPC

Provincial Forest Fire Centre staff cooking bacon for the annual PFFC Klondike Breakfast, 1995

(L to R): Con Dermott, Mike Dubina, Revie Lieskovsky
Alberta Government, AFHPC

25th Anniversary of the Provincial Forest Fire Centre, equipment display, 1995
(L to R): Ben Janz, Mike Dubina, Bob Yost, Revie Lieskovsky
Alberta Government, AFHPC

Forest Officer landed with pilot and charter helicopter during land use inspections, Amadou Lake area, north of Calling Lake, Calling Lake District, Lac La Biche Forest, spring, 1995. Petroleum companies built helicopter-landing pads at a number of sites where costs for building roads through muskegs were prohibitive
Alberta Government, AFHPC

30 Year Service Award, Department of Environmental Protection, 1995
Back Row (L to R): Dale Huberdeau, Cliff Henderson, Howard Gray
Front Row: Minister Ty Lund, Robert Lenton, Elsie Laboucane, Sirkka Kadatz, Deputy Minister Peter Melnychuk
Corinne Huberdeau

Provincial Forest Fire Centre staff dressed up for Halloween, October, 1995
(L to R): Mag Steiestol, Margaret Molinari, Christine Kominek, Cliff Henderson, Rosalia Balogh, Revie Lieskovsky. Front Row: Wendy de Groot
Alberta Government, AFHPC

Slave Lake Forest Leadership Team, mid-1990s
Standing (L to R): Dave West, Land Use Officer; Joe Smith, Chief Ranger Wabasca; Jamie McQuarrie, Forest Protection Officer; Howard Gray, Superintendent; Jerry Sunderland, Forester i/c Forest Management; Jim Stewart, Office Manager; Pat Hendrigan, acting for Hugh Boyd, Smith. Sitting: Russ Stashko, Chief Ranger Red Earth; Wayne Bowles, Chief Ranger High Prairie; Maurice Lavalle, Chief Ranger Kinuso
Jerry Sunderland

Crowsnest Forest Area staff participating in a volunteer stewardship day at Allison Lake, west end of the Crowsnest Pass. The area is within a Forest Land Use Zone where off-highway vehicles are prohibited. Here the staff and volunteers are using mules to pack gravel for trail hardening and restoration work. Crowsnest Forest Area, Prairie Bow Parkland Region, 1995
Darryl Johnson

A role of Land and Forest Service staff was to 'check-scale' a percentage of log loads scaled by industry staff. This picture is of Alberta Pacific Forest Industries (Alpac) staff scaling a load of deciduous timber at their mill north of Grassland, 1995
Alberta Government, AFHPC

Athabasca District staff after the amalgamation of the Smith, Calling Lake and Wandering River Ranger Districts, Calling Lake, Athabasca District, Northeast Boreal Region, 1995
(L to R): Brian Cote, Bruce Mayer, Jeff Scammell, Marvin Pearce, Wes Nimco, Paul St. John, Brian Stanton, Hugh Fritz, Joanne Deren, Ed Barnett, Janice Damgaard, Christine Carter, Dave Callas, Gail Mitchell, Don Podlubny
Alberta Government, AFHPC

Crowsnest Forest Area staff picture, Prairie Bow Parkland Region, October, 1997
Back Row (L to R): Norman Hawkes, Dennis Halladay, Mike Williamson, Mike Alexander, Ian Dunk, Darryl Johnson. Front Row: Rod Houle, Bill Thresher, Ken Snyder, Ken Orich, Doug Nichol
Darryl Johnson

Aerial view of an Alberta Pacific Forest Industries (Alpac) cutblock northeast of Athabasca, Atahabasa District, Northeast Boreal Region, fall, 1998
Alberta Government, AFHPC

Blairmore Public Lands and Forests staff participate in the Crowsnest Pass Food Bank Chili cook-off, 2003
(L to R): Judy Johnson, Rupert Hewison, Ian Dunk, Bill Thresher
Darryl Johnson

Sustainable Resource Development employees presented with a Team Award for work on the Lost Creek fire, 2004
(L to R): Stu Cruikshank, Mike Cardinal (Minister), Mike Dempsey, Donna Babchishin, Stew Walkinshaw, Rick Horne, Andy Gesner, Rick Strickland, Don Livingstone, Darryl Johnson, Quentin Spila, Bob Fessenden (Deputy Minister)
Alberta Government, AFHPC

Fire salvage operation from the Lost Creek fire, Blairmore, Southern Rockies, winter, 2004
Darryl Johnson

Blairmore FireSmart float in the Crowsnest Pass Rum Runner Days Parade, 2004
Darryl Johnson

Forest Guardian training, Hinton Training Centre, Hinton, April, 2005
Ken Snyder

Photo taken at John Benson's retirement party, fall, 1995. All shown were students and/or instructors in Forest Technology School programs
Back Row Standing (L to R): Brydon Ward, Dale Huberdeau, Kelly O'Shea, Ken South, Conn Brown. Front Row Sitting: John Benson, Dick Altmann, Peter Murphy
Corinne Huberdeau

Slave Lake field tour mid-1990s
(L to R): Ken Higginbotham, Howard Gray, Not Identified (hidden), Con Dermott
Jerry Sunderland

Con Dermott retirement party, 1995. Cliff Smith (L) and Con Dermott
Alberta Government, AFHPC

District Manager meeting, Forest Technology School, Hinton, 1996
Back Row (L to R): Butch Shenfield, Drayton Valley; Dennis Cox, Edson; Dave Redgate, Slave Lake; Don Harrison, Hinton; Mike Poscente, Fort Vermilion; Tom Grant, High Level. Front Row: Jim Cochrane, Lac La Biche; John Brewer, Manning; Ray Luchkow, Calgary; Jim Maitland, Valleyview; Rory Thompson, Rocky Mountain House; Dale Huberdeau, Fort McMurray; Russ Stashko, Whitecourt; Don Podlubny, Athabasca; Hugh Boyd, High Prairie, Bob Mazurik, Peace River, Ed Ritcey, Grande Prairie. Missing: Darryl Johnson, Blairmore
Alberta Government, AFHPC

Northeast Boreal Spring Ranger meeting, Fort McMurray, 1996
Back Row (L to R): Tim Burggraaff, Dave Callas, Ursulla Schroeder, Glen Krawchuk, Bob Yowney, Wally Peters, Paul St. John, John Belanger, Dave Scott, Kevin Topolnicki, Brian Stanton, Jeff Scammell, Vince Eggleston, Bob Dunn, Don Podlubny. Middle Row: Jerome Plammondon, Rick Arthur, Bill Black, Bruce Mayer, Mike Symyrozum, Ed Barnett, Noel St. Jean, Barrie Onysty, Dave Lind, John McLevin, Victor Boisvert, Mark Froehler, Gary Schneidmiller, Dan Slaght, Terry Sayers, Pat Rodseth, George Dribnenki, Colin Hardigan?, Brydon Ward, Dale Huberdeau, Howard Herman. Front Row: Collin Williams, Terry Zitnak, Jim Cochrane, Wes Nimco, Edwin Preece, Kurt Frederick, Bruce MacGregor, Mike Dempsey, David Finn, Petter Finstad, Paul Steiestol, Rick Hirtle
Alberta Government, AFHPC

Retired Nordegg Chief Ranger Ted Loblaw attending memorial service for Cliff Brierley, Ranger Creek, Clearwater area, July, 1997
Bob Stevenson

Land and Forest Service Executive retreat, Westridge, July, 1996
(L to R): Neil Barker, Regional Director Lac La Biche; Glenn Selland, Executive Assistant; Craig Quintilio, Director Program Support; Kelly O'Shea, Director Forest Protection; Howard Gray, Regional Director Peace River; Cliff Henderson, Assistant Deputy Minister; Margaret Franklin, Human Resources; Jerry Sunderland, Regional Director Whitecourt; Rick McDonald, Director Land Administration; Pat Guidera, Regional Director Rocky Mountain House; Dennis Quintilio, Director Forest Management
Glenn Selland

Land and Forest Service Executive meeting, Hinton, April 24-25, 1997
Back Row (L to R): Steve McLaughlin, Strategic Alliance Coordinator; Dennis Quintilio, Forest Management; Margaret Franklin, Human Resources; Anne McInerney, Executive Assistant; Rod Stewart, Facilitator – Strategic Alliance; Craig Quintilio, Land Administration; Kelly O'Shea, Forest Protection. Front Row: Jerry Sunderland, Northeast Slopes (Whitecourt); Howard Gray, Northwest Boreal (Peace River); Cliff Henderson, Assistant Deputy Minister; Patrick Guidera, Prairie Bow Parkland (Rocky Mountain House); Neil Barker, Northeast Boreal (Lac La Biche)
Alberta Government, AFHPC

Wildfire Behaviour Officer Kurt Frederick stands beside his 'pride and joy' pumpkin beneath the Bramalea (Great West Life) Building moose, October, 1997. The pumpkin weighed 381 pounds and came in 11th at the Great Pumpkin Commonwealth weigh-off at Smoky Lake, Alberta. Kurt's goal is to grow an Alberta record giant pumpkin. The record weight in 2004 was over 800 lbs. The world record is now approaching 1,500 lbs
Alberta Government, AFHPC

Managers and Executive retreat, Land and Forest Service, Grey Nuns Centre, November, 1997

Back Row (L to R): Jim Maitland, Valleyview; Tom Archibald, Peace River; Rick McDonald, Land Administration; John Brewer, Manning; Craig Quintilio, Program Support; Ray Luchkow, Calgary; Butch Shenfield, Drayton Valley; Pat Guidera, Prairie Bow Parkland (Rocky); Darryl Johnson, Blairmore; Neil Barker, Northeast Boreal (Lac La Biche); Russ Stashko, Whitecourt; Rory Thompson, Rocky Mountain House; Dennis Cox, Edson; Dale Huberdeau, Fort McMurray; Tony Sikora, High Prairie; Dennis Quintilio, Forest Management; Ed Ritcey, Grande Prairie; Kelly O'Shea, Forest Protection. Front Row: Mike Poscente, High Level; Margaret Franklin, Human Resources; Glenn Selland, Executive Assistant; Cliff Henderson, Assistant Deputy Minister; Howard Gray, Northwest Boreal (Peace River); Al Hovan, Slave Lake; Jerry Sunderland, Northeast Slopes (Whitecourt); Jim Cochrane, Lac La Biche. Missing: Bruce Mayer (Acting), Athabasca; Don Harrison, Hinton

Glenn Selland

Prairie Bow Parkland Region Managers, Grey Nuns, December, 1997

(L to R): Patrick Guidera, Regional Director; Butch Shenfield, Drayton Valley; Ray Luchkow, Calgary; Cliff Henderson, Assistant Deputy Minister; Rory Thompson, Rocky Mountain House; Darryl Johnson, Blairmore

Alberta Government, AFHPC

Waterways staff, Fort McMurray, Northeast Boreal Region, December 19, 1996

(L to R): Bert Ciesielski, Dale Huberdeau, Howard Herman, Rick Arthur

Alberta Government, AFHPC

Open House at the Provincial Forest Fire Centre, Great West Life Building, Edmonton, May 9, 1997. Rick Strickland showing Minister Ty Lund and Deputy Minister Jim Nichol infrared equipment used to detect hotspots
Alberta Government, AFHPC

Signing of the Vanderwell Contractors (1971) Ltd. Forest Management Agreement, Slave Lake, July 23, 1997
Seated (L to R): Bob Vanderwell, Minister Ty Lund, MLA Pearl Calahasen, Cliff Henderson. Back Row: Rick Keller, Con Dermott
Rick Keller

Ranger staff from the Northeast Boreal Region participated in refurbishing the Demicharge Cabin, south of Fort Fitzgerald on the Peace River. The cabin had been used as a patrol cabin for the last half century and required new shingles, new chinking and an overall paint job
Back Row (L to R): Rick Arthur, Mike Dempsey, David Lind, David Finn, Bob Yowney. Front Row: Chris Hale, Dale Huberdeau (Captain of the trip), Tim Burggraaff, Bruce Mayer, Wes Nimco
Corinne Huberdeau

Slave Lake Forest Ranger meeting, mid-1990s
Back Row (L to R): Dave Redgate, Ian Whitby, Barry Gladders, Howard Gray (Superintendent), Wes Nimco, Brian Wudarck, Chris Walsh, Wally Born, Rick Goy, Dale Bullock, Leo Forselli, Greg Cariou, Dave Callas, Joe Smith (hat), Ted Cofer, Rick Alguire, Brian Cote. Third Row: Hugh Boyd, Dave Laing, Doug Smith, Forbes Purcell, Wayne Becker, Karl Peck, Doug Ellison. Second Row: Jamie McQuarrie, Shawn Ingram, Ian Johnston, Maurice Lavallee, Shawn Milne, Patti Campsall. Front Row: Darby Paver, Les Weeks, Henry Grierson, Tracey Rosentreter, Wayne Johnson, Sharon Turner, Morgan Kehr, Mark Mill, Wayne Bowles, Jerry Sunderland, Dean Isles
Jerry Sunderland

Prairie Bow Parkland Region staff picture, fall, 1997. Picture includes staff from the Rocky Mountain House Headquarters office, and the Rocky Mountain House, Drayton Valley, Calgary and Blairmore Districts

Back Row (L to R): Dennis Verge, Sirri Poldrok, Not Identified, Rita Stagman, Mike Templeton, Rick Stewart, Maureen Hillaby, Ollie Erdell, Donna Hare, Cindy Timinsky, Dennis Halladay. Fifth Row: Duane Pollock, Jules LeBoeuf, Ted Flanders, Robert Stokes, Bob Lenton, John Augustyn, Brent Davis, Bart Elliott, Pat Guidera (Regional Director), David Irvine (Motivational Speaker), Brian Orum. Fourth Row: Brian Godwin, Rick Smee, Roger Meyer, Peggy Dolen, Andy Gesner, Wes Eror, Mike Kingsbury, Bob Glover, Gerald Sambroke, Bill Thresher, Ken Orich, Sherry Forster (black coat), Dianne Thompson, Butch Shenfield, Ray Luchkow, Tracy Cove, Ken Snyder. Middle Rows: Rod Houle, Wayne Crocker, Don Livingston, Rory Thompson, Darryl Rollings, Jan Simonson, Bart McAnally, Kevin Heartwell, Mike Alexander, Reg Ahlskog, Ed Pichota, Ian Dunk, Mike Duncan, Cheryl Flexhaug, Rick Wolcott, Shirley McGowan (pink), Connie Kadyk, Nancy Chambers, Dollard O'Conner, Frank Vandriel. Front Row: Larae Johnston, Mona Crocker, Angela Burkinshaw, Laurel Koples, Doug Nichol, John Bruce, Len Wilton, Ross Spence, Mike Williamson, Norm Hawkes, Linda Dunn, Rick Moore, Dennis Frisky, Karen MacAulay, Darryl Johnson, Mike Thompson

Rod Houle

Canadian Shield Special Places Project Team, Department of Environmental Protection Award of Excellence recognition banquet, 1998

(L to R): Ron Mossman, Joe Prusak, Ty Lund (Minister), Jim Nichols (Deputy Minister), Dale Huberdeau, Ted Johnson

Corinne Huberdeau

Spruce Budworm spray program, High Level area, AT-302s flying in formation, 1999

Alberta Government, AFHPC

Whitecourt Forest Area staff, 1999

Back Row (L to R): Shannon Wray, Stacey Melanchuk, Brian Wallach, Gordon Graham, Wanda Kuhn, Hazel Zelinski, Margarete Hee, Doug Malmo, Karen Wirtenen, Brad McKenzie, Joyce Paquin, Roger Litke, Dale Asselin, Arlene Pritchard, Diane Gregory, Gary Thompson. Front Row: Herman Stegehuis, Ruth White, Vicki Watson, Kevin Hakes, Cody Fermaniuk, George Robertson (Manager), Martie Jendrick, Mark Ross, Kent McDonald, Wayne Johnson, Jason Proche, Karen Brashko, Evert Smith, Dave Pollard, Gary Dakin

Alberta Government, AFHPC

Northeast Boreal Region Leadership Team, Forest Technology School, Hinton, March 26, 1999

(L to R): Al Hovan, Slave Lake; Terry Zitnak, Operations Coordinator; Neil Barker, Regional Director; Dale Huberdeau, Fort McMurray; Bruce Mayer, Athabasca; Jim Cochrane, Lac La Biche

Alberta Government, AFHPC

Alberta Forest Service friends at memorial gathering for Dale Huberdeau, Lac La Biche, May, 1999

Back Row (L to R): Ralph Woods, Ray Olsson, Collin Williams, Kelly O'Shea, Kevin Gagne, Not Identified, Mike Benedictson, Andy Gesner (hidden), Roy Campbell, Glen Krawchuk, Petter Finstad, Cliff Henderson, Victor Boisvert, Wayne Robinson (hidden), Hugh Boyd, Rick Arthur, Bill Tinge, Kevin Freehill, Bob Mazurik, Bruce MacGregor. Front Row Seated: Craig Quintilio, Dennis Quintilio, Hideji Ono, Bob Young, Jurgen Moll (hidden), Rick Strickland, Al Benson, Howard Gray, Ken McCrae, Ken Orich, Terry Zitnak, Mag Steiestol, Nick Nimchuk

Corinne Huberdeau

Alberta Forest Service friends at memorial gathering for Dale Huberdeau, Lac La Biche, May, 1999

Back Row (L to R): Rick Arthur, Bill Tinge, Kevin Freehill, Bob Mazurik, Bruce MacGregor, Henri Soulodre, Tom Archibald, Cordy Tymstra, Jamie McQuarrie, Bernie Schmitte, Bill Bereksa, Neil Barker, Bob Dunn, Howard Herman, Rob Thorburn, Revie Lieskovsky, Dennis York, John Brewer, Kevin Topolniki, Owen Bolster, Ken Yackimec, Terry Sayers, Ken South, Oliver Glanfield. Front Row Seated: Terry Zitnak, Mag Steiestol, Nick Nimchuk, Sunil Ranasinghe, Dennis Cox, Cinda Lau, Ben Janz, Bill Wuth

Corinne Huberdeau

Aerial view of a stuck truck being pulled out of the soft ground after four-wheel driving, backcountry area west of Rocky Mountain House, early 2000s
Butch Shenfield

FireSmart event in Hinton, summer, 2003
(L to R): Ken Brands (Fire Chief, Town of Hinton), Kevin Freehill, Bertie Beaver, Shelleen Lakusta, Bill Gilmour
Kevin Freehill

Halloween in Slave Lake, Marten Hills Area, 2002
(L to R): Angele Price, Cindy Schellenberg, Doug Ellison (Snow White), Karen MacNeil, Norma Brown
Alberta Government, AFHPC

Cooking up hot dogs for National Forest Week, Whitecourt, May, 2001
(L to R): Mike Poscente, Bertie Beaver, Ken Podulsky
Alberta Government, AFHPC

Wildfire Prevention Officers, Hinton Training Centre, Hinton, December, 2003

Back Row (L to R): Hugh Boyd, Director Wildfire Prevention Branch; Gary Mandrusiak, Rocky Mountain House; Jules LeBoeuf, PFFC; Rick Arthur, Calgary. Middle Row: Darrell Kentner, Whitecourt; Kevin Freehill, Edson; Quentin Spila, Fort McMurray; Rod Houle, PFFC; Owen Spencer, Grande Prairie. Front Row: Chris McGuinty, Peace River; Wayne Bowles, Slave Lake; Herman Stegehuis, PFFC; Wes Nimco, Lac La Biche; John Branderhorst, PFFC
Missing: John McLevin, PFFC; Derrick Downey, High Level

Alberta Government, AFHPC

Wildfire and Air Operations Officers, Hinton Training Centre, Hinton, December, 2003

Back Row (L to R): Roy Campbell, Grande Prairie; Gord Glover, High Level; Morgan Kehr, Edson/PFFC (PB Coordinator); Ray Olsson, Edson; Kent McDonald, Acting FPO Whitecourt. Middle Row: John Belanger, PFFC; Tracey Stewart, PFFC; Trevor Lamabe, Peace River; Ferenc Scobie, FPT Lac La Biche; Jim Maitland, Manager Grande Prairie. Front Row: Paul Steiestol, Fort McMurray; Ken Orich, Slave Lake; Brian Stanton, Lac La Biche; Wally Born, PFFC; Len Wilton, Calgary; Andy Gesner, Rocky Mountain House. Missing: Dennis Driscoll, PFFC

Alberta Government, AFHPC

Forest Protection Division Fire Management Group meeting, Hinton Training Centre, Hinton, December, 2003

Back Row (L to R): Bruce Cartwright, Fort McMurray; Jamie Yee, High Level; Revie Lieskovsky, Wildfire Operations; John Shires, Wildfire Information; Brent Schleppe, Edson; Don Harrison, Wildfire Service; Tracey Stewart, Wildfire Policy and Business Planning (acting for Deanna McCullough). Middle Row: Jean Easton, Human Resources; Jim Maitland, Grande Prairie; Mike Poscente, Whitecourt; Jamie McQuarrie, Provincial Warehouse; Bruce MacGregor, Lac La Biche; Darryl Rollings, Rocky Mountain House; Hugh Boyd, Wildfire Prevention; Tom Archibald, Peace River. Front Row: Chris Churchill, Human Resources; Marilyn McKinnley, Executive Assistant; Anne McInerney, Wildfire Support; John Brewer, Slave Lake; Cliff Henderson, Assistant Deputy Minister Forest Protection; Stew Walkinshaw, Calgary; Rob Thorburn, Hinton Training Centre

Alberta Government, AFHPC

Deputy Minister Bob Fessenden (L) and Minister Mike Cardinal at the 2003 Sustainable Resource Development Staff Recognitions Awards celebration, spring, 2003

Awards Committee

Environmental damage from off-highway vehicles at the Johnson Bog, Ghost-Waiparous, Southern Rockies Area, Southwest Region, September, 2003
Rick Blackwood

Arcview 3.x/Citrix Deployment Team receiving team award during the SRD Employee Recognition Awards ceremony, spring, 2003
(L to R): Stu Churlish, Mike Cardinal (Minister), Brian Fairless, Jim Hammel, Darryl Seeger, Barry Northey, Bob Fessenden (Deputy Minister). Missing: Angela Braun, Christine Kominek, Don Page
Awards Committee

Forest Protection Division Executive, November, 2004
(L to R): John Shires, Wildfire Information Branch; Revie Lieskovsky, Wildfire Operations; Hugh Boyd, Wildfire Prevention; Deanna McCullough, Wildfire Policy and Business Planning; Cliff Henderson, Assistant Deputy Minister; Bruce Mayer, secondment to Consultation and Aboriginal Relations Unit; Don Harrison, Wildfire Service
Alberta Government, AFHPC

Forest Officers conducting a mill inspection at the Colin Ruxton Mill, Wembley, Smoky Area, 2004
Dave Heatherington

Forest Officer Dion Lawrence preparing to winch out vehicle from water hole in road during Commercial Timber Permit inspection, Smoky Area, 2004
Sean Trostem

1993 - 2005

Area Manager Stuart Taylor, with oil and gas company representatives, inspecting a pipeline right-of-way, Smoky Area, 2004
Sean Trostem

Forest Officer Dave Heatherington scaling a deck of tree length logs, Smoky Area, 2004
Sean Trostem

Foresters Dennis Froese and Tracy Douglas conducting a circular plot to check density of tree planting, East Peace Area, 2004
Kelly Boreson

Forester Tracy Douglas measuring diameter of tree at breast height (DBH) for timber cruise, East Peace Area, 2004
Kelly Boreson

Forester Dennis Froese using clinometer to measure height of tree for timber cruise, East Peace Area, 2004
Kelly Boreson

Ross Graham and John Branderhorst receiving their 35-year service pin and award, Sustainable Resource Development Staff Recognition Awards, March, 2004
(L to R): Minister Mike Cardinal, Ross Graham, John Brandershorst, Deputy Minister Bob Fessenden
Awards Committee

Jim Skrenek receiving 40-year service pin and award, Sustainable Resource Development Staff Recognition Awards, March, 2004
(L to R): Minister Mike Cardinal, Jim Skrenek, Deputy Minister Bob Fessenden
Awards Committee

Forest Protection Staff receiving award for design of the Wildfire Threat Assessment Model, Sustainable Resource Development Staff Recognition Awards, March, 2004
(L to R): Minister Mike Cardinal, Rod Houle, Kurt Frederick, Sherra Quintilio, Kevin Keats, Herman Stegehuis, Deputy Minister Bob Fessenden
Awards Committee

Minister Mike Cardinal (L) and Deputy Minister Bob Fessenden at the Sustainable Resource Development Staff Recognition Awards, March, 2004
Awards Committee

Bill 16 Enforcement Committee Team Award, Sustainable Resource Development Staff Recognition Awards, March, 2004
(L to R): Minister Mike Cardinal, Gerry Dube, Doug Amundsen, Dave Bartesko, Doug Luzny, Cec Fagnan, Erika Gerlock, Todd Letwin, Shannon Keehn, Diane Fournier, Diana Brierley, Pat Dunford, Deputy Minister Bob Fessenden
Awards Committee

Sustainable Resource Development Executive Committee, Edmonton, October 26, 2004
Back Row (L to R): Dom Ruggieri, Regional Executive Director Southeast Region; Craig Quintilio, Assistant Deputy Minister Public Lands and Forests Division; Ken Ambrock, Assistant Deputy Minister Fish and Wildlife Division; Neil Barker, Regional Executive Director Northeast Region; David Christiansen, Acting Regional Executive Director Southwest Region. Middle Row: Stew Churlish, Assistant Deputy Minister Strategic Corporate Services; Donna Babchishin, Director Communications; Debra Grainger, Administrative Assistant to DM; Daphne Cheel, Executive Director Policy and Planning; Patti Papirnik, Executive Assistant; Diane Dunn, Executive Director Human Resources; Ken McCrae, Regional Director Northwest Region. Front Row: Cliff Henderson, Assistant Deputy Minister Forest Protection Division; Bob Fessenden, Deputy Minister; Howard Gray, Assistant Deputy Minister Strategic Forestry Initiatives
Alberta Government, AFHPC

Public Lands and Forests Division Manager meeting, Coast Plaza, Edmonton, October 28, 2004
Standing (L to R): Dan Smith, Rangeland Barrhead; Keith Lyseng, Director Rangeland Management Branch; Gerry Dube, Rangeland Lac La Biche; John Laarhuis, Head Rangeland Integration; Dale Willsey, Rangeland Fairview; Ken McCrae, Operations Director; Glen Gache, East Peace; Mark Storie, Yellowhead; George Robertson, Woodlands; Al Hovan, Lesser Slave Lake; Barry Cole, Southeast (Red Deer); Noel St Jean, Waterways; Butch Shenfield, Clearwater; Glenn Selland, Director Land Use Operations. Sitting: Gail Tucker, High Level; Craig Quintilio, Assistant Deputy Minister; Val Hoover, Director Disposition Services; Terry Zitnak, Lac La Biche. Missing: Stu Taylor, Smoky; Bill Irvine, A/Director Business Planning; Rick Blackwood, Southern Rockies; Brian Laing, Rangeland Lethbridge
Alberta Government, AFHPC

Ribbon-cutting, Alberta style
Mayor Stephen Mandel (L) and David Coutts, Minister of Sustainable Resource Development, cut through a Lodgepole pine log — Alberta's official tree — to celebrate the official opening of the Alberta Forest Products Association's new office in the Peace Hills Insurance Building at Jasper Ave and 107th Street, Thursday December 16, 2004. They are assisted by association president Art Lemay, kneeling at front, and George VanderBurg, MLA Whitecourt-Ste. Anne, standing at back
Edmonton Journal, Jason Scott

Forest History Association of Alberta directors and founding members, February 16, 2005
(L to R): Peter Murphy, Neil Shelley, Bob Demulder, Arden Rytz, Bob Stevenson
Forest History Association of Alberta

Forest History Association of Alberta directors elected February 16, 2005
Back Row (L to R): Butch Shenfield, Cliff Henderson, Fred McDougall, Bob Stevenson, Bruce Mayer (Treasurer). Front Row: Bob Udell, David Holehouse (Secretary), Arden Rytz (President), Peter Murphy (Vice President)
Forest History Association of Alberta

Loaded logging truck hauling logs from the Sulphur Lake Woodlot to the Boucher Bros. mill at Nampa, East Peace Area, Northwest Region, March, 2005
Kelly Boreson

2005 Sustainable Resource Development Staff Awards Recognitions, March, 2005
(L to R): Brad Pickering, Deputy Minister; Craig Quintilio, Assistant Deputy Minister Public Lands and Forests Division; Stew Churlish, Assistant Deputy Minister Strategic Corporate Services; Cliff Henderson, Assistant Deputy Minister Forest Protection Division; Ken Ambrock, Assistant Deputy Minister Fish and Wildlife Division; David Coutts, Minister; Jules LeBeouf, Master of Ceremonies
Awards Commitee

40 Years of Service! March, 2005
(L to R): Deputy Minister Brad Pickering, Cliff Henderson, Mike Dubina, Bruce MacGregor, Minister David Coutts
Awards Committee

Ken McCrae receiving his 35 year service pin and certificate, March, 2005
(L to R): Deputy Minister Brad Pickering, Ken McCrae, Minister David Coutts
Awards Committee

Tim Klein and Gordon Graham receiving the Forest Protection Detection Section Team Bright Idea Award, March, 2005
(L to R): Brad Pickering, Tim Klein, Gordon Graham, Minister David Coutts
Awards Committee

Margarete Hee receiving an Individual Achievement Award, March, 2005
(L to R): Deputy Minister Brad Pickering, Margarete Hee, Minister David Coutts
Awards Committee

Members of the Marketing Committee for the Canadian Institute of Forestry, Society of American Foresters conference, March, 2005
(L to R): Brad Pickering, Bob Anderson, Karen Wirtenen, Wendy Mahan, John Shires, Minister David Coutts
Missing: Al Hovan, Donna Babchishin, Duncan MacDonnell, Norman Brownlee, Robert Storrier, David Holehouse
Awards Committee

Health Safety and Wellness Award Recipients, March, 2005
(L to R): Brad Pickering, Marilyn Rayner, Jim Skrenek, Don Harrison, Diane Dunn, Dom Ruggieri, Murray Busch, Sian Swinnerton, Suzanne Hawkes-Gill, Minister David Coutts
Missing: Bill Irvine, Dan Smith, Jamie Yee, Terry Zitnak
Awards Committee

Peace River Fire Manager Tom Archibald presenting a print of appreciation to Woodland Cree First Nation Councilors for past and continued service within the Wildland Firefighting program, April, 2005
(L to R): Kenny Auger, Tom Archibald, Bill Cardinal, Rhonda Laboucan
Alberta Government, AFHPC

1993 - 2005

Customer Service Award Recipients, Public Lands and Forests Division Exploration Unit Team, March, 2005
(L to R): Brad Pickering, Paul Allan, Grant Nieman, Laurice Block, Donna Bambrick, Jerry Riddell, Bev Cormack, Herman Selcho, Blair Stone, Evelyn Finley, Minister David Coutts
Awards Committee

Deputy Minister Award Recipients, Forest Pest Damage Diagnostic System Team, March, 2005
(L to R): Brad Pickering, Sunil Ranasinghe, Linda Joy, Ed Boothman (Cricket Works), Christine Kominek, Cody Crocker, Ena Zefram, Herb Cerezke, Hideji Ono, Ken Mallett, Minister David Coutts
Awards Committee

Mike Cardinal (L), Minister Human Resources and Employment presenting Howard Gray with plaque from the Government of Alberta on Howard's retirement, April 16, 2005
Bob Stevenson

Bob Fessenden (L), Deputy Minister Innovation and Science presenting Howard Gray with plaque from Sustainable Resource Development on Howard's retirement, April 16, 2005
Bob Stevenson

Cliff Henderson (L) presenting Howard Gray with a chrome Pulaski for years of service and dedication to the fire program within Alberta, April 16, 2005
Bob Stevenson

Margaret Franklin (L) and Maureen Lavallee at the Howard Gray retirement function, April 16, 2005
Bob Stevenson

Strategic Forestry Initiatives Division staff, Howard Gray retirement, April 16, 2005
Back Row (L to R): Ziggy Bahde, Howard Gray, Colleen Scott, Glenys Haskins, Gloria Hossinger, Trevor Vegh. Front Row: Jerry Sunderland, Pat Guidera. Missing: Dan Wilkinson, Darren Tapp, Jason Proche, Robert Hendren, Ron Dunnigan and Gordon Giles
Trevor Vegh

Forester Steven Stryde inspecting wood utilization during a waste pile survey in a harvested cutblock, Waterways Area, summer, 2005
Noel St Jean

Information Officer Anastasia Drummond helping Julia Sandford spray a beach ball with a backpack water pump during the Lac La Biche FireSmart community event, July, 2005
Kelly Boreson

Public Lands and Forests Division and Forest Protection Division float in the Lac La Biche Pow Wow Days parade, July 29, 2005
Terry Zitnak

Public Lands and Forests Division and Forest Protection Division float in the Lac La Biche parade, July, 2005
Terry Zitnak

Reviewing aerial photograph of terrain on mountain pine beetle control program, Willmore Wilderness, July, 2005
(L to R): Mike Maximchuk, Erica Lee, Craig Quintilio, Ken McCrae
Duncan MacDonnell

Forest Health Officer Mike Maximchuk pointing out Mountain Pine Beetle infestation areas to Incident Command Team, July, 2005
Duncan MacDonnell

Mountain pine beetle infested area
Andre Roy

Ivan Strang (L), MLA West Yellowhead and Cliff Henderson, Assistant Deputy Minister Forest Protection Division
Bob Stevenson

Aerial view of mountain pine beetle infected trees being burned, Willmore Wilderness, July, 2005
Duncan MacDonnell

Sustainable Resource Development Minister David Coutts providing directions on where trees will be planted at the Grande Prairie Christian School Arbor Day Celebrations, May 5, 2005
Duncan MacDonnell

Sustainable Resource Development Minister David Coutts helping to plant lodgepole pine at the Grande Prairie Christian School Arbor Day Celebrations, May 5, 2005
Duncan MacDonnell

Sustainable Resource Development Centennial Tree Planting Celebration, Legislature Grounds, May 9, 2005
(L to R): David Coutts, Minister; Ernst Klaszus, Chief Warden; Mrs. Phyllis Coutts
Duncan MacDonnell

Sustainable Resource Development Centennial Tree Planting Celebration, Legislature Grounds, May 9, 2005. Minister and Mrs. Coutts with Junior Forest Warden Brennan Christman
Duncan MacDonnell

Sustainable Resource Development Centennial Tree Planting Celebration, Legislature Grounds, May 9, 2005
(L to R): Minister David Coutts, Premier Ralph Klein, Brennan Christman, Art Lemay (President Alberta Forest Products Association)
Duncan MacDonnell

Sustainable Resource Development Centennial Tree Planting Celebration, Legislature Grounds, May 9, 2005. Junior Forest Wardens posing with Premier Ralph Klein and Junior Forest Warden Brennan Christman, Chief Warden Ernst Klaszus, George VanderBurg MLA, Art Lemay, Minister David Coutts and Mrs. Phyllis Coutts

Other participants include Angie (leader), Zach and Maysen Lubbers, Kayla and Morgan Wegernoski, Tyrell Giselbrecht, Anika and Ian Wirtanen, Alexandra Christman, Connor Petrie, Caleb and Chandra Spratt, Keiton Fjoser, Dale (leader), Christy, Geoffrey and Kate Giebelhaus, Ray (leader), Matthew and Sam Sawchuk, Wendy (leader) and Alex McCormack, Savannah and Beaudan Glanz, Elsie (leader), Lane and Lana Matthews, Chelsea Serink
Duncan MacDonnell

National Forest Week SRD picture contest presentation to Gail Greenwood, May, 2005
(L to R): Brad Pickering, Deputy Minister; Kathy Hendren, Junior Forest Warden Regional Coordinator; Gail Greenwood, Junior Forest Ranger Program Coordinator
Alberta Government, AFHPC

Kainai (Blood) First Nation Chief Charlie Weasel Head signing a cooperative agreement with SRD Minister David Coutts for an Aboriginal Junior Forest Ranger crew, May, 2005
Alberta Government, AFHPC

Gordon Fowlie standing behind firefinder stand in Adams Creek Lookout, August 13, 2005. Gordon Fowlie was the first towerperson on Adams Creek Lookout, serving from 1940 to 1945
Alberta Government, AFHPC

Old Adams Creek Lookout in the centre opened up as a tourist attraction at the Grande Cache Interpretive Centre, August 13, 2005. Muskeg Ranger Station on the right
Alberta Government, AFHPC

Ernie Stroebel (L) and Neil Gilliat at the Forest Ranger and Lookout Observer Reunion in Grande Cache, August 13, 2005. Stroebel was a ranger in Calling Lake from the late 1950s to early 1960s before moving to the Muskeg Ranger Station. Gilliat started as a ranger in the Grande Cache area in 1949 and moved to Human Resources from the Superintendent of the Slave Lake Forest in 1968
Alberta Government, AFHPC

Hope and Tim Klein with Simonette Towerman Ernie Basaraba (R) at the Forest Ranger and Lookout Observer Reunion, Grande Cache, August 13, 2005
Alberta Government, AFHPC

Robert Guest (L) and Dave Schenk at the Forest Ranger and Lookout Observer Reunion, Grande Cache, August 13, 2005. A noted artist, Guest is currently lookout observer at Moberly Tower and has served on towers in the Grande Prairie and Grande Cache areas for over 25 years. Dave Schenk was a ranger in the Grande Cache area in the 1950s
Alberta Government, AFHPC

Group photo of rangers and lookout observers attending the Grande Cache reunion, August 13, 2005
Back Row (L to R): Jack Richardson (son of Walt Richardson), Hope Klein, Gordon Campbell, Don Crawford, Vic Fischer, Mansel Davis, Hylo McDonald, Don Lowe, Al Walker, Wayne Bowles, Frank Coggins (son of Tom Coggins). Front Row: Bob Stevenson, Cecil Stollings (son of Earl Stollings), Cyril Lanctot, Ernie Stroebel, Marion Winn (wife of Rex Winn), Jackie Hanington (wife of Bill Hanington), Neil Gilliat, Robert Guest, Gordon Fowlie, Bruce Mayer, Dave Schenk, Ernie Basaraba. Missing from the photo: Tim Klein
Alberta Government, AFHPC

Forest Protection Division Board of Directors, Frying Pan Staging Camp, Smoky Area, August 17, 2005
(L to R): Bruce Mayer, Jamie Yee, Rob Thorburn, Deanna McCullough, Andy Gesner, Roy Campbell, Bruce Cartwright, John Brewer, Brent Schleppe, Patrick Loewen, Darryl Rollings, Hugh Boyd, Cliff Henderson, Revie Lieskovsky, Jim Maitland, Darrell Kentner, Tom Archibald. Missing: Don Harrison, John Shires and Bruce MacGregor
Tom Archibald

A Past to Remember

On September 8, 2005 Sustainable Resource Development celebrated 100 years of Forestry, Lands & Wildlife in a tribute called 'A Past to Remember'. The following are some photographs from that event.

Individuals who have held the post of Minister and Deputy Minister in the various iterations of the Department over the years
Back Row (L to R): Deputy Ministers Peter Melnychuk, Jim Nichols, Doug Radke, Bob Steele, Brad Pickering, Cliff Smith, Fred McDougall. Front Row: Ministers Ty Lund, David Coutts, Mike Cardinal. Unable to attend photo session: Ministers Allan Warrack, Don Getty, Merv Leitch, John Zaozirny, LeRoy Fjordbotten, Brian Evans, Gary Mar and Halvar Jonson; Deputy Ministers Vi Wood and Bob Fessenden
Capture the Moment

Minister David Coutts giving speech during the evening activities
Bob Stevenson

Leo Drapeau (L) and Larry Huberdeau standing in front of a 1/4th replica wooden crawl tower constructed by Tim and Hope Klein
Bob Stevenson

Individuals who have held the post of Deputy Minister in the various iterations of the Department over the years
Back Row (L to R): Jim Nichols, Doug Radke, Brad Pickering, Cliff Smith. Front Row: Peter Melnychuk, Bob Steele, Fred McDougall. Unable to attend photo session: Vi Wood, Bob Fessenden
Capture the Moment

A uniform period parade showing the change in uniforms and ladies dress over the years
Bob Stevenson

Dennis Quintilio (C) and Cliff Henderson (R) preparing packhorse with blanket and pack saddle before loading the packs. Bruce MacGregor looking on
Alberta Government, AFHPC

Water bag pack to the left rear, pack to back right with a diamond hitch tied by Harry Edgecombe and Dexter Champion's saddle and chaps front right
Alberta Government, AFHPC

(L to R): Marg Sutherland, Susan Ward, Fred Sutherland, Brydon Ward, Christie Ward
Alberta Government, AFHPC

(L to R): Grant Sprague, Thomas Booth, Angela Shaver and Tim Freedman
John Laarhuis

(L to R): Bertie Beaver, Norman Brownlee, Deanna McCullough, Karen Wirtanen, Dian Reddekopp and Annette Krum
Norman Brownlee

Organizing Committee for 100 Years of Forestry, Lands & Wildlife, September 8, 2005
Standing (L to R): Kathy Harkin (CEO Inc.), Jules LeBoeuf, Patrick Guidera, Margo Pybus, Jan Patterson. Sitting (L to R): Margaret Blyth, Teresa Stokes, Patrick Loewen
Capture The Moment

(L to R): Doris Braid, Cliff Henderson and Karen Christensen
Patrick Loewen

Singing of the national anthem at the opening ceremonies
(L to R): Patrick Guidera, Mike Underschultz, Bob Fessenden, Pat Wearmouth
Patrick Guidera

(L to R): Hope Klein, Tim Klein and Tanya Birnie-Browne standing in front of the 1/4th scale replica wooden crawl tower constructed by the Klein's
Norman Brownlee

Display of a bush camp and tools used for early forestry work
Butch Shenfield

Communication technicians Mike Peebles (L) and Gerald Carlson at the communications display
Norman Brownlee

Bertie Beaver memorabilia display, with wooden carving done by Howard Townsend in the left foreground
Norman Brownlee

Cribbed fire pit for cooking in foreground, canvas tents and equipment from a fire camp in background
Dave Braa

Paul St. John (L), Instructor at the Hinton Training Centre and Don Pope standing in front of the forestry training display
Dave Braa

Ed Gillespie and Don Fregren standing in front of Bob Young's restored 1971 Chevrolet Long Rider Alberta Forest Service truck
Ken Dutchak

Oliver Glanfield (blue shirt, suspenders) talking with Peter Murphy (R)
Jean Lussier

United Way Contests

Alberta Forest Service employees have volunteered and supported United Way and Canadian Cancer Society contests since the early 1990s. Rick Strickland, Wildfire Information Officer, Provincial Forest Fire Centre, was the main coordinator for the Chicken Wing Eating Contest, from 1993 to 2005, and the head shave contest in 2004. Dollars raised over those years has surpassed $100,000.

United Way Chicken Wing Eating contestants, 1993
(L to R): Lou Foley, Revie Lieskovsky, Con Dermott
Alberta Government, AFHPC

United Way Chicken Wing Eating contestants, 1993
Mike Dubina (L) and Debra McIntosh in the 'heat of the contest'
Alberta Government, AFHPC

Assistant Deputy Minister Ken Higginbotham riding tricycle in United Way program. Evelynne Wrangler at right, Edmonton, mid-1990s
Alberta Government, AFHPC

United Way Chicken Wing Eating contestants, 1993. Lou Foley receiving a 'hug' from Revie Lieskovsky
Alberta Government, AFHPC

Sustainable Resource Development staff participating in the 2004 United Way 'Shave a Head' contest, April 30, 2004. SRD raised $32,014.41 during this 2004 United Way contest
(L to R): Rick Strickland, Wildfire Information Officer PFFC; Cliff Henderson, Assistant Deputy Minister Forest Protection Division; Bob Fessenden, Deputy Minister. Haircuts provided by Cindy Ferrari, Eliza Young, Laurie Murowchuk from Jagged Edges
Alberta Government, AFHPC

Alberta Forest Service

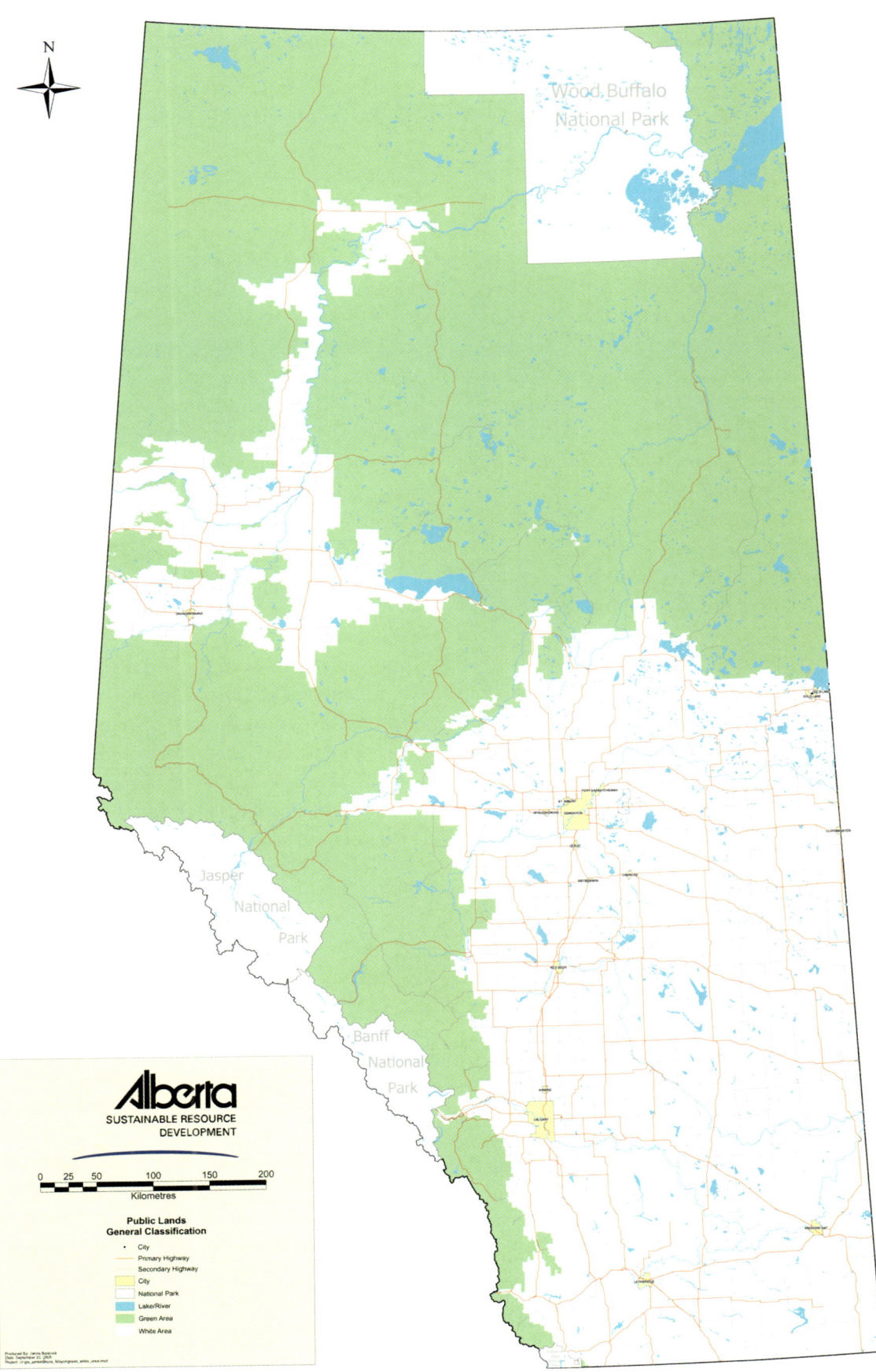

CHAPTER 10

Fire in Alberta

Since the time that trees returned to the post-glacial landscape about 10,000 years ago, forests evolved under the influence of fire.

Fires in the forest were common and part of the natural pattern of disturbance until modern fire suppression efforts were introduced in the latter half of the 20th century. Indigenous forest residents, and later the tide of European settlers, also burned forest areas for their own economic and cultural reasons.

During the first 75 years of the Alberta Forest Service, nearly 47,000 fires and almost 12.3 million hectares of burned area were recorded – an average of about 625 fires and 164,000 hectares each year. The fire record, however, is incomplete, especially during the first 30 years (1930 - 1960), so the average number of fires may well have been closer to 1,000 per year. The majority of these fires were small. A few became very large, some one million hectares or larger. Several of these fires were notable, some because of their size, and others as a result of their effect on the ecosystem, people, values, economy or their intensity or newsworthiness.

Forest fire prevention signs at Outfitter/Guide hunting camp, Rocky Mountains Forest Reserve, 1912
Dominion Forestry Branch, AFHPC

Legends, aboriginal stories, and historical journal descriptions written by early explorers, fur traders and settlers describe some of the early fires. Analysis of forest stand ages and fire scars also provide accounts of these fires.

Fires set the stage for growth of new forests and different species, providing a mosaic of diverse forest communities. Fires also resulted in tangled blowdown of dead trees that greatly impeded the travel of explorers and by horse.

The cumulative effect of recurrent fires was well described by the early foresters who surveyed the Dominion lands of the Northwest Territories for the Dominion Forestry Branch. J.A. Doucet reported on his trip in the Smoky River area that: "out of 8,000 square miles of land examined in the Peace River region in 1913 that was well wooded even during the last 100 years, an area of only 1,350 square miles, or not quite 17% is found today bearing forest cover 75 years of age and over. It has been said also that during the last 50 years fires have run over 65% of the entire territory, or 4,240 square miles."[1] This rate of burn is typical of the results of more recent studies. Early explorers, mountain climbers, and surveyors had very strong opinions of fire, were excellent observers of fire history and contributed valuable information through their reports and journals.

The following stories provide some historical insights into fire events in Alberta. These stories stood out by virtue of legend, historical interest, records set and involvement of people. It is interesting that they are dispersed over two centuries, although there are clusters around some of the more active periods of fire. This is a reflection of the persistence of forest fires and the determination of people to try to control them.

346 Alberta Forest Service

1792 and 1793 Buckingham House [near Elk Point] to the Rocky Mountains

Explorer Peter Fidler wrote about his experiences in the southern part of what is now Alberta in his *'Journal of a Journey over Land from Buckingham House to the Rocky Mountains in 1792 &3.'* In his journal dated Tuesday December 18, 1792, Fidler noted:

"Grass all lately burnt the way we have passed this Day towards the Mountain, but not to the South of us, but at a good distance in that direction the Grass is now burning very great fury, supposed to be set on fire by the Connen na hew Indians. Every fall & spring, & even in the winter when there is no snow, these large plains either in one place or other is constantly on fire, & when the Grass happens to be long & the wind high, the sight is grand & awful, & it drives along with amazing swiftness, indeed several Indians I have heard being burnt in this manner to death, the fire coming upon them in the night when asleep. The flames roars along like the waves in the ocean in a storm. The only way they avoid these fires, when aware of it, is by immediately setting fire to the Grass they are at, & when a little space is burnt themselves, Horses, &c. go upon the burnt part & when the fire comes to this place just burnt, it becomes dark in an Instant when this happens in the night."[2]

Peter Fidler's last description indicates that local native people understood the value of burning out a piece of ground and moving into the 'burn' (or black) to save themselves.

1812 Fort Edmonton Area, Hudson's Bay Company

Among the early journal accounts of fire was the Hudson's Bay Company record of Fort Edmonton for October 12, 1812: "the Plains are, and have been the several days past, burning in a most dreadful manner. Fires are raging in all directions, and the sun obscured with smoke that covers the whole country, and should the remarkable dry weather which has now continued so long, not change very soon, the

Forest Ranger Ferguson warning campers about fire, Cypress Hills Forest Reserve, 1908
Dominion Forestry Branch, AFHPC

Plains must be burned to such an extent as to preclude all hopes of our getting a large supply of dry provisions, for which appearances on our arrival here were very flattering."[3] The Sarcees reported later that from Fort Edmonton to the banks of the South Saskatchewan there was not a buffalo bull to be seen nor a bit of dry ground unburned. The Hudson's Bay Company was compelled to send men to Lac Ste. Anne to fish for food. The coordinates suggest the fire may have burned up to 6.5 million hectares.

1858 Southern Rockies, Palliser Expedition

The Palliser Expedition into southwestern Alberta, led by Captain John Palliser with James Hector, is well documented and has numerous references to fire and the difficulty of travel as a result of fallen timber. One of the first references described by I.M. Spry was along the Red Deer River: "At last they got their first view of the Rocky Mountains . . . a less attractive sight, which they saw over and over again, was trees destroyed by the ravages of fire (almost invariably due to wanton carelessness and mischief, as a result of signal fires, often used for trivial messages), which brought starvation and misery to the Indian tribes themselves by spoiling their hunting grounds." Palliser later recorded, as they made their way up the Kananaskis valley: "The obstacle which a burnt forest presents to the traveler is of all others the most arduous; sometimes we were in a network of trees, lying at all angles the one to the other, and requiring no small amount of skill to choose which should be removed first." Spry noted that, "After working hard with their axes almost all day, clearing the fallen timber, the party was exhausted. They saw mountain trees burnt in places so precipitous that no human hand could ever have reached them, which convinced them that lightning had started fires in the mountains."[4]

1858 Upper Saskatchewan, James Hector (Palliser Expedition)

Hector, traveling in September in the vicinity of Mount Forbes [now in Banff National Park] in the Upper Saskatchewan wrote: "During the night we saw a great glare of flame down the valley at the lower end of the lake, and we rightly conjectured that the fire we left at our last halting place among the fallen woods had set the forest on fire ... the first passage of fire is rapid, but it often remains smouldering for months, in spots."[5]

1858 Prairie Grasslands, Henry Youle Hind

Henry Youle Hind reported an extensive fire in the prairie grasslands, the size reflecting the potential for large fires in the west, given the continuity of fuels coupled with effects of weather. "From beyond the south branch of the Saskatchewan to Red River all the prairies were burned last autumn, a vast conflagration extended for 1,000 miles in length and several hundreds in breadth. The dry season had so withered the grass that the whole country of the Saskatchewan was in flames… we traced the fire from the 49th parallel to the 53rd, and from the 98th to the 108th degree of longitude. It extended no doubt to the Rocky Mountains."[6] The coordinates of this burn suggest the fire could have been as large as 50 million hectares.

1859 Brazeau River Area, Earl of Southesk

The Earl of Southesk traveling down the Brazeau River notes: "Fires which are the curse of this region have destroyed the beauty of those noble valleys, ruining the magnificent forests that ages had matured and leaving in their stead endless track of charred and decaying remains amidst which wretched seedlings struggle up as best they may…."[7]

Stony soil exposed after removal of forest cover by repeated fires, 1908
Dominion Forestry Branch, AFHPC

1863 McLeod River Area, Lord Milton and Walter Cheadle

Lord Milton and Walter Cheadle were also adventurers who sought a "north-west passage by land" to the Pacific. They also contributed to the fire history of the area. Camped one June morning in the vicinity of the McLeod River while their guide was scouting a trail, they had a smudge fire smouldering for the horses and were cooking their pemmican by the tents when: "suddenly a louder crackling and roaring of the other fire attracted our attention, and, on looking round, we saw, to our horror, that some of the trees surrounding the little clearing had caught fire." The smudge fire had escaped. "Cheadle, seizing an axe, rushed to the place and felled tree after tree to isolate those already fired from the rest, whilst Milton ran to and fro, fetching water in a bucket from a little pool and poured it on the thick, dry moss through which the fire was rapidly spreading. We were by this time nearly surrounded by blazing trees, and the flames flared and leapt up from branch to branch, and from tree to tree, in the most appalling manner." Fortunately, in the midst of their struggle, their guide returned and helped to save the horses and camp. They quickly packed up and left upwind to the west. They later remarked: "Clouds of smoke visible during this and the following day behind us, showed that the fire was still burning furiously."[8] It was neither the first nor the last accidental fire in Alberta.

1866 Rocky Mountain House, Henry Stelfox

Indian Agent Henry Stelfox describes the fire that originated from the charred remains of the Hudson Bay Fort (Rocky Mountain House) near the confluence of the Clearwater and North Saskatchewan River and burned west to the mountains, north to Jasper and east to Cooking Lake. Chief Jim O'Chiese had no choice but to move his band out of the region. They traveled to the Peace River country, Saskatchewan and Manitoba and eventually returned through Beaver Indian territory to the Baptiste and Nordegg Rivers.[9]

1880s, 1890s Rocky Mountain and Foothills Fires

The large and destructive fires along the East Slopes of the Rockies from the U.S. border to north of Grande Prairie during the 1880s and 1890s are legendary. Recent studies have provided more details about the years and sizes of burns, but more work is needed to understand what was happening during those two decades. For example, the age class data compiled by staff of the Foothills Model Forest suggests that 36 per cent of the burnable area of Jasper National Park and the Weldwood Forest Management Agreement area burned between 1887 and 1896, an average of 3.6 per cent a year. As much as 40 per cent may have burned during the two decades of the 1880s and 1890s. Even over 20 years, that represents an average of two per cent of the area burned per year.

Age class maps of the area south of Grande Prairie clearly show the influence of burns during that period. Albert Hanson, retired lumberman, gave an interesting insight when he told about Mrs. Campbell, an elderly aboriginal woman, who lived through those fire years. She explained to him that by 1889 it had quit raining, that it had not rained for three years. When it quit raining, the forest started burning. "All this country was all heavy forest. The fire burnt for three years. It would start up in the spring, after winter snow had left, igniting from smouldering muskegs, and continued to burn all summer. When winter came it slowed down, the muskeg still burning and smoking, then the following year start up again. So it burned constantly from 1889 to 1890 and 1891. The way the forest burned ... was that first the needles burnt off, then the ground started burning, burnt the roots, the trees fell down and then the trunks burnt up. By the time it was done burning, there was nothing left."[10] To survive the fire, they spent the three days of the worst of the fire in Lake Saskatoon - holding their children above their heads, their materials and supplies buried along the shore. They could come ashore at night, when the fires went down, but went back in the lake in the daytime.

There is also evidence of major forest fires during those two decades in Banff National Park, Clearwater and Bow River Forests, and the C5 Forest Management Unit in the Oldman River area of the Crowsnest Forest.

Fire killed timber and debris, Crowsnest Forest, Rocky Mountains Forest Reserve, 1910
Dominion Forestry Branch, AFHPC

1894 Great Northern Alberta Fire, J.G. MacGregor

The great fire of 1894 followed a summer drought that persisted into late fall and affected much of northern Alberta. As the geese gathered to fly south a campfire along the Athabasca Trail blew into tinder dry fuel and developed into a giant forest fire eventually spreading into eastern Saskatchewan. The fire burned from the Smoky River area to the Saskatchewan River through over-mature forest stands and then across prairie grasslands until the snow finally arrived.[11] Similar fires had occurred along the east slopes of the Rockies between the 1840s and the 1880s. By the turn of the century, travel in the extensive area of burned forests was difficult and required many hours of laborious trail clearing and subsequent maintenance.

1898 McLeod River Burns, James McEvoy, Dominion Land Surveyor

James McEvoy, Dominion Land Surveyor, observed, "beyond the McLeod River most of the valuable timber has been destroyed by fire."[12]

1903 Upper Brazeau River Burns, Stutfield and Collie

Mountain climbing adventurers Hugh Stutfield and Norman Collie traveling along the upper Brazeau River stated: "The frequent, and often wanton destruction of forests in the Canadian Rockies by fire is deplorable. Sometimes they are set alight on purpose by prospectors in order to clear the ground, but nine times out of ten the fires are the result of sheer carelessness. There are severe penalties attached to the offence, but, as evidence is very difficult to obtain, much fine timber is wasted, and trails rendered almost useless for years to come."[13]

1908 Jasper Area Burns, Mary Schaffer

Mary Schaffer, traveling near Jasper, noted the effect of forest fires in her diary: "for the next two hours the trail led us down a fire swept valley where chopping was incessant and heavy."[14]

1909 Fires of Spring: Burning in Jasper Area, Edward Moberly

In contrast to threats posed by wildfires, First Nations and Métis people had long used controlled burns on selected sites. Edward Moberly offers another perspective about the use of fire by his family and others in the Jasper area until their eviction in 1910. "Then in the spring that's the first thing everybody does is burn the meadows. If there's one [important] thing, it's going to need help. Well, everyone goes and helps [with the burn]. But burn [at the time and place] when he wants to burn. This way the meadow doesn't grow in - willows and things doesn't come in - it's always the same size and it's always clean."[15]

1910 "Big Blowup" Burns in the Southern Rockies

Countess Alice Grey, wife of Earl Grey, the Governor General of Canada, was camped near the Gap on the Oldman River and reported "terrific fires" in the vicinity. In the same year fires swept through north of Coleman and Blairmore and on past Lille. On August 1, 1910, while miners and mill hands fought fire north of Blairmore the local hotel keepers sent out a few kegs of free beer which resulted in fights and a stabbing and a subsequent three year jail sentence.[16] Meanwhile, other fires were burning in the Elbow, Sheep and Highwood valleys that summer. The notorious "Big Blowup of 1910" in western Montana and northern Idaho culminated on August 20-21 when the many widespread persistent fires blew up under gale-force winds. In total about 1.2 million hectares burned and 85 lives were lost in the United States. In Alberta an early snowstorm August 22 or 23 stopped the 1910 fires.

1911 Dominion Forestry Fire Prevention among aboriginal people

The Dominion Forestry fire prevention program included aboriginal people. In northern Manitoba, the region's population was mostly Indians, freighters, and canoemen for the Hudson's Bay Company. To interest the Indians in fire prevention, the rangers distributed a

Aboriginal people take pledge to aid in preventing forest fires
Dominion Forestry Branch, AFHPC

small fire ranger's badge and pledge, with Indian lettering. The pledge translated as follows:

"We Indians appreciate the work the government is doing to prevent forest fires in our district. We pledge ourselves to do all we can to help. We promise to put out our own campfires every time before leaving camp. We accept the badges given by the government as a pledge."

Chief Fire Ranger A. McLean reported on the reception of the pledges by the Indians: "With few exceptions the Indians have all been very careful this summer. Not a single report of fire has been received by me. We had no trouble getting the Indians to sign the enclosed pledges. Before I would let them sign I made sure they thoroughly understood the nature of their pledge. One and all they were much pleased with the badges given them. They pinned them in all sorts of places on their clothing, where each one's fancy thought the most conspicuous place. One man would not take his badge until he had first washed himself and changed his shirt, and then the badge was used as a collar button or brooch. Hats and shirts were the favourite spots for adornment but not a few - after carefully polishing them - would fold them in cloth to keep them for some special occasion. A great per centage of them feel they are thus, having received the badge from the government, constituted minor chiefs and guardians of the forest."[17] Alberta photos show similar ceremonies taking place at Fort McKay and Hay River, suggesting the pledge quickly became common practice throughout the Dominion Forests.

1915 Smoky River Fire, Caroline Hineman

Caroline Hineman, traveling south from Grande Cache and guided by Curly Philips, observed a fire in progress. "In the upper reaches of the Smoky River a fire had burned down the right bank of the Smoky River from Calumet Creek for a distance of seven miles in a band two miles wide. Jackson the Chief Warden stated it had been started by four Indians, now in jail, who had been hunting goats in Moose Pass."[18]

1919 Lac La Biche and the Great Fire of 1919

The Village of Lac La Biche burned on May 19, 1919, destroying all but three buildings - the railway station, church and railway tycoon J.D. McArthur's hotel. The hotel served as a post-fire nursing station. Since the telegraph lines also burned, word could not be sent to Edmonton until the next day when a locomotive made it across the burn enabling the conductor to make a call. The Province and Red Cross quickly responded with a relief train from Edmonton. This seems to have been the first recorded wildland-urban interface fire in Alberta. The

Crew lays first round of logs for the Moberly cabin, a ranger station on the Lower Trail north of Entrance. Note fire-damaged forest in background, 1939
Alberta Government, AFHPC

Livingstone (Gap) Ranger Station, aside the Livingstone River, in the foreground. Background hills and mountain slopes show snags from the 1919 summer burns. The photo was taken by National Air Board (later RCAF) pilot A.A. (Ack Ack) Leitch who was posted at the High River DFB airbase. He enjoyed getting out with the rangers on his days off. Ranger in charge of the District was Harry Nash. Mid 1920s
Peter Murphy photo and story

smoke was so thick that residents thought there was an eclipse of the sun. Children were loaded into wagons and left in the shallows of the lake to keep them safe.

The Annual Report of the Canada Department of the Interior for the year 1919 referred briefly to a large fire in the Battleford Fire Ranging District in Saskatchewan that burned over 2,740,000 acres (1.1 million hectares). However, there were actually several fires, many of which burned together, during the last two weeks in May. They covered an estimated area extending 450 kilometres west to east from near Boyle, Alberta to north of Prince Albert, Saskatchewan. This fire was up to 150 kilometres wide. The area within this roughly determined perimeter was 3.1 million hectares; actual area burned may have been well over two million hectares. The fire burned out McArthur's logging camp south of Lac La Biche and the extensive stands of spruce east across the Saskatchewan border. There were many close calls among settlers, and lots of stories about saving their homes by putting children on the roofs with wet sacks to put out falling embers while the parents cleared lines from which to backfire. A band of 21 Cree camped at Big Island Lake was overrun. Despite taking refuge in the lake, 12 died and of the nine survivors many were badly scarred. Two Cree hunters also died on Wolf Mountain.

Along the eastern slopes in the summer of 1919, many fires were reported in the Crowsnest Forest, the larger ones in the Porcupine Hills and Racehorse Creek. There are also references to two big fires in the Bow River, Clearwater and Brazeau Forests.

1931 Rocky Mountain House Fire, Eric Huestis

In 1931, Eric Huestis was Supervisor of the Clearwater Forest at Rocky Mountain House. He described this fire in February of 1931:

"The fire started about nine miles west of Rocky pushed by an 80 mile-per-hour wind - it moved very fast, headed straight for the town. We gathered up every able-bodied man, we told the women to be ready to evacuate the town and we headed down to the riverbank thinking that the fire wouldn't jump the Saskatchewan. The fire came down to within about a mile of

Youth Forestry Training Program enrolees at Highwood Camp learn how to use surveying equipment for road construction purposes
Dominion Forestry Branch, AFHPC

the river. I was tearing up and down the town bringing people I found down with the Model T. The last guy I found on the street was an Anglican Minister - so I picked him up. I didn't know what I'd do with him once I got down there, but he was the key man in that whole fire. What happened was that as we drove down and got a few of the men across the river, he said, My God, what are you going to do? I said - Well sir, as far as I'm concerned the only thing we can do is start praying. So I stopped the car, he got out on his knees in the ditch and started to pray. Fifteen minutes later the wind came in from the north and it started snowing. I don't know to this day whether this was on the way, or whether the praying did it. But he rode high, wide and handsome for several years after that. It saved the town and saved the bacon. This was the middle of February so anything can happen in this country, when you start getting into fires. Don't think you're out of the woods because anything can happen at anytime, as long as the conditions are right - and all the equipment in the world won't stop it. The Lord has to help you, always keep that in mind."[19]

1933 15 cents an hour and grub: Grande Prairie Trail Fire

The 1930s became known as the "Dirty Thirties" because of the Great Depression and, on the Prairies, because of the extended drought that led to soil drifting and forest fires. As former firefighter Rod Gregg recounted: "The summer of 1933 was hot and dry: by mid August the air was heavy with the smell of burning conifers, the sun a dark red orb. 1933, too, witnessed the very depth of the great depression - in the community only the very few railway men who possessed what they called 'whiskers,' and a small number of government agents continued in regular employment. A few established merchants, on a cash only basis, managed to keep afloat. Meanwhile the greatest number of us picked up what few jobs we could find - anything at all that would generate a few cents." Forestry was offering 15 cents an hour and grub so he was quick to volunteer to fight a fire north of Edson beyond the Athabasca along the old Grande Prairie Trail. Reporting early the next morning, they walked for about four days with a team and wagon to carry supplies. They crossed the Athabasca in a rowboat, floated the wagon and swam the horses. When they finally reached the fire they found it to be "too large for our small party to do anything other than return and report."[20]

This story is similar to the situation that Eric Huestis was told about by the ranger at Carrot Creek at the time. His district ran south as far as Rocky Mountain House. As he explained to Huestis: "I had a team of horses and a democrat. I would get word by CN telegram that there was a fire down at Breton or Winfield. I would hook up the team and away we'd go and four days later I'd arrive. Either the fire was all over hells' half acre or it was out. So then I'd get on the phone or go to the nearest town and phone in and they would tell me there's another fire up near Edson. So I'd head back and the same thing would happen, either it was all over hells' half acre or it was out, depending on conditions." Huestis commented: "It was a hopeless proposition to do anything as far as fire protection was concerned, with only about a dozen rangers in the whole north country."[21]

1934 Castle River Fire, Joe Kovach, Ranger

The 1934 Castle River fire, as Huestis explained, came down the Castle River on both sides of the main branch. "We had 160 men in there fighting it. All of a sudden a wind took off and I ended up with the cook and packer and myself outside the fire with 160 men inside but no way of getting them out! The packer had a brain wave - he had the cook get a bunch of supplies and he packed them up as close to the fire edge as possible. There was only one trail in the valley. We dug a hole in the middle of the trail and covered it with canvas, and then earth. And then when the boys followed the fire down they found the grub! We finally got them out through the river at night when the fire died down."[22]

1936 Highwood River Fire and Phillips Fire in B.C.

The Phillips fire in British Columbia escaped initial attack and headed north up Elk River while the B.C. Forest Service held it at the lower end. The fire headed north into the Elk River valley. Surrounded by mountains, it was too hot and dangerous for anyone to get in front, so it was abandoned on the B.C. side. It was observed from the Alberta side in early July and as the month progressed the fire made major runs in the Elk Valley and eventually spotted into the Highwood Valley in Alberta. [23]

Author R.M. Patterson writes that in 1936, the Highwood River went dry for a short distance above the Sentinel (Highwood) Ranger Station and the north fork of Sheep Creek was dry where it crossed the Calgary Trail. The summer of 1936 was supposed to have been the driest since 1886; even though there had been years in between that were so wet one could bog a wagon in the middle of a rocky flat. "You could see that fire roaring up the Elk Valley at a most ungodly speed and it would not be too long that an up-draught will fling glowing coals over the Divide and into the green forests of the Highwood." [24] On July 23 at the Continental Divide all hell broke loose.

Alberta rangers could see the smoke billowing over the mountains so they sent two crews from the Highwood with packhorses and hand tools; one each to two high mountain passes on the Continental Divide. Their objective was to build a fireline across the passes to stop any threat to Alberta. Late one afternoon the fire blew up under tremendous winds, burned to treeline and spotted well over the mountains to start fires along the Highwood and Cataract Creek behind them. The crews quickly packed up and headed back that night to try to hold the fire from spreading east across the rivers. At Lookout Ridge the upper Highwood valley was a terrible sight to see. All of the Divide and the Misty Range were covered in thick smoke. In one day the fire traveled 15 miles from Weary Creek in B.C. to Sheep Creek in Alberta. The AFS put 100 men on the Highwood fire. Patterson wrote: "One bunch of men got cut off the trail to Fording Pass by a switch in wind that sent the fire racing through young pine in a sudden flurry of flame; the bunch scrambled and hacked their way through old forest down to Lost Creek and crossed the dry Highwood River to reach safety in the valley bottom. Another bunch of men and horses scrambled for dear life up Storm and Mist Creeks leaving their camp in flames. Ashes were falling in Calgary and High River." [25]

Saskatchewan River fire burned into mature spruce, Banff National Park, 1940
Dominion Forestry Branch, AFHPC

Fish caught by Ranger Dexter Champion and other firefighters on the Elk River fire, Bow River Forest, Rocky Mountains Forest Reserve, 1936
Jay Champion

In the meantime a third crew went over Elk Pass from the Kananaskis side and built hand line to try to stop the fire on the B.C. side. They lost line in one blowup but stopped it again about a kilometre north and held it. Most of the Highwood Valley west of the river burned out, and the Lineham Lumber Company closed its sawmill at High River. [26]

1936 Galatea Fire, Kananaskis Valley

While the Phillips and Highwood fires were burning, lightning started another fire near the head of Galatea Creek. Its smoke was not noticed among the heavy billows generated by the fires in the Elk and Highwood Rivers. When Ranger Griffiths first assessed it: "... due to its location ... high up on Mount Galatea (he) did not consider it serious..." However, he sent in a crew the next day and on August 5, 25 men arrived at the Boundary Cabin and headed for Galatea Creek. Strong southwest winds created extreme fire conditions and the fire spread steadily until August 9 when it literally exploded at about 3 pm. The crew took refuge in the Kananaskis River as the fire headed north and east. "Had the wind continued blowing on Monday from the same direction and with the same intensity as on Sunday, the fire would most certainly not have stopped south of the Bow River." Fortunately the wind did change and it stopped at Mount Lorette. This burn was later salvage logged for mine props for the underground coalmines at Canmore and, during World War II, was logged by German prisoners-of-war. [27] The Kananaskis Village and the Kananaskis golf course are located on areas burned in this fire.

Long climb to fire tower cupola, 1956
Alberta Government, AFHPC

The *Calgary Herald*, in a 1937 article, summarized the conditions for Ranger Dexter Champion's wife Louise and son Jay. Champion was stationed, with his family, at the Kananaskis Lakes Ranger Station, near Boulton Creek. "Last summer, when fires were raging in the mountains, the blinding smoke, falling ashes and burning cinders caused the ranger's family to be moved to safety by forestry truck, to Wasootch Camp. Later, removal to the Dominion headquarters camp near Seebe was necessary. When the fire was subdued and the Lakes district known to have escaped the destroyer, the little party returned home."

1936 West Castle Fire, Eric Huestis and Joe Kovach

Eric Huestis explained that the B.C. Forest Service told him they wouldn't fight the fire, so he sent a crew of "damn good miners" in from Alberta to build fireguards on the B.C. side. "We had our crew on our side part way up a mountain where the river ran. And all of a sudden a hell of a wind blew the thing completely over the top and the crew was inside. The fire jumped right over them!" [28]

Ranger Joe Kovach and 56 men were surrounded by fire and retreated to a high cliff where they were busy extinguishing sparks that ignited their clothing as the fire overran their position and spread down the valley. They then moved up to a glacier and spent a cold but safe night with a second crew who had food with them. Huestis said he had arranged for a plane to fly over the burn for over an hour before he finally spotted where they had camped. "It was in an open place, and the table was still there and I could see a mound of earth." He said at the time: "Those miners have buried the whole camp, and maybe they will get by." Then he saw where they had escaped into the rocks. He learned later that as the fire came over they had to tie up four of the younger generation with ropes to keep them from running out. The fire spread on a 25-mile front to the Beaver Mines area where it finally ran into grasslands. One hundred and eighty men spent the summer on the fire earning 15 cents an hour; horse packers working on the fire earned

Looking southwest towards Wedge Mountain from the Boundary Cabin, Kananaskis District. Photo shows where the 1936 Galatea fire burned into the Evans Thomas drainage. November, 1938
Alberta Government, AFHPC

Fire in Alberta 355

50 cents a day. The fire started on July 17 and was declared out on November 18.₂₉

1941 Spring Fires, Rocky Mountain House

During the winter of 1940-41 there was little snow in the Rocky Mountain House District and spring came early. J.R.H. Hall, Acting Forest Superintendent, Rocky Mountain House prepared a story for the Pulp and Paper Magazine of Canada which stated: "The morning of April 26th broke clear, fairly calm, and very warm. Lookouts reported the odd smoke from settlers brush burning operations in the settlement away east of the forested area. At about 11 AM a high wind started to blow from the southeast and several fires, probably smoldering spots from settlers early April clearing fires, sprang up and were soon fanned into roaring infernos headed for the timbered areas. The annual battle was on." In the story Hall summarizes the spring events, where "… in less than an hour one of these fires had traveled three miles and had spread to the north and south. Near noon the spearhead of the fire hit the McTighe lumber camp and by this time the blaze was so hot small conifer trees burned like tinder and shrubs and grass hundreds of feet ahead of the main fire broke into flames as burning bark and other debris hit the ground." This careless burning of brush piles by settlers resulted in the death of an old age pensioner who died of smoke inhalation, the destruction of three sawmills and the loss of large quantities of logs and lumber. "… a beautiful forest was turned into an area covered with charred and blackened dead trees, driving out wildlife, and taking away the living of a large number of woodsmen."₃₀

1944 Bad Heart River Fire, Western Producer Newspaper

Firefighting is fraught with hazards. The fire hazard was very high in the spring of 1944 as a result of low snowfall the previous winter and

(L to R): Rangers Harold Dunlop, Harold Enfield, Ben Shantz and Bob Diesel at firefighter training camp, Whitecourt Forest, spring, 1966
Alberta Government, AFHPC

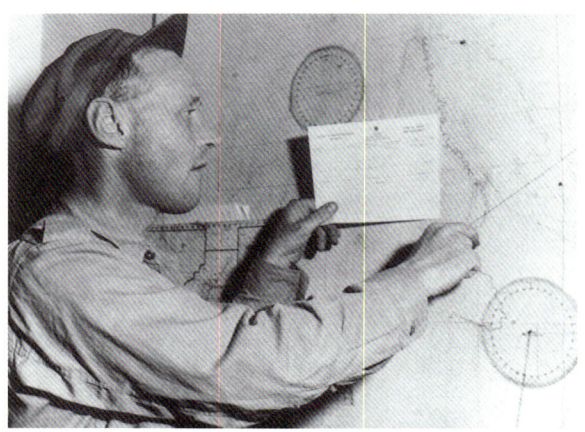

Jack MacGregor plotting smokes, Whitecourt Division, 1956
Alberta Government, AFHPC

A powerful prevention message: Forest Officer Ernie Stroebel at the Calling Lake Airstrip cabin, Lac La Biche Division, 1960s
Alberta Government, AFHPC

warm, dry and windy weather. This was during World War II when manpower and resources were scarce. As a result, many fires burned out of control in the Peace River area. As reported in the *Western Producer*, a ranger called on Stanley Krzyczowski and sent him to a sawmill site with a number of others. He was age 20 at the time. There were about 12 men on the fire, including the owner of the sawmill, who was in charge. They focused on trying to save the mill buildings from embers falling from the approaching fire.

Very strong winds suddenly caused the fire to 'blow up', burning and spreading rapidly as a crown fire. It quickly became apparent that the

Ranger Sam Sinclair (L) and Fire Control Officer Ben Shantz at native firefighter training, Alexis, Whitecourt Forest, April, 1965
Alberta Government, AFHPC

Alexis firefighter training graduating class, Glenevis, Whitecourt Forest, April, 1965
Back Row (L to R): Sam Sinclair (AFS), David Bird, Ralph Adams, Jacob Rain, Alvin Bearhead, Mike House, Paul Kootenay Sr., Peter Bird, Paul Kootenay Jr., Mike Burke (AFS), Philip Cardinal, Lazarus Rain, Lawrence Rabbit, Art Peter (AFS). Front Row: Ben Shantz (AFS), Peter Rain, Steve Bearhead, Sam Bearhead, Frank Bearhead, William Rabbit, Bob Diesel (AFS), Harold Enfield (AFS)
Alberta Government, AFHPC

mill could not be saved. After turning horses loose from the barn, the mill owner led what he must have believed was the entire crew to a beaver dam on the Bad Heart River where they survived, although all were affected by the heavy smoke. Unfortunately, in the chaos of events, reduced visibility and noise, Krzyczowski and two others found themselves alone with fire burning all around them. They tried to escape through the fire, but two were overcome by the dense smoke and perished. Stanley Krzyczowski made it to the Bad Heart River getting badly burned on the way, then followed it upstream to Webster, about eight miles, where he was taken to the Grande Prairie hospital. He spent three painful months in hospital recovering from severe burns to his hands, face and feet.[31]

1949 Upper Hay River and North Peace Fires

The fire record for 1949 shows that fire starts were above average with over a half-million hectares burned. One of the larger and more visible of those was one that burned along the Little Smoky River valley between Triangle and Valleyview and across the old gravel highway that linked Edmonton to Grande Prairie and the Alaska Highway. However, in 2004 Sustainable Resource Development forester Darrell Kentner followed up on stories about a forest fire in the Upper Hay River area around 1949 that had "burned all the way to Saskatchewan." Since that country was north of the protection area at that time, there were no fire reports. Kentner pieced together evidence of the fire events using the Phase I inventory maps that were based on the 1950 aerial photography. Candidate areas were digitized to create a new composite map and ages were checked with contemporary age-class data as a guide. The map presents a reasoned estimate of the area that appeared to have burned in 1949. Rather than a single fire, there was a dispersed mix of large and smaller fires extending at least between the British Columbia border and Wood Buffalo National Park. In total there may have been around 560,000 hectares burned, which would have doubled the area of record for 1949. Kentner also noted that had the communities of

Planning session during the spring, 1968, Lac La Biche fires.
(L to R): Len Allen, Bill Bereksa, Bert Varty (seated), and an Ontario Lands and Forests employee
Alberta Government, AFHPC

Goose Mountain tower site, overrun by fire, 1998. The site burned over at 0200 hours May 4, 1998
Kent McDonald

High Level, Bushe River, Meander River, and Paddle Prairie existed at the time they would have witnessed wildland-urban interface fires.

1950 Chinchaga River Fire

This fire started June 1 about 18 miles northwest of Fort St. John, British Columbia. Documents show that the fire started in logging slash from an escaped smudge fire from a seismic crews' horse camp. When discovered, the fire was estimated at almost 100 acres. This was one of many fires burning at the time, and was in a zone where fire permits were not needed. So, in a triage-type decision, no direct action was taken by the British Columbia government. It flared up periodically over the following months as it headed northeast into the Chinchaga River valley in Alberta.

Keg River Ranger Frank LaFoy continued to be concerned and requested permission to attack the fire during its quiet spells, but because the fire lay outside the protected area he was turned down. It blew up on September 4-5 then again on September 20-21. Smoke from this and other fires in the region drifted over Ontario, the northeastern United States, Atlantic provinces, England and Holland before dispersing over Germany. When the Chinchaga River fire was finally stopped by rain and snow at the end of October it had spread 150 miles and burned 3.5

Talbot Lake fire Overhead Team, 1981
(L to R): Terry Van Nest, Dave Bartesko, Bob Hilbert, Darren Fanton, Frank Lewis (Fire Boss), Lyall Gill, Ed Dechant
Terry Van Nest

million acres. Of the final area, about 70 per cent was in Alberta, and the balance in British Columbia. This became the first fire event to effectively change Alberta government fire policy. In 1951 the protected area, in which fires could be attacked, was extended from the mid-province line north of Slave Lake to include the entire forested area. The Chinchaga fire was the first time a backfire was successfully used in the province to save a town, mainly with the perseverance of LaFoy and the townsfolk of Keg River. Frank LaFoy later transferred to the Fish & Wildlife Division, mainly over his dissatisfaction with the AFS's inattention to this fire and saddened that so much of his district had burned.[32]

1956 Gregg River Fire, Robert J. Adams

The 1956 Gregg River fire and two others burned approximately 61,000 acres of the North Western Pulp and Power Ltd. FMA (almost 3% of the area) before the mill was operational. This event challenged foresters to consider fire in subsequent forest management planning. The fire started in early July and was described by Bob Adams in his best-selling book *Beyond the Stump Farm*. Of particular note was the episode about the muskeg coffin where Bob and his Métis companion Chief were dropped off at the fire by the forest ranger and were directed to set up a base camp while he rounded up additional resources. The fire blew up and overran their camp and they debated their fate. "I heard once that you can escape a fire by burying yourself in a muskeg… the Chief was already heading for the muskeg with a Pulaski and a shovel."[33] They survived by digging a trench deep into the muskeg and pulling the wet moss layer over them until the fire front passed. After three days of walking in the burn they arrived at the Gregg Cabin and reported their location on the forestry radio.

Company response was immediate and critical. Huestis mediated the ensuing discussions that resulted in greatly increased government support for fire control, as well as greater government-industry cooperation in fire management. This was the second major fire event to directly influence government fire policy.

1968 Seven Days in May (May 18-25 Fires)

In May of 1968 a blocking high-pressure area developed over Saskatchewan that brought warm, dry air into Alberta along with persistent SE winds. Windrow burning during the winter months near Rocky Mountain House, Whitecourt, Slave Lake, Athabasca and Lac La Biche resulted in many holdover fires and in the third week of May strong gusty winds pushed over 30 fires into the

Associated Helicopters pilot Harvey Trace training firefighters on safe operation in and around helicopter, Fox Creek, Whitecourt Forest, 1967
Alberta Government, AFHPC

Aerial view of the 1972 Judy Creek fire (DW5-05-72) at Swan Hills
Alberta Government, AFHPC

Okanagan Helicopters Bell 212 releases load of water from belly tank
Alberta Government, AFHPC

Fire in Alberta

Conair DC-6 airtanker drops fire retardant slurry
Alberta Government, AFHPC

Map showing 1968 Vega fire
Jamie Badcock

protection zone. "The fires established two new records – the greatest number of outbreaks ever recorded so early in the season, and the largest acreage ever incinerated so early in the season. The fires burned over 900,000 acres in total, with timber loss estimated to exceed two billion board feet. One sawmill was destroyed and several timber quotas were severely disrupted. Suppression costs ran over $5 million. Fortunately, no lives were lost and property loss was not high, although several settled areas had to be evacuated during the peak of the fires."[34]

On May 23 the Vega fire, which was driven by low-level jet winds, burned from the Athabasca River to just south of Lesser Slave Lake in 10 hours, a distance of 60 km. "Fire intensity was extreme with spruce trees, 24 inches in diameter, snapped off by the force of fire wind turbulence at peak periods."[35] The fire was influenced by drought conditions, late spring green-up, and a winter with very little snow. Following the run of May 23, however, moisture arrived and saved the town of Slave Lake from burning. Interestingly, 33 years later on May 23, 2001, the Chisholm fire began and, under very similar conditions, ran towards Lesser Slave Lake. The Chisholm fire slowed up just south of the town of Slave Lake, stalled once again by a cold front arriving with moisture. Both fires were precedent-setting events in Alberta. As of 2004, the Vega fire spread rate and Chisholm fire intensity values were the highest on record in the province. The impact of the Chisholm fire, however, was significantly greater than the 1968 Vega fire as a result of residential and industrial development in the area over the intervening 33-year period.

1972, 1981, 1998 Swan Hills Fires

The town of Swan Hills was evacuated three times in less than 30 years because of the threat of wildfires. The first was in 1972 when the Judy Creek fire ran just south of the town and across Highway 33. As bulldozers and crews pulled back to the highway the fire spotted across in two places and despite immediate action a fire front established on the east side and continued to run.

The Moosehorn fire forced a second evacuation in 1981 and as the fire made a run for the Swan River a last ditch backfire operation was implemented. Two aerial ignition torches fired a line just over the crest of the river valley and with up-slope conditions and in-draft from the fire front, spotting was eliminated and the fire run contained. A second line was then ignited adjacent to the river edge and all fuel on the slope was eliminated. The resulting up-drafts created a perfect tunnel as descending smoke on the south side of the river wrapped back to the north side allowing helicopters to patrol through the tunnel, confirming that the fire had remained on the north side of the river.

The third evacuation of the town of Swan Hills in 1998 was again a close call as a fire front to the west of Goose Mountain overran Goose Mountain Tower in the early morning of May 4. The fire impacted the forest industry, oil and gas operations, power distribution and more, and although Whitecourt was not evacuated, businesses were impacted for several days. [36]

1980 Cold Lake Air Weapons Range Fire (DND-4-80)

In 1980, most of east-central Alberta was free of snow by the first of April, adding to the impact of low precipitation levels from the previous summer and fall. The spring became the forerunner of three consecutive record fire seasons in Alberta that resulted in new fire training and policy manuals. On May 2 a fire (DND-4-80) in the Cold Lake Air Weapons Range ran 18 kilometres in five hours and eventually burned 137,313 hectares in Alberta and 40,500 hectares in Saskatchewan. The fire was documented as a case study of 'blow-up' fire conditions that included observations of lower and upper atmosphere temperature and wind profiles. This analysis of the influence of jet winds contributing to extreme fire behaviour has alerted fire managers to the importance of including upper atmosphere monitoring as standard practice in fire weather forecasts. [37] The "flying drip torch" was used operationally for the first time in Alberta on the Department of National Defence fire. The backfire tactic is now routinely applied on most large fires. As the DND fire moved into Saskatchewan a new fire start in the Peace River Forest prompted a shift in fire suppression priorities and crews and overhead

Air Spray B-26 airtanker makes a drop
Alberta Government, AFHPC

Skidder burned in Chisholm fire, 2001
Alberta Government, AFHPC

Fire camp south-east of Mariana Lake, 1995
Alberta Government, AFHPC

Fire in Alberta 361

personnel were moved to fight the Notikewin fire. The Winefred airstrip was a busy area as 600 army firefighters set up camp to work on the DND fire while Alberta crews packed their camps to fly to Peace River where timber values were threatened. Ironically, as the organization geared up on the Notikewin fire, Ken South was assigned as Division Boss on the north half of the fire and Jim North was Division Boss on the south half of the fire. This understandably caused confusion with helicopter pilots ending up at the wrong camp on more than one occasion.

1981 Keane Tower Fire

Wind, wind, wind - the common denominator of the large fires in Alberta. Either high velocity or change of direction has driven the most innocent fire into an uncontrollable inferno. In 1981, following a slow fire season, winds in late August and early September resulted in the largest recorded fire in Alberta. The Keane Tower fire burned 420,000 hectares in Alberta and, spreading east into Saskatchewan, burned at least another 100,000 hectares under the influence of strong, cold winds. The total area of 1.36 million hectares burned in 1981 is still an Alberta record and, interestingly, occurred late in the year when a record was certainly not predicted. The area of the larger Chinchaga fire of 1950 (1.4 million hectares) was not recorded at the time; the area was calculated from maps in 1986.

1982 120 fire starts in one day

After the busy 1980 and 1981 fire seasons, the spring of 1982 was uneventful and fire managers were relieved as forest green-up progressed. On June 12, however, there were 120 new fire starts in the province, mostly in the Lac La Biche, Slave Lake and Peace River Forests. Escaped fires were priorized and for the next month the majority of AFS staff fought fire with no break in the weather. The impact of the three consecutive fire seasons was significant and resulted in several enhancements, among them helitack crews, a presuppression preparedness system and an advanced training initiative that is the foundation of today's fire management program.

Alberta government CL-215 aircraft dropping load of water and foam
Alberta Government, AFHPC

Chisholm Fire Boss Rob Thorburn presented with a thank-you cake from residents of the Chisholm, Flatbush and Fawcett area, 2001
Alberta Government, AFHPC

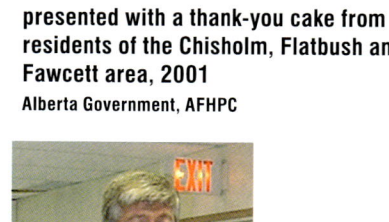
Chisholm Fire Boss Rob Thorburn, 2001
Alberta Government, AFHPC

1990 A day of fire starts

On August 6, 1990 an upper ridge breakdown occurred during severe burning conditions. A total of 114 fire starts was recorded, mainly in the Lac La Biche and Slave Lake Forests. Over the next few days, eight E class fires (greater than 200 hectares in size) started, some as a result of the August 6th fire starts, and others as a result of holdover and new lightning fire starts. The most notable were two fires in the Lac La Biche Forest, the May Tower fire (DL4-40-90), in the Wandering River District, that burned almost 10,000 hectares; and the Round Hill fire (DL2-66-90), in the Beaver Lake District, that burned almost 12,000 hectares.

Virginia Hills Firefighter Training Camp, Whitecourt Wildfire Management Area, 2001. Tents are used for classroom instruction
Alberta Government, AFHPC

1995 Foggy Mountain and Mariana Lake Fires

Although these two fires were over 400 kilometres apart, the fire behaviour characteristics were similar. They signaled the beginning of the Buildup Index breaking the 100-level on a routine basis in Alberta. Both fires started during the morning of May 28, 1995. There were many challenges for the fire management teams over the next month. The high Buildup Index, reflecting deep drying of forest fuels, meant that considerable additional fuel was made available to the fire fronts. Both fires made major runs as cold fronts moved through northern Alberta along with an associated low level jet wind – causing instant changes in wind direction and much greater wind speeds. The results were a clear demonstration of how large-scale weather systems can profoundly affect fire behaviour. Early on in both fires the frontal winds created simultaneous runs, first on the Foggy Mountain fire and, just hours later, on the Mariana Lake fire. Of particular concern at Mariana Lake was the constant threat to the Algar Tower base camp and the extended closure of Highway 63, the main supply route in and out of Fort McMurray.

Highway 63 leads into the Mariana Lake inferno, 1995
Rick Arthur

The base camp ultimately burned after a timely evacuation ahead of a predicted blow-up at the front of the fire. The last picture taken before the fire overran Algar tower site appeared in magazines across North America for the rest of the summer. In recent years Buildup Indexes have approached 200.

1998 Mitsue, Virginia Hills and Tolko-High Prairie Fires

The years 1996 and 1997 experienced above-average precipitation that resulted in a lush growth of grass and other quick-drying fuels. That was followed by low snow pack and a dry spring in 1998. Early in May these cured, easily ignitable fuels were dry and plentiful in central Alberta. Following a weekend of unrelenting winds, fire starts in both the protection and non-protection zones quickly spread unchecked under the influence of erratic and gusty winds. The Mitsue fire near Lesser Slave Lake threatened the Mitsue industrial properties and eventually burned into Slave Lake Pulp Corporation's log decks. The Virginia Hills fire northwest of Whitecourt burned over Goose Mountain Tower and ran towards Swan Hills through the Blue Ridge and Millar Western Forest Management Agreement (FMA) areas. Forested areas and regenerated blocks were burned on both FMAs, which significantly impacted the forest age class and future harvest sequencing options. The other major concurrent fire was the Tolko-High Prairie fire that started along the railway line to McLennan and ran about five kilometres east before a cold front drove it south into the Tolko mill yard. It burned the log decks, full with the winter haul, and it was only through heroic efforts that the mill itself was saved.

Bruce MacGregor's Alberta Incident Command Team, Montana, 2000

Standing (L to R): Dave Heatherington, Andrew McWhirter, Margarete Hee, Marc Gamache, Brian Orum, Bill Cooper, Paul St. John, Wayne Werstiuk, Mike Templeton (white shirt), Billy Tchir, Don Podlubny, Bill Bereska, Brad Bailey, Wayne Bowles, John Belanger, Bob Yowney, Frank Kachuk, Paula Gray, Lori Vance, Heidi Ahlefeldt, Tim Klein, Norm Quilichini, Marc Freedman, Shawn Zwerzinski, Joe Wagenfehr (Montana Liaison). Kneeling: Evert Smith, Doug Smith, Chad Morrison, Butch Shenfield, Sheri Larsback, Bruce MacGregor, Hugh Boyd, Tara Lee Manwaring, Mark Donovan, Owen Spencer, Kurt Frederick, Chantel Ritcey
Bruce MacGregor

2001 Chisholm Fire

This year was a high-profile fire season as a result of unprecedented community impact and homes being lost in fast-spreading fire fronts. The number of structures lost in central Alberta in May, 2001, totalled 129, of which 22 were homes. There were a total of 12 homes and 58 outbuildings destroyed outside the Forest Protection Area, with 524 people being evacuated. Within the Forest Protection Area, 10 homes and 49 outbuildings were destroyed and 84 people evacuated. The 10 homes were burned in the hamlet of Chisholm during the May 28 run of

Lost Creek fire Incident Commander Andy Gesner (L) with Premier Ralph Klein, Blairmore, Southern Rockies Wildfire Management Area, 2003
Alberta Government, AFHPC

Bruce MacGregor thanking the Salish and Kootenai tribes for the hospitality given to the Alberta overhead team, Montana, 2000
Alberta Government, AFHPC

the Chisholm fire, which was driven again by a low-level jet wind. This fire is well documented and is acknowledged as the most intense fire in Alberta's history, based on the fuel consumed and rate of spread. Interestingly the convection column was analyzed from a U.S. satellite - two super-cell thunderstorms created by the heat from the fire transported smoke and particulate matter into the stratosphere 16 kilometres above the fire. The 2001 fire season was the longest on record for Alberta, extending from March 1 to November 30. At the end of the season a total of 153,573 hectares had burned and expenditures totaled $174,000,000.

2002 House River Fire

For the third consecutive year the fire season commenced on March 1, one month earlier than the normal first day of April. The early date was consistent with the three-year drought in central Alberta and the development of extreme spring fire hazards as a result of delayed green-up. The House River fire started on May 17 and when it was declared controlled on June 17 it was the second largest recorded fire in Alberta's history, burning 238,867 hectares. At the same time, record fire events were occurring in Saskatchewan, Quebec, Arizona and Colorado. Fire impacts on communities and industry operations and facilities were unprecedented in North America. The House River fire was fought for a full month under extreme weather conditions, including no green-up of minor vegetation. One thousand firefighters were working on the House River fire at its peak. Province-wide the area burned totaled 496,397 hectares and expenditures were $255,385,000 by the end of the fire season.

2003 Lost Creek Fire

The Lost Creek fire started in the vicinity

Air Spray Lockheed L-188 Electra dropping load of retardant, 2004
Gail Greenwood

of the '1936 fire' south of the Crowsnest Pass that had spotted over the mountains from B.C. This man-caused fire was ignited on July 23, 2003 and for the next 33 days threatened the communities of Coleman, Blairmore, Bellevue, Hillcrest, Burmis, and Beaver Mines as well as acreages, ranches and the West Castle Ski Resort. The extreme fire behaviour associated with the mountainous terrain of the Lost Creek fire had not been experienced in Alberta since 1936. The firefighters of 2003 certainly gained a respect for their predecessor's efforts as they fought the adjacent 1936 West Castle fire with packhorses and hand tools. The mountain terrain created unique overnight fire excursions, upslope and downslope fire runs and hourly wind changes depending upon valley and slope configurations. All this spectacular day and night fire behaviour was in full view of community residents, contrasting with the majority of fire events described earlier in this chapter.

2004 Another 120 starts in one day

July 15, 2004, saw 120 new fire starts. Sustainable Resource Development attacked these fires vigorously and succeeded in controlling over 80 per cent of them by 10 a.m. the next day. Of those that escaped initial attack only four, or three per cent of the fires grew to E Class, larger than 200 hectares in size. A total of 1,178 personnel were working fires or man-up activities on July 15th, supported by 100 helicopters, 70 dozers and 29 airtankers. The majority of the fires occurred in the Slave Lake, High Level, Fort McMurray and Lac La Biche Wildfire Management Areas.

The major contributor to the high number of fire starts was lightning generated by the passage of a cold front. Events of this magnitude may have happened before, but detection and record keeping systems were never as good as they were in 2004. This impressive statistic is a credit to the integrated network of fixed, mobile and lightning detection systems – an impressive performance.

Fire Years – Implications

Each fire season is evaluated, both for its

Darwin Lake, north of Fort Chipewyan, played host to a number of studies and experiments as part of the Canadian Forest Service prescribed burning trials. The Darwin Lake and Steen River (north of High Level) burns documented initial spread rates and intensities in two major forest (fuel) types, namely jack pine and black spruce, and helped to calibrate the Canadian Forest Fire Danger Rating System. The burning trials were also useful for training of fire control personnel and the assessment of fire suppression equipment and techniques, including the use of hand tools and retardants, for fireline construction. These burns were generally a collaborative effort between fire researchers from the CFS Northern Forest Research Centre in Edmonton and Alberta Forest Service HQ and District staff. Darwin Lake burn team, summer, 1974

(L to R): Not Identified, Bert Clinton (CFS photographer), Peter Arnold (cook), Edward Whiteknife, Not Identified, Fred Burbidge, Ed Stechishen, John Muraro, Charlie Van Wagner, Jorn Thomsen (AFS), John Walker, Dale Huberdeau (AFS), Dennis Quintilio, Jack Bell (PFES)

Terry Van Nest photo; Dave Kiil story

impact and perceived effectiveness of the fire control capability. The above-average fire years in terms of area burned or number of fires started were given particular scrutiny to try to determine the reasons why and to effect improvements. Most major changes in policy and levels of support followed the most serious fire seasons. Among these periods were the 1880s and 1890s that led to formation of the Dominion Forestry Branch in 1899; the 1910 and 1919 periods resulted in increased staff; the 1936 fire season was the first major test for the new Alberta Forest Service, which performed well although there were no funds available during the 1930s for additional support. There were extensive fires in 1944 but this was wartime and no additional help was available. In 1949 there were large fires in the Peace River country, then the 1950 Chinchaga fire came in from British Columbia, which caused the government to declare the entire north as an area to be protected from fire. Three fires in 1956 burned on the forest management lease of Alberta's first pulp mill at Hinton, which was the catalyst for the first major increases in support in the history of the AFS. This kind of response followed the fire seasons of 1968, 1980-82, 1989-91 and, most recently, 1998 and 2001-2004. In each case the response included a reassessment of, and changes in, procedures. In the spirit of adaptive management this will likely continue.

Prescribed burning and research

Preventing forest fires and controlling those that invariably escaped was the aim of the DFB and AFS from the start - "to check this evil," as stated so eloquently in proceedings of the Council of Assiniboia in 1832. [38] However, by the 1950s serious questions were raised about the wisdom of trying to eliminate all fires. As described by professor Hank Lewis of the University of Alberta in his 1979 film *Fires of Spring*, one of the traditional aboriginal practices was to burn meadows when they were flammable between spring break-up and "green-up." Forest ranger Sam Sinclair helped to reintroduce spring burning in the Lesser Slave Lake area when he started his "Native firefighter training program" in 1960 and practiced on meadow burning. Dale Huberdeau did the same in the Footner Lake Forest in the 1970s. These were conducted to reduce the fire hazard of cured fine fuels, to sustain the meadows and to train fire crews. In 1983, Fire Control Officer Ben Shantz took prescribed burning a step further when he proposed to burn some high-elevation forests on Ram Mountain, west of Rocky Mountain House, to improve winter grazing for Bighorn sheep. This was contentious but Shantz accepted responsibility and conducted a successful burn using a combination of aerial and ground ignition. It set a precedent that was followed by a range of prescribed burns for such purposes as habitat improvement and hazard reduction.

Research scientist Dave Kiil, with the Canadian Forest Service (CFS), worked on the prototype Fire Danger Rating system, which involved setting small (about one metre square) plots to test ignition and initial spread in conifer needles. In the 1960s he worked with North Western Pulp & Power Ltd. staff to test the feasibility of prescribed fire to burn deep duff to prepare seedbeds for spruce on cool moist sites.

The result of this experiment was that repeated burns were required to reach mineral soil and create a receptive mineral seed bed. The cost of the controlled burns exceeded the cost of mechanical scarification, therefore this practice was not suitable for operational use.

Research forester Dennis Quintilio, also with the CFS, took the next step by setting fires under prescribed and instrumented conditions to study fire behaviour as influenced by fuels and weather. In 1969 he ignited a series of test fires in lodgepole pine logging slash in the Kananaskis valley under a range of intensities that clearly showed the value of that approach. This was followed by a cooperative agreement with the Alberta Forest Service to study fire behaviour in aspen and jack pine in Slave Lake Forest. In 1974, a more comprehensive project was initiated at Darwin Lake in the far northeastern corner of Alberta and a series of burns in jack pine stands north of Lake Athabasca was completed. This was the first major multi-agency project and its results set the stage for ongoing trials in northern Alberta and adjacent Northwest Territories. These trials were designed to be burned under increasingly high burning indices to document crown fire behaviour.

An advantage of a prescribed burn is that it can be ignited under optimal conditions with favourable weather forecasts. However, there is always an element of risk. For example, on the 1969 Kananaskis burns the series was completed, including burns in the extreme fire danger class, and post-fire measurements were in progress.

Paramount Cub pump, early firefighting pump, Clearwater Forest, Rocky Mountains Forest Reserve, 1950
Jim Clark

Mop-up had been completed, however as a precaution a pump and gas tank were left at the perimeter of the burn plots adjacent to a beaver dam and creek. Two weeks after the last burn a strong wind developed in the night and as the research crew traveled to the project site the next morning they observed a smoke column suspiciously close to the burn plots. On arrival the story unfolded - the winds had "discovered" a holdover fire in the plot next to the standby pump. The wind blew embers across the fireline and into the small stand of pine where the pump and gas tank were located. The surface fire, which would normally have stopped at the beaver dam, blew up the gas tank and spotted across the creek. The wind was still raging and the research crew rushed to the creek to set up the backup Paramount pump, which had a spark arrest setting and was notorious for being very temperamental. All eyes were on Stan Lux, as he carefully set the spark arrest and pulled the start cord with all his strength and when the pump roared to life the fire never had a chance. The Paramount was an old dinosaur pump but could outdo all the modern competition – if it started.

Darwin Lake was the site of the second close call. After a long stretch of cool damp weather ended and a welcome drying trend established over the experimental project area, the series of burns were started at the low and moderate hazard levels that increased as the drying days continued. The big day was approaching for the first anticipated crown fire in the jack pine stand known as Plot 6. A Bell 206 helicopter with bucket was scheduled for this high hazard burn but the burn was cancelled at the last minute. Then, after a full summer of waiting for the ideal conditions for Plot 6, Dale Huberdeau and Dennis Quintilio decided to go ahead with the ignition and as predicted the plot crowned immediately.

Aerial view of the International Crown Fire Modeling Experiment (ICFME) site located near Fort Providence, NWT. This project ran from 1995 to 2001 and involved more than 100 participants representing some 30 organizations from 14 countries
Gary Dakin

Aerial view of the ICFME site showing test plots that were burned in 1997, Fort Providence, NWT, 1998
Gary Dakin

House constructed within one of the ICFME test plots to study the effects of radiant heat on structures and the fuel-free zone required for structures to survive wildfires in a forested setting, Fort Providence, NWT, 1998
Gary Dakin

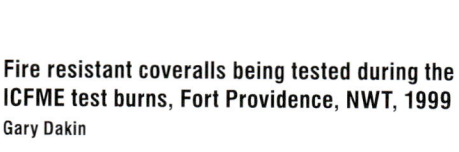

Fire resistant coveralls being tested during the ICFME test burns, Fort Providence, NWT, 1999
Gary Dakin

Crown fire developing in International Crown Fire Modeling Experiment test plot, Fort Providence, NWT, 1999
Gary Dakin

Numerous coverall samples adjacent to ICFME test plot, Fort Providence, NWT, 1999
Gary Dakin

Full crown fire in International Crown Fire Modeling Experiment test plot, Fort Providence, NWT, 1999
Gary Dakin

Aerial view of developing crown fire in International Crown Fire Modeling Experiment test plot, Fort Providence, NWT, 1999
Gary Dakin

The pair headed for the downwind boundary of the plot, which had a hand line supported by a sprinkler system. After listening to the head fire aggressively heading in their direction, they decided to backfire from the sprinkler line. The challenge of hand lighting is all in the timing, so the backfire is usually lit early and does not immediately pull into the headfire. This was the case and spot fires were jumping the full length of the sprinkler line. Huberdeau had a nozzle in his back pocket and proceeded to disconnect the sprinkler line hose and attach the nozzle and disappear into the smoke for 30 minutes of serious firefighting outside the plot boundary. The helicopter came back the next day but so did the rain and the experimental burning was over for the summer.

Following up on the successful experimental burns in the 1970s and early 1980s, Marty Alexander, of the Canadian Forest Service (CFS), led a joint CFS and AFS experimental burn project at Big Fish Lake, 100 kilometres northeast of High Level. The experimental burning took place from 1984 to 1989 and involved CFS staff from the Northern, Great Lakes and Pacific Forestry Centres and the Petawawa National Forestry Institute as well as AFS staff including Dale Huberdeau, Lyall Gill, Gerald Sambrooke, Ken Orich and Bill Lesiuk. The purpose of the Big Fish Lake project was to systematically document the behaviour of experimental fires in the black spruce-Labrador tea-cladonia fuel complex. This fuel type was considered a major problem because when dry it was easily ignited, spread fast and burned hot, so it was difficult to control. This fuel type was a problem in not only northern Alberta but other western Canadian provinces, the Yukon and Northwest Territories. The experimental burns strengthened the reliability of the Fire Behaviour Prediction System, and were also used to assist in assessment of discovery times from a fixed detection point. Ponton Tower was able to detect smoke columns of the initiating fires within three to nine minutes of ignition.

The International Crown Fire Modeling Experiment (ICFME) focused on the behaviour and implications of high-intensity crown fires. A site was selected near Fort Providence, NWT, where there were uniform stands of jack pine with a black spruce understory, representative of northern forests. The main project ran from 1995 to 2001 and involved more than 100 participants representing some 30 organizations from 14 countries. A total of 18 successful experimental high-intensity crown fires were conducted and documented. ICFME provided valuable new data and insights into the nature and characteristics of crowning forest fires. The concept of the ICFME evolved as the result of a number of converging issues: the recognition that the United States and Canada could not continue separate approaches to fire behaviour model development, the opening of Russia to the western world, increased communication, and the formation of international associations to facilitate collaboration. While the initial impetus for ICFME was the desire to improve the physical modeling of crown fire propagation and spread, the project also created the opportunity to examine many other aspects and impacts of crown fires. The feasibility of igniting and studying crown fires in this way had been clearly demonstrated in the previous collaborative studies, which also enabled refinement of the techniques. The *Canadian Journal of Forest Research* featured a special issue of papers on ICFME studies in their August 2004 edition (Stocks, Alexander and Lanoville 2004.)[39] Marty Alexander (CFS Edmonton), Brian Stocks (CFS Sault Ste. Marie, Ontario) and Rick Lanoville (GNWT Forest Management Division) were the co-coordinators for the project. The principal Alberta SRD representative was Gary Dakin who worked jointly with Mark Ackerman from the Department of Mechanical Engineering at the University of Alberta.

Forests, Fire and People

The forests of Alberta evolved under the influence of recurrent fires. Fires created a mosaic of forest stands of different ages and species mixes, while individual plant species developed survival and renewal strategies to perpetuate a variety of forest communities. Aboriginal people used fire in selected areas to create habitat conditions important for sustaining their ways of life, but they could also be adversely affected by wildfires. When European settlement began, these newcomers created new values in and around the forest that were at risk from uncontrolled fires. This led to organized fire control efforts after 1899. Since 1930 the Alberta Forest Service has continued its dedication to refining forest protection practices. The AFS has developed increasingly innovative means by which to improve the effectiveness of its fire management efforts and at the same time is striving to enable fire to play a role in the forested ecosystem in balance with inherent risks. Challenges have been met by a succession of committed, knowledgeable and experienced individuals whose capabilities are recognized worldwide.

Fire on edge of city limits, Palangkaraya, Indonesia, November, 1997
Canadian Expert Fire Instructors train Indonesians in use of Canadian fire equipment and fire tactics. Dale Huberdeau (L), Canadian Instructor and an Indonesian trained military firefighter
Tom Laxdal

International Export Assignments

A long history of assisting fire management agencies outside of Canada began in 1973 with an urgent request from Bogotá, Colombia as eleven mountain fires threatened villages, plantations, and watersheds. The Province of Alberta was asked by the Canadian International Development Agency (CIDA) to supply fire control assistance to Colombia. Confirmation was forwarded to CIDA and a team was appointed and advised to leave immediately for Colombia via Ottawa, New York, and Miami. The team consisted of Stan Hughes, Carson McDonald and Bill Wuth. The request was made at 1530 hours February 27 and the team arrived in Bogotá at 1830 hours, March 1st without having slept for 40 hours enroute. The team spent a week analyzing the fire situation and filed their report titled *Recommendations on the role of fire control in a resource management program for Colombia.*

In 1983, the province of Alberta was again asked to assist internationally, this time in China where the current fire detection system was being evaluated. Joe Niederleitner, a detection expert with AFS, traveled to China numerous times over the next five years and contributed much wisdom and technical help in re-vamping their tower network. He evaluated tower structure, firefinder technology, data recording systems and training, and produced operational manuals at the conclusion of the project. During one of his visits to northern China, he observed a major fire outbreak, which burned villages near the Mongolian border and took 40 lives.

Following a disastrous 1997 fire season in Southeast Asia, Canada was asked to provide a team of experts to train Indonesian firefighters. Alberta Lands and Forests Service selected Dale Huberdeau and he joined Mark Heathcott (Parks Canada), Tom Laxdal (Saskatchewan) and Jim Motishaw (British Columbia) in Indonesia in November and December of 1997. Successive years of drought had caused severe jungle wildfires to burn out of control and smoke and haze from ground fires created social, health, and economic issues for the Indonesian people and the neighbouring countries of Singapore, Malaysia, Brunei, Thailand, Philippines and Vietnam. The Canadian team was tasked with training local Indonesians in fire suppression tactics, safety, and efficient equipment use.

Canadian Forest Service fire research crew, Big Fish Lake Experimental Burning Project, 1984
(L to R): Tim Lynham, Bryan Lee, George Dalrymple, Gilles Delisle, Gary Hartley, Brad Hawkes, Jack Bell, John Mason, Marty Alexander, Charlie Van Wagner, Gyula Pech, Kelvin Hirsch, Brian Stocks, Mike Weber
Marty Alexander

Dale Huberdeau was an integral part of the training exercise, passing on his fire experience in education, prevention, detection, and suppression. Huberdeau was assigned the responsibility of equipment inventory management as the team traveled throughout the remote jungle forests of Borneo. A four-day basic training program was delivered with equipment inventories necessary to support an initial attack or sustained action operation for military and government firefighters. Huberdeau worked with local Indonesian officials to organize equipment units and to manage the inventory on the fireline. Huberdeau was an authoritarian individual and his size and stature, compared to the average Indonesian, commanded immediate attention and respect. Dale Huberdeau or "Boone" as his colleagues referred to him became the focus of many inquisitive eyes, particularly during a haircut day as the Indonesians marveled at his red hair, moustache, and height.

Over 325 firefighters were trained throughout the Provinces of East and Central Kalimantan. The Indonesian Government was very appreciative of the Canadian team's contribution and the mission was concluded with a high level of formal thanks.

In August 2000, Incident Commander Bruce MacGregor and his Alberta team were assigned to the Clear Creek Divide Fire Complex on the Flathead Indian Reservation in Montana. This was the first export operation where a Canadian team managed a fire complex independently. The Alberta firefighters were admired for both their aggressive firefighting and their respect for the tribal lands. Bill Bereska, a fire behaviour specialist with the team, noted a serious wind shift one afternoon and advised branch director Butch Shenfield to move his men off the line. The crews had just gathered at a safe location when a red flag weather advisory arrived at the base camp from the headquarters meteorologists. Bereska had read all the warning signals in the field well ahead of the weather alert. There were many highlights during the fire assignment including a visit from President Clinton but MacGregor recalls the most significant event was the ceremonial thank you from the Salish and Kootenai tribes as the team prepared to leave for home. Ken Orich's team arrived and carried on the suppression operations and when a New Zealand/Australian crew was scheduled to replace them, the Flathead Nation intervened and the fire was fought from start to finish by Alberta.

Treasured Memory

Leadership and teamwork are defining characteristics of successful programs. Among the many individuals who exhibited these qualities are two former rangers who always went that extra mile and are paid special tribute for their contributions to forest fire management.

Harry Edgecombe was raised in Fort Vermilion and joined the AFS in 1946 after wartime service. Starting as assistant ranger in the Clearwater Forest he served in Fort McMurray, High Level and Grande Prairie before moving to Hinton where he later led the fire management training program. He was widely respected for his savvy about the forest and fires, for innovative approaches to problems and particularly for his ability to work with people anywhere. As he often said: "When it's too tough for everyone else, it's just right for us" and got others to share that point of view. Harry retired in 1979.

Dale Huberdeau was a remarkably perceptive individual. Fire was certainly his major interest, not just for the complexities of fires themselves, but for the intellectual challenge and the opportunity for action to get people and machines together. He was raised in Colinton, near Athabasca. Harry Edgecombe introduced him to fires when he was a grade 10 student. As he remarked: "I spent about 40 of those 60 days in July and August on fires for 55 cents per hour and I thought that was a pretty good lifestyle." He joined the AFS in 1965 as assistant ranger at Fox Creek, moving to Fort McKay the next year. He spent his career largely in the Fort McMurray and Footner Lake forest areas – becoming area manager at Fort McMurray where he passed away in 1999. He was especially respected for his skills in fire management, from the fireline to planning and organization. He was quoted as saying: "When we're out there fighting fires, I don't want to see burn holes in the back of anyone's shirt. I want 'em on the front." After what he called the "disastrous" fire year of 1981 at Footner Lake, he commented: "… we pretty well shook the organization upside down and built a new one in terms of forest protection." He also reintroduced the concept of aboriginal spring burning in the High Level area, involving the original practitioners in the process. These were typical of his many contributions.

Harry Edgecombe
Alberta Government, AFHPC

Dale Huberdeau in 1995
Alberta Government, AFHPC

Firefighting training in Slave Lake, Lesser Slave Lake Forest, April, 1925
Dominion Forest Service, AFHPC

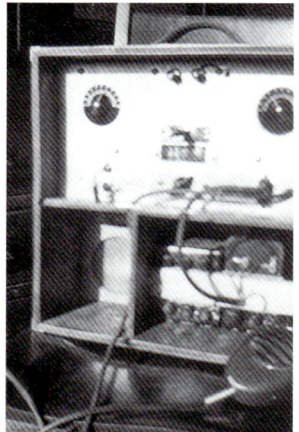

Three watt ultra high frequency radio, early 1940s
Ron Linsdell

Portable radio from the 1950s and 1960s
Ron Linsdell and Tim Klein

Collection of forestry radios from the 1980s to present, 2005
(L to R): Kenwood TK270G VHF (128 Channels), Kenwood TK270 VHF (32 Channels), Yaesu FTH-2008 VHF (32 Channels), GE MPD UHF (20 Channels), Motorola PT500 VHF (8 Channels), Motorola MX-350 UHF (20 Channels)
Mike Peebles

Overhead team for Fire 2 in District 23B, 1961 (Fire 23B-02-61), Edson Forest
(L to R): Al Needham, Don Lowe, Bill Lee, Vic Fischer, Doug Quinnell
Don Lowe

Fire guard constructed on the south-east corner of Fire 23B-02-61, Edson Forest, 1961
Vic Fischer

Rangers Bill Lee (L) and Al Needham at Fire 23B-02, Edson Forest, 1961
Don Lowe

W.J. (Jock) McLean was appointed as Fire Weather Officer in November, 1963. This first major advance in applying weather to fire concerns was brought about by the urgings of Frank Platt. McLean was a forester by training, had a valid commercial pilot's licence and was checked out on both fixed-wing and helicopters. McLean's interest in meteorology and his forestry and flying background made him the perfect candidate to study the relationship between forest fires and weather, providing guidance to field staff
Alberta Government, AFHPC

Firefighter training Assumption, Habay District, Peace River Division, 1964
Back Row Standing (L to R): Larone Ferguson, Jim Hamelin, Clarence Ferguson, Murray Mitchell, George Wanuch, Jim Belrose, Alphonse Ducharme, Adolph Auger, Dick Mitchell, Cliff McGillivary, Absolam Laboucan. Middle Row: Larone Nanooch, Tommy Didzena, Germain Pastion, Joseph Chambaud, John Schasee, Walter Kazonie, John Chalifoux. Front Row: Paul Cousta, Paul Courtoreille, Andrew Deedza, Clement Fournier, Joe Bulldog, Alex Meneen, William R. Cardinal
Corinne Huberdeau

Refueling Wajax Pump at Mayberne Airstrip, Edson Forest, 1960s
Alberta Government, AHFPC

Harry Edgecombe in the 1970s
Alberta Government, AFHPC

Painting on a rock outcropping near Prairie Creek, done by an Indian member of a minimum-security crew, doing trail and road improvement work for AFS, spring, 1966
Alberta Government, AHFPC

Slave Lake Fire Control Officer Bernie Simpson instructing at native firefighter course, Slave Lake Forest, 1968
Alberta Government, AFHPC

Members of Algar Lake Fire Overhead Team, Fort McMurray, Athabasca Forest, 1971
(L to R): Frank Lewis, Dale Huberdeau (Fire Boss), Fred Schroeder
Frank Lewis

Lac La Biche Forest Superintedent Bert Varty (plaid shirt) pointing to location on map during planning exercise, Lac La Biche Forest, May, 1968 fires. Seated to the back left is Ranger Harry Jeremy
Alberta Government, AFHPC

Meander River Type III Emergency Firefighter Squad after winning firefighter competition, Footner Lake Forest, 1973
Corinne Huberdeau

Algar Lake Fire Overhead Team, Fort McMurray, Athabasca Forest, 1971
(L to R): Bernie Brouwer, Fire Control Officer; Fred Schroeder, Dale Huberdeau, Fire Boss; Frank Lewis, Bob Plankenhorn
Frank Lewis

Staff from the Alberta Forest Service were chosen by the Canadian government to assist and advise the Colombian government on nearly a dozen wildfires that were threatening the capital city Bogotá.
(L to R): Bill Wuth, Fire Control Technician, Footner Lake Forest; Carson McDonald, Fire Control Officer, Edmonton Fire Control Centre; and Stan Hughes, Head Forest Protection Branch
"At eleven a.m. on Tuesday, February 27, 1973, the Canadian International Development Agency (C.I.D.A.) in Ottawa received a telephone call from the Canadian Embassy in Bogotá, Colombia, South America; relaying a request from the Colombian government for provision of professional and technical advice to further their efforts to combat the wildfires near the nation's capital city of Bogotá."
The team had two primary objectives:
1. To provide professional and technical fire control advice to the Government of Colombia in order to assist in the containment of fires in the Bogotá area, and
2. To provide long term fire control recommendations that might assist the future development of resource protection in the Republic of Colombia.
March, 1973
Photo courtesy of the Canadian Forest Service; reproduced with permission

Canadian Forest Service staff meeting to discuss the introduction of the Drought Code into the Canadian Forest Fire Danger Rating System, February, 1973
Back Row (L to R): Dennis Quintilio, Bruce Lawson, Dick Silversides, Jim Kayll, Not Identified, Not Identified, John Muraro, George Chrosciewicz. Front Row: Charlie Van Wagner, John Turner, Peter Paul, John Walker, Brian Stocks, Luc Pouliot?, Bob Vines
Dave Kiil

Fire in Alberta

Graduates from Initial Attack Leader training, Forest Technology School, Hinton, April-May, 1977
Alberta Government, AFHPC

Dale Huberdeau with drip torch conducting hazard reduction burning, Footner Lake Forest, spring, 1977
Corinne Huberdeau

Overhead team at Edra Airstrip on Fire DA2-30-79, Athabasca Forest, July, 1979
Back Row (L to R): Mike Dubina (Fire Boss), Bob Petite, Terry Wagner, Scott Mayston
Front Row (L to R): Tom Trott, Jamie McQuarrie, Peter Nortcliffe
Rick Keller

Ponton Tower with fire DF2-37-84 in the background, Footner Lake Forest, 1984
Henry Grierson

Edra Tower and airstrip with base camp for DA2-30-79, Athabasca Forest, July, 1979
Rick Keller

Forest Protection Technician Len Allen (on phone) and Forest Protection Officer Howard Gray (on radio) in the Slave Lake Forest Headquarters Radio Room, Slave Lake Forest, April, 1980. Slave Lake staff were dealing with a number of fires at this time, when an Air Spray B-26 airtanker crashed while working on a forest fire near the Town of Slave Lake. Pilot Flight Lieutenant Teras (Terry) Kitzul died in the crash
Edmonton Journal, Jim Cochrane

Buffalo Head Hills fire camp, Footner Lake Forest, April, 1980. Photo shows the 'kitchen' fire and storage tents
Rob Thorburn

Camp kitchen supply tents at fire camp, 1980s
Alberta Government, AFHPC

Almost 'all the luxuries of home', 1980s
Alberta Government, AFHPC

Canvas tents used for accommodation at fire camp, 1980s
Alberta Government, AFHPC

Graduates from Initial Attack Leader training, Forest Technology School, Hinton, May, 1980. Instructor Ken South in top right corner
Alberta Government, AFHPC

Firefighters using hardhats to get water from waterhole for use on surface fire, 1980s
Alberta Government, AFHPC

Firefighters working on a surface fire in an old burn, 1980s
Alberta Government, AFHPC

Slave Lake Superintendent Carson McDonald showing media the map location of fire threatening the Town of Swan Hills, August, 1981. The Town of Swan Hills had previously been evacuated in 1972 when a fire came within 3 kilometres of the town, and was threatened with an evacuation in 1967 when a tree fell on a power line and started a fire that destroyed 6,000 hectares of forest 16 kilometres to the northeast.
Edmonton Journal photo and story (August 11, 1981)

Aerial view of forest fire that forced the evacuation of the Town of Swan Hills, August, 1981. Swan Hills Chief Ranger Kelly O'Shea said the evacuation of the town of 2,200 was "prompted by a wind shift Monday afternoon which turned the fires south toward the town". Slave Lake Forest Superintendent Carson McDonald said the fire "is very definitely out of control. It's in a very difficult area to fight. The land is hilly which makes it difficult to put men and bulldozers on the ground near the fire."
Ken Orr photo, Edmonton Journal stories (August 11, 1981)

Alaska presented an award to the Footner Lake Forest staff following the successful import of Alaska firefighters to fires within the Footner Lake Forest in 1981. This was a historical event in that Alberta had never imported that many firefighters from out of province before. Rob Thorburn said "we were impressed with their ability to camp on the line progressively in pup tents and eat c-rations with only supplementing them with a steak and a few vegetables every several days. It was this event that eventually led to Alberta's consideration for individual 8 person strike teams so to speak." Jamie McQuarrie said, "The Alaskans were great firefighters – hard workers, no complaints and very, very disciplined. I'm not sure if it was a first for Alberta, but I know the Alaskan gals were the first female firefighters I had seen."
(L to R): Jamie McQuarrie, Forest Protection Technician; Rob Thorburn, Air Attack Coordinator; Carl Leary, Superintendent Footner Lake Forest; Dale Huberdeau, Forest Protection Officer; Warren Sinclair, Dispatcher
Corrine Huberdeau

Graduates from Initial Attack Leader training, Forest Technology School, Hinton, April-May, 1983
Alberta Government, AFHPC

Helitack training graduates, Forest Technology School, Hinton, May, 1983
Alberta Government, AFHPC

Helitack Rappel crew member training, Forest Technology School, Hinton, April-May, 1984
Alberta Government, AFHPC

Ranger Lyall Gill (white hardhat) instructing firefighters correct procedures on the use of prescribed fire, Footner Lake Forest, early 1980s
Alberta Government, AFHPC

Crew preparing mixture for use in aerial drip torch, north of Wabasca, Slave Lake Forest, 1982. Dennis Quintilio and Lyall Gill were the two aerial ignition specialists, with mixing crew assistance provided by Brent Simmonds and Gulf Canada employees
Alberta Government, AFHPC

Helicopter at mixing location to get a new ignition drum for backfire operations north of Wabasca, Slave Lake Forest, 1982
Alberta Government, AFHPC

Forest Officers involved in firefighter training program, Cold Creek District, Whitecourt Forest, spring, 1985
Standing (L to R): Fred Paget, Len Stroebel, Ian Hancock, Mark Froehler, Rick Arthur, Lloyd Seedhouse. Kneeling: Dennis Halladay, Norm Olsen
Fred Paget

Sketch of firefighters in silhouette
Artist Lorna Bennett

Boston Bar Overhead Team – Fra Fire, 1985. Dennis Cox described the export as "not any small task" … "where drought codes were in the 600-800 range; terrain was extremely steep; fuel types were challenging and varied from recently harvested cut blocks to old growth timber that was in excess of 200 feet tall; temperatures were consistently in the 30s and RHs were recorded as low as 8% at our mud (retardant) mixing pit one day." In addition "access was very poor and no heavy equipment was readily available since the only way across the Fraser at Boston Bar was an aerial tram that could only take two regular sized vehicles at a time. It took three or four days to get any heavy equipment, as it had to be transported to Lillouet, put on a train and then sent to Boston Bar. Manpower was in as short a supply, as was equipment. The BC Fire Boss handled the political issues, while Dale handled the practical issues."
(L to R): Dave Hargreaves, Line Boss (BCFS); John Hall, Fire Boss (BCFS); Preston ? (BCFS), Pete Valk, Air Operations Coordinator (BCFS); Rob Thorburn, Air Operations Coordinator; Bill Davidson (BCFS); Dale Huberdeau, Fire Boss; Terry Van Nest, Plans Chief; Dennis Cox, Line Boss; Bob Glover, Service Chief
Terry Van Nest photo; Dennis Cox story

Fra fire, Boston Bar, British Columbia, July 18, 1985; 1730 hours
Terry Van Nest

Fra fire, Boston Bar, British Columbia, July 19, 1985; 1600 hours
Terry Van Nest

Graduates from Initial Attack Leader training, Forest Technology School, Hinton, April-May, 1986
Alberta Government, AFHPC

Stan Clarke (L) and Dale Huberdeau while on export to Oregon, 1987
Bruce MacGregor

Forest Protection Officer spring workshop, Kananaskis, 1988
Back Row (L to R): Owen Bolster, PFFC – Fire Operations Officer; Jamie McQuarrie, Slave Lake; Gary Dakin, Fort McMurray; Darryl Rollings, Rocky Mountain House; Kelly O'Shea, Calgary; Hylo McDonald, Edson; Jurgen Moll, Whitecourt; John McLevin, Lac La Biche (Acting for Bill Bereska). Front Row: Mag Steiestol, PFFC – Fire Prevention Coordinator; Lou Foley, PFFC – Wildfire Operations; Gordon Bisgrove, Manager Wildfire and Air Operations; Don Law, Grande Prairie; Ken South, Peace River
Corinne Huberdeau

Sketch of *'Start of the Corral Creek Fire'* south of Grande Cache, 1987
Artist Robert Guest

384 Alberta Forest Service

Forest Officer Jim Lunn and crew complete hazard reduction burning, Footner Lake Forest, late 1980s
Corinne Huberdeau

Kakwa Staging Camp, Grande Prairie Forest, 1991
Alberta Government, AFHPC

Instructor Gordon Baron conducting Straw Boss training, 1992
Alberta Government, AFHPC

Overhead team from DND-003-93 in front of the La Corey Ranger Station sign, Lac La Biche Forest, May, 1993 (L to R): Len Wilton, Ken Porter, Ralph Woods, Rick Stewart
Bruce MacGregor

Fire in Alberta

Land and Forest Service staff work with Alpac staff to install a prevention sign north of the Alberta-Pacific millsite, Northeast Boreal Region, mid-1990s
(L to R): Kurt Frederick, John McLevin, John Belanger, Ed Barnett, Don Pope (Alpac)
Bruce MacGregor

Phoenix Helicopters C-FHLF with aerial heli-torch, Mariana Lake fire, May, 1995
Alberta Government, AFHPC

Phoenix Helicopters C-FHLF with aerial heli-torch, Mariana Lake fire, May, 1995
Alberta Government, AFHPC

Sketch *'Fire in an Old Burn'*, 1996
Artist Robert Guest

Land and Forest Service Overhead Team meeting, Swan Hills Fire Base, Whitecourt Forest, spring, 1996

(L to R): Bob Mazurik, Wayne Bowles, Bill Bereska, Howard Herman, Tom Grant, Gary Schneidmiller, Chris McGuinty, Bruce MacGregor, Cordy Tymstra, Conrad Gray, Don Podlubny, Bruce Cartwright, Dennis York, Ray Olsson, Butch Shenfield, Russ Stashko, Bob Yates, Rob Thorburn, Dave Lind, Ken Porter, Leon Graham, Gary Mandrusiak, Jim Cochrane, Hugh Boyd, Mike Dubina, Kurt Frederick, Larry Warren, Mark Froehler, Dennis Driscoll

Alberta Government, AFHPC

John Belanger (L) and John McLevin (R) conducting Type III firefighter training, Northeast Boreal Region, Lac La Biche, mid-1990s

Bruce MacGregor

Land and Forest Service conducting a prescribed burn at Yarrow Creek, north of Waterton Lake National Park, Crowsnest Forest Area, Prairie Bow Parkland Region, 1997

Darryl Johnson

Fire overhead team in the Caribou Mountains, Northwest Boreal Region, High Level District, mid-1990s

(L to R): Butch Shenfield, Bill Lesiuk, Ken McCrae (Forest Protection Officer, Peace River), Bruce MacGregor (Forest Protection Officer, Lac La Biche), Barry Gladders, Ken Bisgrove

Bruce MacGregor

Canadian team that provided training to Indonesian firefighters, November-December, 1997
(L to R): Mark Heathcott, Parks Canada; Dale Huberdeau, Alberta; Tom Laxdal, Saskatchewan; Jim Motishaw, British Columbia
Tom Laxdall

Firefighter applying foam to hotspots, 1990s
Alberta Government, AFHPC

Aerial view of dozers constructing fireguard, 1990s
Alberta Government, AFHPC

Ministerial tour of fires north of Lac La Biche, Lakeland District, Northeast Boreal Region, May, 1998
(L to R): Bruce MacGregor, Regional Forest Protection Officer; Kelly O'Shea, Director of Forest Protection; Jim Cochrane, District Manager Lakeland District; Ty Lund, Minister of Environment; Neil Barker, Regional Director
Bruce MacGregor

Aerial view of
Canadair CL-215
dropping foam and
water mixture on fire,
spring, 1998
Alberta Government, AFHPC

Aerial view of Agnes Lake fire, Slave Lake
District, Northeast Boreal Region, spring 1998
Alberta Government, AFHPC

Jesse Lake fire, Lakeshore
District, September, 1999
Alberta Government, AFHPC

Base camp at the Round Hill
Tower airstrip, Lakeland District,
Northeast Boreal Region, 1999
Bruce MacGregor

Fire in Alberta 389

Land and Forest Service staff on export to British Columbia. Photo taken at Moyie Base Camp, August 25, 2000
(L to R): Jeff Henricks (Alberta Liaison), Jim Hennessey, Dave Mireau, Jamie Sylvester, Vance Soto, Jeff Swingler, B.C. Minister of Forests Jim Abbott, Robert Anderson, Cranbrook MLA Erda Walsh, Trevor Lamabe, British Columbia Premier Ujjal Dosanjh, Darren Janvier, Mike Lutz, Keith Fickling, Garth Kowalyk, Derrek Harrison, Ambrose Jacobs, Max Mathon
Jeff Henricks

Airtractor AT-802 dropping retardant to build a fireguard on the Gorge Creek fire, Crowsnest Pass, 2000
Darryl Johnson

Bruce MacGregor's Alberta Overhead Team enroute Montana. This team was the first Canadian team in charge of their own fire complex in the United States, 2000
(L to R): Kurt Frederick, Hugh Boyd, Don Podlubny, Paul St. John, Bill Bereska, Bruce MacGregor, Wayne Bowles
Bruce MacGregor

Members of the Bruce MacGregor Incident Command team on export to Montana, 2000
(L to R): Hugh Boyd, Bruce MacGregor, Butch Shenfield, Joe Wagenfeur, Homer Courville
Bruce MacGregor

Ken Orich's Alberta Incident Command Team, Montana, 2000
Standing (L to R): Brent Korolischuk, Paul Malysh, John Bruce, Darrell Beam, Brian Lopushinsky, Terry Sayers, Gordon Crowder, Kent McDonald, John Brewer, Chuck Spencer (hidden), Morgan Kehr, Wayne Becker, Ken Orich, David Finn, Rick Horne (hidden), Russ Difiore, Rick Stewart, Ollie Erdell. Kneeling: Michael Weyer, Kevin Hakes, Gordon Graham, Shawn Milne, Sherra Quintilio, Robin Barnes, Darlene Belleau, Doreen Leichnitz, Patti Campsall, Cheryl Flexhaug, Cindy Schellenberg
Brian Lopushinsky

Air Spray's "Longliner", Lockheed Electra L-188 working on the Sunpine Forest Products log yard fire, Rocky Mountain House, May 6, 2001
Photo credit unknown

Operations field meeting in Montana, 2000
(L to R): Bill Bereska, Bruce MacGregor, Hugh Boyd, Butch Shenfield
Bruce MacGregor

Conair Convair dropping retardant on fire east of Whitecourt, Whitecourt Wildfire Management Area, spring, 2001
Alberta Government, AFHPC

Fire in Alberta 391

Fire behaviour on the Chisholm fire May 28, 2001
Bill Bereska

Dennis Driscoll and Mike Dubina at the Virginia Hills Firefighter Training camp, dining area under plastic tarp in the background, May, 2001
Alberta Government, AHPFC

Type III Emergency Firefighter training at the Virginia Hills Firefighter Training camp. Here firefighters learn the proper way to sharpen fireline tools – pulaskis and shovels. Whitecourt Wildfire Management Area, May, 2001
Alberta Government, AFHPC

Aerial photo showing the Hamlet of Chisholm after the Chisholm fire ran through the hamlet, May, 2001
Stew Walkinshaw

Aerial view of the Chisholm fire base camp, May, 2001
Alberta Government, AFHPC

Firefighter checking pump and water supplying water to the sprinklers set up on house for structure protection, Chisholm fire, Lac La Biche Wildfire Management Area, 2001
Rick Moore

Whitecourt-Ste. Anne MLA George VanderBurg presenting the joint Whitecourt and Edson Wildfire Management Area Helitac Type I crew the first place trophy in the firefighter competition at the Virginia Hills Firefighter Training site, August, 2001
Back Row (L to R): Kurt Richter, Ryan Beniuk, Andrea Schlender, Forrest Barrett, George VanderBurg, Geoffrey Driscoll. Front Row: Robert Anderson, Dustin Thatcher, Tyler Hauck. Mike Dubina standing to the right at the prize table
Alberta Government, AFHPC

Map showing the 2001 Chisholm fire in relation to the Vega fire of 1968, as well as other E Class fires, by date in the vicinity
Jamie Badcock

View of smoke column from the Dog Rib fire, west of Sundre, October, 2001
Alberta Government, AFHPC

Fire in Alberta 393

Medium helicopters at Mariana Lake base camp, House River fire, Highway 63 near Mariana Lake, Lac La Biche Wildfire Management Area, 2002
Barry Alexander

Alberta government's Canadair CL-215 scooping water, Amoco Kirby Lake area, Lac La Biche Wildfire Management Area, 2002
Alberta Government, AFHPC

Reunion of old friends, Nanaimo, British Columbia 2002
(L to R): Neil Gilliat, Wally Harrison, Rex Winn
Neil Gilliat

House River fire, northwest of Lac La Biche, Lac La Biche Wildfire Management Area, 2002
Barry Alexander

Photo taken from back seat of helicopter that was using aerial heli-torch, House River fire, 2002
Shawn Milne

Running crown fire on House River fire, Lac La Biche Wildfire Management Area, 2002
Barry Alexander

Aerial view of backfire started along dozer cleared torch line, House River fire, 2002
Shawn Milne

Airbrush version of Bertie Beaver
Artist Lorna Bennett

Aerial view of backfire on fire SWF-237-2002, north of Slave Lake, Slave Lake Wildfire Management Area, 2002
Alberta Government, AFHPC

Photo taken from back seat of helicopter using aerial heli-torch, House River fire, 2002
Shawn Milne

Fire in Alberta 395

Start of the Rock Creek prescribed burn, Willmore Wilderness, Foothills Wildfire Management Area and Jasper National Park, July, 2003
Herman Stegehuis

Rock Creek prescribed burn, Willmore Wilderness, Foothills Wildfire Management Area, and Jasper National Park, July, 2003
Herman Stegehuis

Rock Creek prescribed burn, Willmore Wilderness, Foothills Wildfire Management Area and Jasper National Park, July, 2003
Herman Stegehuis

Provincial Wildfire Information Officer Rick Strickland explaining to media current status and updates of suppression efforts on the Lost Creek fire, Blairmore, Southern Rockies Wildfire Management Area, 2003
Alberta Government, AFHPC

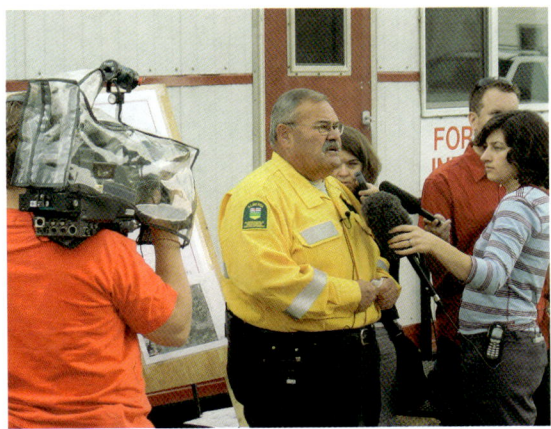

Information Officer Brydon Ward conducting interview with media on current status and updates of suppression efforts on the Lost Creek fire, Blairmore, Southern Rockies Wildfire Management Area, 2003
Alberta Government, AFHPC

Livingstone-Macleod MLA David Coutts and Premier Ralph Klein talking to media during briefing of suppression activities on the Lost Creek fire, Blairmore, Southern Rockies Wildfire Management Area, 2003
Alberta Government, AFHPC

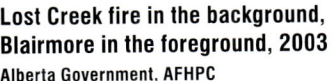

Lost Creek fire in the background, Blairmore in the foreground, 2003
Alberta Government, AFHPC

Fire in Alberta

Sketch of firefighter with fire hose
Artist Lorna Bennett

Thank you sign from the residents of the Crowsnest Pass to all the firefighters (volunteers, Municipal and Government) who worked on the Lost Creek Fire, 2003
Darryl Johnson

Wes Nimco's Incident Command team on export to Armstrong, Ontario, August, 2005
Back Row (L to R): Petter Finstad, Adam Sobchuk (Ontario), Bernie Schmitte, Chad Morrison, Ed Pichota, Mike Williamson, Chris McGuinty, Emile Desnoyers, Corey Davis, Herman Stegehuis, Ken Dutchak, Ed Edwards. Front Row: James Fix, Doreen Leichnitz, Sharron Purcell, Wes Nimco, Brian Wotton (Manitoba), Grant Forster, Ciara MacRory (Ontario), Adina Arcand, Laurie Brouzes (Manitoba), Tim O'Brien (Ontario). Team is along side a Coulson Helicopters Sikorsky S-61
Ken Dutchak

Incident Commander Wes Nimco taking notes during daily briefing session, Armstrong, Ontario, August, 2005

Bertie Beaver Fire Danger sign, Lac La Biche Airtanker Base, Lac La Biche Wildfire Management Area, 2005
Alberta Government, AFHPC

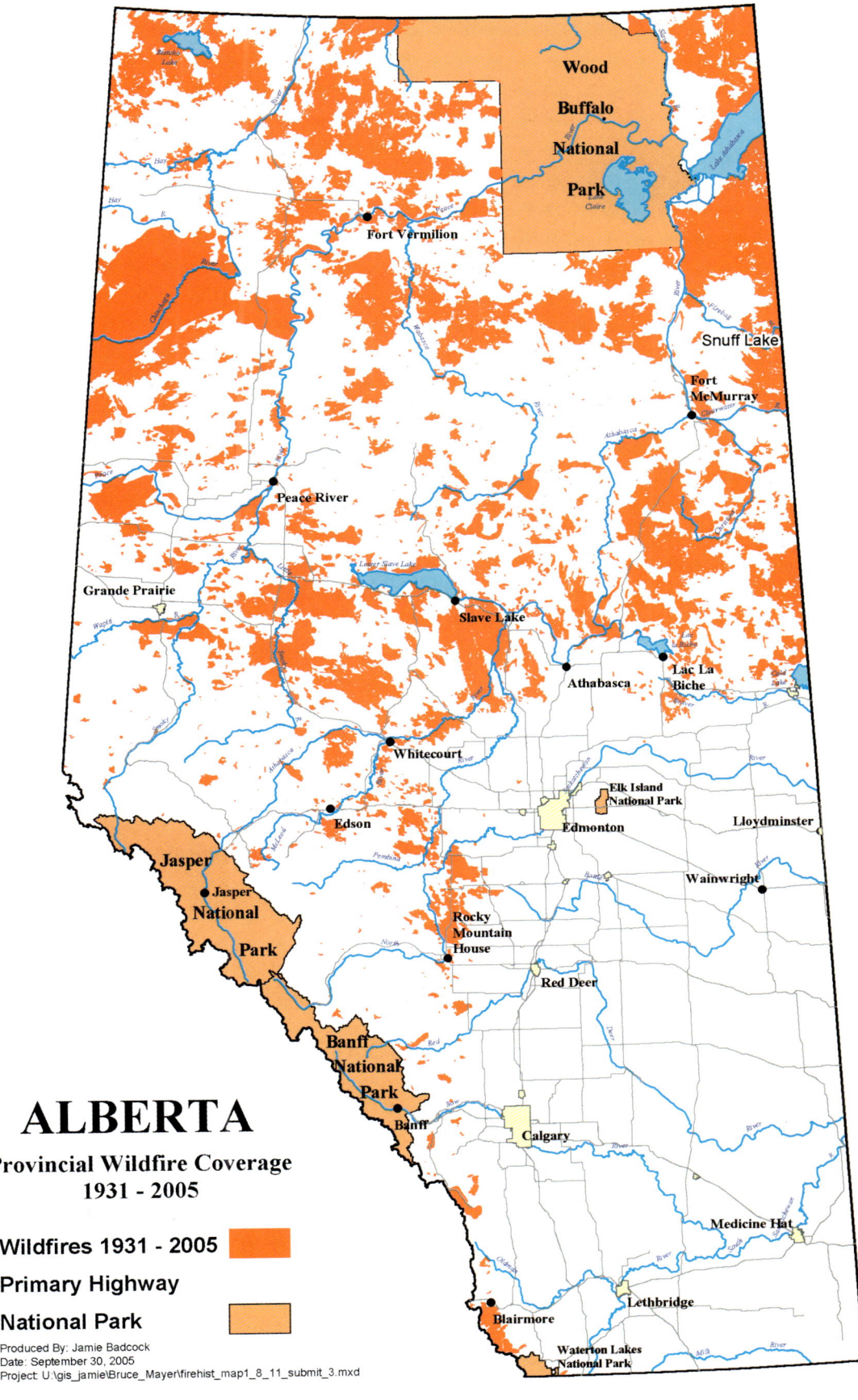

Fire in Alberta 399

CHAPTER 11

Alberta Forest Service Leaders Reflect

The Directors of Forestry have been leaders in bringing foresight, stewardship and innovation to protecting and managing the forests of Alberta. Their mandate has encompassed a complex package of responsibilities, from developing well-managed forests that are adequately protected from fire, to supporting a viable forest industry along with wildlife, fisheries and healthy watersheds, to meeting the cultural, spiritual and recreational needs of Albertans. Each Director faced a distinctive array of challenges, each had his own management style, and all were effective in their own way. The story of the Alberta Forest Service is reflected in their unique recollections. Unfortunately, there is no record of retrospective views from Blefgen, Huestis or Brennan, but some of their views are conveyed through comments they made while in office.

Ted Blefgen, 1930-1948

Ted Blefgen, 1930-1948

Ted Blefgen was the first Director of Forestry for Alberta, taking up the reins upon the transfer of resources in October, 1930. Eric Huestis, his successor, neatly reviewed Blefgen's career in 1948 after Blefgen's early retirement for health reasons.[1]

1930   AFS created October1, 1930
1931   Oliver Farm near Edmonton chosen for forest nursery site
1932   Work relief camps started
1936   Major fires in foothills
1939   World War II begins, many AFS staff enlist
1942   Increased lumber production for war effort
1947   Eastern Rockies Forest Conservation Board

**Alberta Forest Service leaders, November, 1970
(L to R): Ted Blefgen, Eric Huestis, Bob Steele**
Edo Nyland

"The Director of Forestry, Mr. T.F. Blefgen, has risen from the ranks to head the Forest Service. He started as an Assistant Ranger in the Crowsnest Forest on May 1, 1911, received various promotions to Ranger, Assistant Supervisor, Supervisor, Assistant District Forest Inspector under the Dominion Government, and with the transfer of the Natural Resources from the Dominion to the Province in 1930, became Director of the new Alberta Forest Service. His service was broken only by the period spent in the air force in the First World War. A great deal of credit is due to a man who rises from the lowest in an organization to the highest. It gives him, in his experience in the field, a wider and more lasting knowledge of the work of the Forest Service and gives him an intimate knowledge of the problems from the very lowest to the highest. It also gives him an intimate knowledge of the various Forests, making it possible for him to give better decisions on details in view of his close relationship to these details over a period of years."

Blefgen began his work as Director in a positive vein, resolved to maintain the level of programs started by the Dominion. However, he immediately faced the economic consequences of the Depression and forest fire loads during the drought years of the 1930s. Then he was confronted with the impact of the Second World War and its increased demand for wood and a shortage of staff. His most significant achievement was to keep the Forest Service functioning during this time. In fact, through innovation, the AFS continued to build lookouts, trails and cabins, built a forest nursery at Oliver and introduced radios to lookouts and ranger districts.

That Blefgen respected his staff was revealed in his annual report for 1942-43. After describing the great difficulties created by the war effort and staff shortages, he stated: "In the face of this definite increased demand for products of the forest, and the great difficulty in assisting inexperienced operators in starting, the majority of the Forestry staff have put forth Herculean efforts in keeping up their work as well as possible, and are to be heartily commended as a result. They are to be especially commended in such times as these when every hand is urgently required to help with the view of keeping this country free of Europe's tyrants."

In his reports during the early 1940s, Blefgen wrote about future needs for reforestation, the contributions of rangers to fish and wildlife work, introduction of bulldozers to firefighting, trials by the RCAF of aerial photography, preparatory surveys to distinguish agricultural and forested areas, and interest by pulp and paper companies in Alberta forests. But, even in his last report, for 1946-47, he repeated the main constraint imposed on him: "… during the depression years we were definitely informed that no money could be made available, and during the war years the necessary labour could not be secured."

He did, however, articulate his vision and lay the groundwork for achieving it. He would have gained great satisfaction later as his ideas became reality through the establishment of the Green Area in 1948, the aerial photography and forest inventories of 1949, the revised *Forests Act* of 1949, negotiations for pulpwood lease areas, and the hiring of new foresters in 1949. He prevailed during a time of great challenge and, with Eric Huestis, set the stage for a rapid unfolding of the forestry sector in Alberta.

Eric Huestis, 1948-1963
(Deputy Minister 1963-1966, Civil Service Ombudsman 1966-1970)

Eric Huestis also started his career with the Dominion Forestry Branch. He worked in 1923 and 1924 as a summer student during his forestry program at the University of British Columbia, first in Slave Lake under Ted Blefgen, then in the Brazeau and

Eric Huestis, 1948-1963

1948	Green Area set up
1949	New *Forests Act*, forest inventory begins
1951	Forestry Training School opens at Kananaskis
1952	Rabies control program
1954	First pulpwood lease agreement for NWP&P
1956	Large fires in west and north
1957	First AFS aircraft purchase – Helio Courier
1960	Forestry Training School opens in Hinton

Cypress Hills Forest Reserves. He started work full-time in 1925, and transferred to the AFS in 1930. He had worked in most of the Forests by the time he moved to Edmonton in 1948 to become Acting Director of Forestry.

Like Blefgen, he coped with hard times, commenting during an interview in 1983: "The basic problem of doing anything in Alberta in forestry, and particularly firefighting, was lack of money."₂

Oil revenues began to flow to the government after 1947 following the Leduc Number One discovery. However, it was not until after the AFS became part of Lands and Forests under Norman Willmore as Minister in 1955 that ministerial interest and levels of support really began to increase. As a reminder of the humble beginnings of the AFS, Huestis often referred to a dog-eared list of the names of the eight rangers in 1939 who were the only full-time permanent staff in the AFS staff outside the forest reserves.

Huestis was particularly pleased when in 1948 he got the minister of the day to budget for aerial photography to be taken across the entire province to produce a series of base maps. Huestis argued that these maps were needed to support oil exploration activities, and he was able to tack onto the project a forest inventory for the southern half of the boreal forest. This was also the year in which the Green Area was established. These achievements were followed in 1949 by his hiring of Reg Loomis to take charge of the forest inventory and his obtaining approval to hire nine new foresters. As he said: "I went over to UBC. I had enough money to hire nine foresters so I went through a graduating class of 21 and picked out my nine!" Also in 1949 he added a clause to his revised *Forests Act* to enable negotiated leases "for the purpose of growing continuously and perpetually successive crops of forest products – adjusted to the sustained yield capacity of the lands." These events set the stage for advancing the technical aspects of forest management. In the meantime, Huestis lobbied for support for fire control. Capitalizing on the serious fire year of 1956, he succeeded in winning industry, government and public support to start to beef up the AFS fire control capabilities.

In 1962-63, his last year as Director of Forestry, Huestis noted that the fire season had been "a welcome relief" from that of 1961; a combination of favourable weather along with the build-up in fire-control strength he'd been able to initiate. He would also surely have been proud when he relinquished command of the AFS, looking back at well established programs and good people in forest management, silviculture, land use, construction of roads, ranger stations and lookouts, and expanding training efforts at his new Forestry Training School at Hinton with its adjacent Cache Percotte Forest area set up for education and study.

Huestis had also served as Commissioner for Fish and Game from 1940 to 1958. He was especially pleased to have had Forest Rangers appointed as Fish and Wildlife Officers in 1940 and to get them involved throughout the forested area. Forestry staff had worked closely with trappers to restore beaver populations by defining trapping areas and sealing beaver pelts to manage trapping. They were rewarded in the mid-1950s by a resurgence in the numbers of beaver. As well, forestry staff were deeply involved in the extensive rabies control effort of the early 1950s in northern Alberta.

His term as Director enabled Huestis to initiate many ideas, both his own and those of his staff, that had been so long thwarted by adverse circumstances. He continued his career for another six years as Deputy Minister of Lands and Forests and four more as Civil Service Ombudsman.

Bob Steele, 1963-1973
(Deputy Minister 1973-79 and Deputy Minister of Telephones & Utilities 1979-1982)

Bob Steele, 1963-1973

Bob Steele presided over a number of major developments, perhaps most notably in forest management and the Quota System, along with effective use of the growing aircraft fleet. Among his comments about his decade as Director of

1963	Second Forest Management Agreement, Grande Prairie; the 990-kilometre Forestry Trunk Road officially opened
1966	Quota System – full commitment to forest regeneration after harvest
1968	Fires in May, worst in recorded history
1969	AFS reorganized, Ranger Districts centralized
1972	Report on effects of timber harvesting

Forestry, he said.₃

"The fact that we developed (forest) management plans that were fairly comprehensive and enabled the province to become involved in a greater development of the timber resources is probably the best thing of that period. I took a great deal of interest in it. I used to work extra hours all the time on management planning because I thought it was essential to do the job properly. I've always believed that you shouldn't get involved in things unless you know what you're doing or know what you're talking about. I think that was probably the greatest thing, seeing the development of the forest industry to a significant extent in Alberta.

"I worked with Al Saunders on development of the Forest Technology program at the Northern Alberta Institute of Technology. We got along very well and everything worked out very satisfactorily with the Forest Technology School at Hinton and NAIT.

"The other thing that I thought worked out very well was the organization of our field offices. We set it up so the offices would be more efficient and more self-sufficient. We allowed them more scope to manage the business out there on their own without being interfered with too much, because I felt there were too many people in head office wanting to know every little thing that was going on out there and it was not an efficient way of handling it.

"There's no doubt that the use of aircraft made a tremendous difference. Pretty well every provincial government elsewhere in Canada eventually ended up with a lot of aircraft capabilities, but I think Alberta almost pioneered a lot of it and developed the ideas of how the system should be administered and what type of aircraft to use."

Fred McDougall, 1974-1978
(Deputy Minister 1978-1989)

Fred McDougall, 1974-1978

"I was appointed Director of Forestry in the Department of Lands and Forests on February 1, 1974. By that time, Huestis and Steele had built the Forest Service from modest beginnings into a large and effective

1974	Computerized silvicultural record system set up, first in Canada; *Forest Development Research Trust Fund Act*
1975	New Department of Energy & Natural Resources, major reorganization
1976	Pine Ridge Forest Nursery at Smoky Lake approved
1977	Eastern Slopes Policy; Phase III inventory approved
1978	Mountain Pine Beetle outbreak in Crowsnest Pass area vigorously attacked with a variety of control measures, controlled in 1988

organization, with over 600 permanent staff.₄

"The breadth of activity was staggering. Through the Forest Land Use Branch, the Forest

Service was the land management agency for the Green Area of 388,500 square kilometres, or almost two-thirds of the province. It administered grazing, oil and gas operations, forest recreation, watershed management, reclamation and land use planning. The Forest Management Branch supervised all aspects of the growing forest industry and was rapidly expanding its silviculture programs. There was a Construction and Maintenance Branch, which built and maintained 5,800 kilometres of forestry roads plus bridges, airstrips, fire bomber (airtanker) bases, 170 fire lookouts and towers, 41 ranger stations and regional offices. There were 1,200 buildings in the Forest Service inventory. There was a mechanical equipment section doing maintenance and developing new equipment (primarily for forest fire control).

"The Training Branch, based out of a new school at Hinton, offered the second year of the two-year Forest Technology course in conjunction with the Northern Alberta Institute of Technology, an in-service Forest Technician program, and also offered a series of nationally recognized Forest Fire Control training programs. A well-organized Junior Forest Ranger Program was also administered from the Training Branch. And last, but certainly not least, was the Forest Protection Branch, which managed all eight Alberta government aircraft plus a seasonal fleet of 18 leased aircraft (helicopters and fire bombers), a major communications system, a weather forecasting section, a fire research program, and a very large and effective forest firefighting system supported by a huge inventory of firefighting equipment and a small army of seasonal firefighters.

"But the most impressive thing about the organization was its people. They were knowledgeable and well trained. The exceptional characteristic was their total dedication. The Forest Service was a huge family, as years of working together, often living together in remote locations, had resulted in an unusual degree of interdependence and many lifetime friendships.

"The Director of Forestry in those days carried an intimidating responsibility, and during my first year, 1974-75, I often felt overwhelmed by the challenges. The Forest Service continued to grow rapidly (67 positions were added in 1974-75). Approval was obtained to proceed with the Pine Ridge Forest Nursery, funded by the Heritage Trust Fund. The Forest Development Research Trust Fund was established under its own Act and administrator with $90,000 allocated to 10 research projects in the first year. A forest management agreement was executed with Alberta Aspen Board, establishing a poplar waferboard plant near Slave Lake. In September, 1974, the Environment Conservation Authority reported on the results of their hearings into Land Use and Resource Development in the eastern slopes. The Forest Service was to be heavily involved in implementing nearly all of the recommendations.

"A major, though unheralded, achievement in 1974 was the development and implementation of a comprehensive silvicultural record system whereby every cutblock in the province from 1966 onward was located, numbered and entered

Investment in the 1980s, increased Pine Ridge Forest Nursery capacity
Bob Stevenson

into a computerized record system. The system enabled complete and accurate enforcement of the provincial reforestation regulations. This system was carefully updated for many years. Alberta was the only province in Canada to maintain accurate records of all areas cut with up-to-date information on their reforestation status.

"Following the 1975 provincial election, Don Getty was appointed Minister for the newly created Department of Energy and Natural Resources. The Department of Lands and Forests became the Renewable Resources group in the new Department. Although the Forest Service was retained as a major division in the new Department, it was diminished as the Construction and Maintenance Branch was eliminated with responsibility for roads, bridges, and airports transferred to the Transportation Department. The administration and management of government aircraft was transferred to Government Services. Along with these changes, the historic title of Director of Forestry was lost as the head of the Forest Service was now to be titled Assistant Deputy Minister, a change I disliked and unsuccessfully resisted. I did not want to be the last Director of Forestry.

"In May, 1978, I was promoted to the position of Deputy Minister, Renewable Resources. Looking back on my four years as Director of Forestry (and subsequently as Assistant Deputy Minister) I am most proud of the fact that I was able to maintain positive momentum and continued growth and improvement in the Forest Service through a difficult period of change. In spite of government re-organization, and the loss of important program elements, the Forest Service continued to improve its effectiveness in such core areas as forest fire control, reforestation and forest management. We made a major contribution to provincial land use policies, and implemented new eastern slopes policies quickly and effectively. Our mountain pine beetle control program from 1977 to 1988 was a major success, avoiding huge timber losses. And forest fire losses from 1974 to 1978 were small, averaging less than 13,000 hectares per year."

Al Brennan, 1978-1985
(Executive Director Forest Industry Development 1986-1993)

During his seven-year term as ADM of Forestry Al Brennan led and presided over a number of significant events. These were gleaned from a review of the annual reports for the years during his tenure. Those issues that he seemed to feel

Al Brennan, 1978-1985

1979	Pine Ridge Forest Nursery officially opened, Maintaining Our Forests program begins
1979-83	Series of major fire years led to major investments in fire management
1980	AFS 50th Anniversary
1981	1.3 million ha burned in major fire season
1983	First Oriented Strandboard (OSB) mill, Edson
1984	Pine Ridge ships 100-millionth tree seedling; Eastern Slopes Policy revision; Phase III forest inventory completed
1985	Position Paper on Forest Industry Development

were the most important were forest protection, forest management planning, forest regeneration and forest industry development.[5]

The forest protection system was severely tested during the major fire years 1979 to 1983. These seasons also highlighted the problem of fires in the wildland-urban interface. The experience led to reassessment and strengthening of forest protection, including the Presuppression Preparedness Rating System (PPRS) and Helitack programs, and the purchase of CL-215 water bombers.

The Phase III forest inventory was completed. Data from the inventory were used to prepare the publication *Location of Future Forest Industry in Alberta*, setting the stage for the industry expansion which started in earnest at the end of his term. The integrated resource management plan for the eastern slopes area was revised through REAP and released in 1984 as *Policy for*

Resource Management of the Eastern Slopes. This continued to be a major guiding policy. As well, preparations for the renewal of quota certificates got underway, supported by fieldwork and planning to reassess the quality and quantity of timber available.

The major event in silviculture was opening of the Pine Ridge Forest Nursery in 1979. Designed to produce 9.2 million seedlings per year, by 1984 it had been increased to 32 million and in that same year Pine Ridge celebrated cumulative production of the 100-millionth seedling. A Silviculture Improvement Advisory Council was formed to further collaboration among industry and government foresters. Through the Maintaining Our Forests (MOF) program, $25 million from the Heritage Savings Trust Fund was invested in silviculture programs during the seven years of his term.

In 1985, during Brennan's last year as Assistant Deputy Minister, he assisted in writing the government's Position Paper outlining its proposed initiative to encourage investment in the forest products sector. The paper, which reflected his faith in Alberta's tenure system, noted that the existing softwood industry utilized only 60 per cent of the available annual allowable cut, and the use of hardwoods was significantly lower at only six per cent of its AAC. The conclusion was that there was tremendous potential for further growth and that the tenure system would continue to support a strong and stable forest industry in the province: "A healthy and vibrant forest products industry, strengthened by the Alberta forest industries development initiative, will continue to be a significant source of strength and diversification to the Alberta economy", the paper said. [6]

Brennan helped bring this vision to reality in 1986 when he moved to take charge of what became the government's new Forest Industry Development program. He held that position until ill health caused him to retire early, in 1993.

Cliff Smith, 1985-1989
(Deputy Minister 1989-1993)

Cliff Smith, 1985-1989

1985	Government White Paper identifies forestry as a potential major contributor to economic diversification
1986	Department of Forestry, Lands & Wildlife, Forest Industry Development Division (FIDD); first Medium Density Fibreboard (MDF) mill at Blue Ridge; CL-215 airtankers purchased
1988	Alberta Forest Research Advisory Council (AFRAC); Alberta Newsprint Company mill announced
1989	Expert Panel on Forestry (Dancik) appointed

"I had the privilege of leading the Alberta Forest Service from 1985 to 1989. This was a most exciting period in the history of the AFS; one that led to a considerable number of new activities and new initiatives. [7] In 1985 we were fortunate to have John Zaozirny as the Minister of Energy & Natural Resources. John not only highly valued the Forest Service and its people, but also recognized the potential of Alberta's fledgling forest industry sector. Through his leadership the government adopted a White Paper that identified forestry as a major force, capable of providing the much-needed diversification to Alberta's economy, which to that point was over-reliant on its agriculture and energy sectors. The White Paper paved the way for a number of new challenges for the Forest Service. Among them was the development of Alberta's forest industry, which entailed a major expansion on all fronts, including pulp and paper, lumber and panel products. The development opportunities included Alberta's first newsprint mill at Whitecourt, Canada's first oriented strandboard plant at Edson, and Canada's first medium density fibreboard plant at Blue Ridge. Industrial forestry had truly arrived in Alberta.

"One of the most rewarding spin-offs from this development within the industry was the emerging interest in aspen and balsam poplar; of which Alberta had a bountiful supply and which until the 1980s were not considered much more than weeds. Several of the new developments were based on these poplar resources, either in whole or in part. At last we were able to carry out truly integrated harvesting, which in turn provided for more deliverable silvicultural prescriptions.

"New Free to Grow standards were developed to strengthen Alberta's already strong track record in reforestation. The conceptual design of ecosystem management was something that most of us subscribed to but may not have really understood, but we recognized that there was a very strong need for some different approaches. In timber harvesting we've gone away from the idea of blocks and rectangles and squares, which at the time were thought to be good and didn't do a bad job silviculturally, but certainly didn't address a lot of the other needs of the landscape. I think now there is a much softer approach in terms of impact.

"The AFS also made major gains in its fire management program, in terms of both technology and staff expertise and enjoyed an enviable record throughout the latter half of the eighties. None of this would have been possible without a strongly dedicated, focussed and capable staff, all proud to wear the AFS badge. Most of them viewed their vocation more as a way of life than a job."

Ken Higginbotham, 1989-1995

Ken Higginbotham, 1989-1995

Ken Higginbotham taught forest ecology at the University of Alberta forestry program before starting full-time work with the Alberta Forest Service in 1988 as Director of the Forest Research Branch. He became Acting Assistant Deputy Minister for

- 1990 Four new mill projects completed at Peace River, Whitecourt, Blue Ridge and Hinton, Al-Pac approved
- 1991 Free to Grow standards for regeneration, public liaison committees for FMAs, Daishowa court case hearings about the issue of perpetual sustained yield
- 1992 Land and Forest Service in Department of Environmental Protection, Daishowa court case decision
- 1994 Special Places initiative begins

Forestry in 1989, ADM in 1990.$_8$

"This all happened at a time when the expansion of the forest industry was going at full swing, so my early days in that office [1990] were involved with FMA negotiations with Cliff Smith, Deputy Minister. The expansion of the forest sector seemed to coincide with a growing environmental presence. I think the uproar that occurred around Alberta-Pacific and Daishowa Marubeni was prompted in large part by the 1985 White Paper that encouraged use of the forest and its opportunities as a way to diversify the provincial economy. This led to a huge focus in my early days in trying to deal with environmental and public expectations.

"For instance, it led to the development of a policy that required FMA holders to formally seek public involvement in creating detailed forest management plans. It also became very clear to us pretty quickly that the forest management plans were too obscure or high-level for people to care about - they were much more interested in annual operating plans. That's one of the things that led to AOPs (Annual Operating Plans) in Alberta becoming three- to five-year plans. I think that it was the public interest that led us to realize that people were going to insist that they have some up-to-date knowledge, for example, on where road systems were going to be developed and where logging was going to occur and how wildlife interests were going to be looked after.

"With the advent of major government downsizing, the Forest Industry Development Division was moved into the Economic Development portfolio, and the Forest Service was moved into Environmental Protection. Not only did Forestry, Lands and Wildlife get wrapped into Environmental Protection, but ultimately the organization of the Forest Service with its 10 forests or 10 regions was destined to disappear. It was reduced first to six regions with complete universality of the regional structure for forests, environment, wildlife and so on. I think that the Forest Service budget, apart from firefighting of course, went from about $100-$110 million annually down to about $85 million. A major change which occurred during this time was combining AFS and the Public Lands Division to create a new entity, the Land and Forest Service (LFS). Unfortunately, these changes led to a lot of people leaving. I personally supported and believed strongly in the benefits of what the government was trying to do with the downsizing. But it also meant that some of the wonderful people that had been so important to the Forest Service over the years chose it as a time to leave - some of those people that in my mind embodied the spirit of the Forest Service. To me, that was the important part of being in the Forest Service - the spirit of that organization.

"One time, when the Heritage Savings and Trust Fund Oversight Committee visited Pine Ridge Forest Nursery, Grant Mitchell, who was then the Liberal forestry critic, asked: 'Why in this day and age would people actually be willing to wear a uniform?' I said, 'Mr. Mitchell, these people wear the uniform proudly. It's part of the tradition of this as an agency that is different than any other government agency.' I also remember clearly the time when Jim Kitz passed away. He was a young man with high potential in the Forest Service and with great respect from everybody in the organization. At his funeral, the most emotionally powerful thing to me was seeing 180 Forest Service people, everyone who had been able to get there, in uniform, forming an honour guard for Jim's casket and for his family. It really embodied for me what the Forest Service was all about.

"The Forest Service and forestry in general was also greatly increasing its focus on First Nations issues through that period. Premier Klein was admired by many among the aboriginal groups and he undertook to work towards some enhancement of aboriginal rights. The whole business of aboriginal rights and participation in co-operative management was developing at the time when Tolko came into the province and a wood supply was being put together for their High Prairie OSB mill. We took part of the S9 wood supply and put it in the hands of the Whitefish band, for example. There were agreements that the wood was going to end up going to the Tolko mill but it gave some opportunity there for the Whitefish band to have an ongoing revenue source. Similar kinds of agreements were made with the Little Red River Cree Nation, Tallcree First Nation and the Dene Tha' First Nation in the High Level area. The province had also done some work around Métis settlements with the aim of establishing a land base for them, so we became heavily involved in working with Métis settlements, helping them with the calculation of AACs and management systems for their lands."

Cliff Henderson
(1995-2001 ADM Land and Forest Service, 2001-present, ADM of Forest Protection)

Cliff Henderson was raised on a farm near Ponoka, Alberta and graduated with a Bachelor of Science in Forestry from the University of Idaho.9 His first posting with the AFS was to Fort McMurray in February, 1966. After holding

Cliff Henderson, 1995 - present

1995	Major reduction in all government services
1996	Forest Management Science Council
1997	Foothills Model Forest Phase II
1997	Designated Administrative Organizations (DOAs) reflect privatization initiatives; Alberta Forest Conservation Strategy; Award-winning Fire Simulator training CD-ROM
1999	Integrated Resource Management (IRM) Division created
2001	"Forestry" now led by three ADMs in new Department of Sustainable Resource Development
2002	FireSmart program introduced
2002	Wildland-urban Interface program introduced
2004	Municipal Wildfire Assistance program

various positions in the field and in Edmonton, Cliff became ADM of the Land and Forest Service in 1995.

Cliff recalled after Premier Klein took over leadership of the province in 1991, work began to downsize the provincial government and develop a business planning process for program funding. This process was instrumental in creating a framework for change over the next 10 years. During the early 1990s the Service was reduced from 1,800 to 1,150 staff and the field organization, which had been 10 Forests and 42 Districts, was reconfigured to four Regions and 18 Districts. During the 90s, the Forest Service was a Branch within the Department of Environmental Protection. Ty Lund was the Minister for over five years, and under his watch the protection of the environment was carefully balanced with resource development. Significant change was under way, but the Forest Service maintained responsibility for protecting and managing lands and forests.

It was exciting times in the 1990s as the Forest Service worked to keep up to the government changes. The downsizing prompted creation of Delegated Authority Organizations. These were independently governed organizations, which assumed responsibility for program delivery. The two providing support to the Forest Service were the Forest Resource Improvement Association of Alberta (FRIAA) and Alberta Conservation Association (ACA).

The 1990s also witnessed rapid development of timber resources. The forest industries sawn products increased by three times to production exceeding three billion board feet. The market-rate system of timber dues encouraged mill upgrading to increase efficiency and utilization, resulting in some of the most modern and efficient mills in North America. A component of the dues was set aside under FRIAA to provide funds for research and forest improvement programs. This system provided opportunity for companies to participate in the Foothills Model Forest and Sustainable Forest Mangement Network at the University of Alberta. Significant operational research results now help guide the management of Alberta's forests.

The Forest Service had another major reorganizational change in 2001 when it moved to a new department called Sustainable Resource Development. The new Department included Fish and Wildlife, Public Lands, Lands and Forests, Forest Protection and Strategic Forest Initiatives. Cliff Henderson became the ADM of Forest Protection.

The Forest Protection Division grew out of the core of the Forest Service and came about as a result of catastrophic fire seasons that occurred in 1998 and in some of the years that followed. The division has become a leader in North America. One of the most significant programs developed, is the FireSmart program to implement efforts to reduce fire risk along the urban interface with the wildland landscape.

Building on the past and striving for continuous improvement, Henderson maintains the 'can do' attitude of his predecessors. He is taking the Forest Protection Division to new levels by developing and implementing long-term strategic plans.

Henderson is also recognized as being an ardent supporter and advocate of recording and preserving forest history within the province. This passion has led to the development of a historical photo CD, the production of this history book of the 100 years of forest service, and the support of the Forest History Association of Alberta.

Howard Gray, 2001-2005
(2001-2003 ADM Land and Forest Division, 2003-2005 ADM Strategic Forestry Initiatives)

Howard Gray, 2001 - 2005

1998	Appointed Executive Officer of Forest Industry Development Division
2001	"Forestry" now led by three ADMs in new Department of Sustainable Resource Development
2002	Canada-US Softwood Lumber dispute is ongoing concern
2003	Minister Cardinal announces that "forestry" is the primary industry in over 45 communities

Howard Gray's career began in 1966 as a warehouseman in Footner Lake, becoming assistant ranger at Meander River in 1968. In 1972 he left to work in the Northwest Territories where he became head of forest protection. Returning to Alberta in 1976 he alternated between positions at the Fire Centre and field offices until 1986 when he was appointed Forest Superintendent at Slave Lake, later becoming head of the Northwest Boreal Region at Peace River. In 1998 he returned to Edmonton, first as director of the Forest Industry program then in 2001 as ADM Land and Forest Division and in 2003 ADM Strategic Forestry Initiatives.[10]

His memorable achievements include a fire costing system still in use today, designed to measure operational success against fiscal responsibility; and helping develop the concept of up-front preparedness by spending more money at the front end in order to save considerable losses and expense at the back end of a fire event. He developed a helitack and rapattack program, along with specialized training courses at the Hinton Training Centre. As well, he initiated an advanced mock scenario of a major fire in Kananaskis Country to help officials in southern Alberta appreciate the potential danger of wildfires to communities.

Gray encouraged development of First Nations fire crews, and in his later career also found ways to include aboriginal communities in forest management and manufacturing opportunities, starting with the agreement with Tolko at High Prairie that enabled the Whitefish First Nations' community to partner with the forest company. "It was highly successful and is still working today. It was a real opportunity for the aboriginal people to have a bargaining chip when they sat at the table to cut a deal that would bring them a share of the business."

The ongoing softwood lumber trade dispute with the United States took a lot of effort. "The U.S. has identified the policies and practices it believes Alberta should change if it wants to achieve free trade." At the time of writing, the provincial government was faced with the decision of whether and how to go along with some of those changes. Gray envisaged a possible need for revisions to the *Forests Act* to resolve some of the trade issues. "It would be an interesting challenge that would allow us to fix a lot of things in the Act and the regulations. They need overhauling anyway to enable us to achieve the kind of forest management we should be doing in the 21st century."

He has also placed a high priority on the

promotion of good forest management. "I put a lot of emphasis on reforestation . . . having our staff go out and find out what's going on – are we really establishing forests or are we just planting trees? I always wanted to move on the issues of sustainability. We're here to look out for the public good of Alberta. The legacy we leave is good, healthy forests."

Gray, who is proud of a career that developed without aid of a university education, has supported the introduction of ever-increasing levels of science to the management of public forests, in the belief that better science builds better policy. Over his varied career, Gray says, he has tried to stay "on the leading edge, not fall into complacency."

"For me, it was a continual state of wanting to do it better, make it work better, to challenge the organization and never accept anything just on face value. This is especially true of forest sustainability, making sure we're managing sustainably, that we're moving forward and being as creative as we can to find solutions for problems."

He recognizes that the world of the forester has changed dramatically since the 1930s and 1940s. "I think we've done a hell of a good job in the first 75 years, considering what we did and what we had to work with," he says. "But now we're moving into a time where we really have to get focused and start to be more integrated. The fish and wildlife issues and the environmental issues and the market demands all have to start coming together for us in government. We've got to deal with biodiversity and landscape management and cumulative impacts and all those things, which are hugely complex. And they require a lot of expertise. We have to become a profession that's much more visible, much more articulate about what we're doing and why we're doing it. We also have to be more inclusive in listening to different opinions."

Craig Quintilio, 2003-present
(ADM Public Lands and Forests 2003-present)

Craig Quintilio is Assistant Deputy Minister of the Department of Sustainable Resource Development in charge of Public Lands and Forests.[11] Born and raised in the Crowsnest Pass, he developed an interest in forestry,

Craig Quintilio, 2003 - present

- 2001 "Forestry" now led by three ADMs in new Department of Sustainable Resource Development
- 2003 Forest Resources Improvement Association (FRIAA) mandate extended to 2007
- 2004 Bighorn Access Management Plan completed and a committee formed to implement the plan
- 2005 Softwood Lumber dispute remains unresolved; Mountain Pine Beetle outbreak

attended the University of Montana, graduating in 1972. After working in the coast forests of British Columbia he joined the Alberta Forest Service in 1973, working in a variety of field and head office positions. In 1995 he was appointed Director of Program Support, then Forest Protection and Public Lands before his present appointment in 2003. He emphasized two major aspects: integration of land uses and integration within the Department.

"First, as a Department we are the ones who are charged with proper management of the public land base. It's a huge mandate. Our job as the public land manager is to try to get a smooth integration of all those forest-based activities. This can't be just a government-run process. There are huge industrial players out there, including the oil and gas sector and the forest industry, the agriculture and range cattle folks – the big industrial sectors. So among the challenges will be to get a host of players to understand there has got to be some cooperation on the land base. And we need to recognize a whole range of other values, such as biodiversity, recreation and whatever values citizens want. So it's a matter of trying to come up with a kind of landscape approach to planning and management.

"Everybody has to take some responsibility for proper management of that land base. At the end of the day it's going to require some kind of a social licence to operate on it – regardless of what the operation is. For example, people out there on a quad or grazing cattle or a company harvesting timber, or an oil and gas company trying to put in a pipeline – they are all connected. If we don't do this right, everybody's going to lose.

"At the same time, we have to recognize that this province is focused on its developing economy. The very good lifestyle that we enjoy in Alberta is largely the result of the industrial base, so it's a huge job to enable those activities to occur on the public land base but to get a balance so that it is done sustainably.

"The second point for us within government is to effectively integrate management for sustainability. Since we are constrained by budgets, we have to be creative in how we deliver programs and this will mean more partnerships. I believe that if we want this public land base to be sustainable, we must all contribute to managing it. The foundation is still that the government sets the rules and is still ultimately responsible for what goes on out there. There is still a big role for government to play, less in doing the work and more in managing and being a catalyst for change. Part of the job is to get reasonable people in a room together, and a lot of times when you do that you can get out of the way and they'll come up with some pretty creative solutions. We are also working on new ways of doing business. If we expect integration on the land base, if we're going to be the catalyst, then our staff has to get into that mind-set, too.

"What has impressed me is that despite all these problems, the inherent strength of the organization has kept things going in timber and wildlife and forestry and fire. Throughout that period of downsizing, changes and reorganization we lost a lot of good people – but we also kept a lot of good people. It is a credit to the people who remained with this organization that despite the turmoil and the constant reorganization, the staff got the job done. We continue to get the job done and we deliver. We have done amazing things. The positive results are due to the will of the people who are here."

Helicopter tour of Mountain Pine Beetle infested sites, August, 2005
(L to R): Craig Quintilio, Assistant Deputy Minister Public Lands and Forests Division; Mark Storie, Area Manager Foothills Area; Darryl Johnson, Incident Commander
Duncan MacDonnell

Display of the many badges worn by Forest Rangers and Forest Officers over the last 100 years

Epilogue

The mandate of the Alberta Forest Service in 1930 was straightforward. As stated by Director of Forestry T.F. Blefgen in his first annual report in 1931, the AFS became responsible for the major forestry activities within the province. These included protection and administration of the Forest Reserves and forest protection on the Edmonton Fire Ranging District, which essentially comprised the rest of the provincial forest area. However, the job was made difficult by a combination of vast areas, long distances, lack of access, few staff, little funding and a sparse provincial infrastructure such as telephone and telegraph service. Once started, the task was made even more complicated with the onset of the Great Depression that further reduced funding and staff. Added to that was the beginning of the extended drought period during the 1930s, followed by the war effort until 1945. It is amazing how they coped. Their story is a tribute to Blefgen and his dedicated AFS forest rangers and staff. This tradition of dedication and commitment has characterized the AFS throughout its history.

When Professor J.H. Morgan reported in 1886 on his preliminary review on the forests of Canada, he was impressed with their extent but appalled by the losses through uncontrolled logging and forest fire. He also lamented not knowing the extent and nature of the forest, the need to organize a system of forest management, and absence of any forestry schools to train qualified staff. Thirteen years later, in 1899, what was to become the Dominion Forestry Branch was created to start to address these concerns on the Dominion lands of the west.

Then, fast-forward to 1946, when the Alberta Post-war Reconstruction Committee recommended that Alberta expand fire protection services, make a physical inventory of the forest, institute a long-range program of reforestation and inaugurate a training program in forestry.

In 1990, 44 years later, the report of the Expert Panel on Forest Management in Alberta contained 133 recommendations, among which were several dealing with forest fires, forest inventories, reforestation and education.

Despite the persistence of these concerns, successes were being achieved at every stage. AFS staff worked within a philosophy of "continual improvement" in which they strived to do better despite constraints imposed by limitations of physical and economic resources. But, while making progress, the task of "protecting and managing" the forest itself was becoming increasingly complex through greater demands for forest products, increased activities in the forest (both commercial and recreational) and rising public expectations of what the forest could provide. It became an ongoing race between these driving forces, on one hand, and the generation and application of energy, ingenuity, experience, research, knowledge, education and technology by the forest managers on the other.

Consider forest inventory, for example. In 1886 the actual extent of the forest was not well understood. By 1946, the boundaries of the forest reserves had been surveyed, additional prospective reserves had been mapped and a tentative border between lands best suited respectively for forestry or agriculture was being defined. To get more precise estimates of timber values and rates of growth the entire forest was mapped during the 1950s using the new technology of aerial photography. By 1990 estimates of growth and yield had become more precise. Satellite imagery and remote sensing made it possible to monitor changes and computers could be used to store and analyze the results. However, by 1990 it was also clear that inventory data were also needed for other forest-related values, such as wildlife and their habitats, watersheds and fisheries habitat, soils and landforms, cultural and historic sites and viewscapes.

The point of this example is that while the need for inventories has persisted, the scope of the needs has grown. Research and knowledge have led to greater understanding and technologies have been applied to facilitate them. Increased populations and economic activities have resulted in greater public concerns about impacts on the forest which, at the same time, led to greater efforts to increase knowledge about the forest ecosystem and how best to ensure its sustainability. As concerns became more complex, managers of the forests became more skilled in addressing them.

The three-year business plan for the Alberta Sustainable Resource Development, released on March 2, 2004, says:

"Alberta's natural resources contribute to the high quality of life, and the high level of education, health and social programs Albertans enjoy. Alberta's dynamic economy and the ability to maintain it over the long-term are the direct result of the sustainable management of the province's natural resources. It requires a balance among environmental, economic and social benefits that Albertans receive from these resources. Key to achieving success in sustainable resource management is viewing natural resources and environment as interrelated parts of a single system."

The plan identified a number of emerging challenges and opportunities, including industrial footprint and access management, sustainable resource management and climate variability. Some of the more specific issues affecting the forest included:

- Sustainable forest management and measuring progress through generally acceptable criteria and indicators
- Certification – choosing between rules- and objectives-based systems or finding a balance between them; threats of boycotts and marketplace or other consumer actions
- Impacts of energy and other sectors on achievement of sustainable forest management and sustainability of wood supply to forest industries
- Economics – sharing of costs of sustainable forest management among forest users and beneficiaries
- Softwood lumber countervail and anti-dumping actions by United States interests
- Climate variability and/or climate change and possible influences on forest fires, insects, diseases and adaptability of tree species, as well as implications for biodiversity
- Forest fires – finding a balance among costs of protection, the natural role of fire in the ecosystem and values-at-risk
- Integration of uses on forest lands as well as integration of government administration agencies and incorporation of changing public values
- Roles, responsibilities and accountability of professional forest managers

This may appear to be a formidable list, but in a sense these elements represent variations of the issues faced by Ted Blefgen, Director of Forestry, in October 1930 when the Alberta Forest Service was established. In the same spirit that he and his AFS colleagues tackled their problems with innovation, ingenuity, persistence and intelligence, so will the present forestry community. One of the major differences now is that more people are involved. The processes of collaboration, consultation and networking make available the combined talents and intellect of the many people interested and involved in the forest and its sustainability. As well, there are growing capabilities in forest research, education and training that are extending our knowledge and thereby increasing abilities to apply it. That, combined with the inherent dedication, energy and creativity of those involved, bodes well.

If the past is prologue, the future looks encouraging, indeed.

Conversions

1 hectare	2.4711 acre	1 gallon	4.5461 litres
1 kilometre	0.6214 mile	1 acre	4046.9 m²
1 square km	0.3861 mile²	1 inch	2.54 cm
1 kilogram	2.2046 pounds	1 pound	0.4536 kg
1 cubic metre	233 foot board measure		
1 foot board measure	12" x 12" x 1" thick board		

References

Chapter One

1. P.J. Murphy, *History of forest and prairie fire control policy in Alberta* (Edmonton: Energy and Natural Resources Report Number T/77) 1985.
2. Canada 1885, "Department of the Interior Annual Report for the fiscal year 1885," Sessional paper, Ottawa, 1885.
3. R. Huth, *Horses to Helicopters, Stories of the Alberta Forest Service* (Edmonton: Energy and Natural Resources) 1980.
4. Canada 1902, "Department of the Interior Annual Report for the fiscal year 1900-1901," Sessional paper, Ottawa, 1902.
5. Canada 1911, "Department of the Interior Annual Report for the fiscal year ending March 31, 1910," Sessional paper, Ottawa, 1911.
6. E.H. Finlayson, "Instructions and specifications for trail construction," *Alberta Inspection District* (Calgary: AFS Archives) 1916.
7. Canada 1911, "Department of the Interior Annual Report for the fiscal year ending 31 March, 1910," Sessional paper, Ottawa, 1911.
8. A. Knechtel, "The Forest Reserves," *Forestry Branch Bulletin No. 3*. (Ottawa: Department of the Interior) 1910.
9. R. Huth, *Horses to Helicopters, Stories of the Alberta Forest Service* (Edmonton: Energy and Natural Resources) 1980.
10. G.H. Edgecombe and P.Z. Caverhill, "Rocky Mountains Forest Reserve. Report of boundary survey parties," *Forestry Branch Bulletin No. 18*. (Ottawa: Department of the Interior) 1911.
11. R. Huth, *Horses to Helicopters, Stories of the Alberta Forest Service* (Edmonton: Energy and Natural Resources) 1980.
12. Canada 1916, "Department of the Interior Annual Report for the fiscal year ending 31 March, 1915," Sessional paper, Ottawa, 1916.
13. Canada 1923, "Director of Forestry Annual Report for the fiscal year 1921-22," Sessional paper, Ottawa, 1923.
14. R. Huth, *Horses to Helicopters, Stories of the Alberta Forest Service* (Edmonton: Energy and Natural Resources) 1980.

Chapter Two

1. Alberta, "Department of Lands and Mines Annual Report 1930-31."

2. Ibid.
3. Ibid.
4. Ibid.
5. Sam Fomuk, *Telegraph keys in the Alberta Forest Service* (Edmonton: AFS Archives) 1970.
6. Alberta, "Department of Lands and Mines Annual Report 1930-31."
7. Alberta, "Department of Lands and Mines Annual Report 1944-45."
8. Alberta, "Department of Lands and Mines Annual Report 1945-46."

Chapter Three
1. Robert Diesel, interview by P.J. Murphy, Lac La Biche, AB., 10 September 1998.
2. Jack Grant, interview by P.J. Murphy, Edmonton, AB., 26 January 1999.
3. Chuck Rattliff, interview by P.J. Murphy, Grande Prairie, AB., 26 March 1999.
4. Land Forest Wildlife magazine, date unknown.
5. Bill Balmer, interview by P.J. Murphy, Grande Prairie, AB., 6 June 1993.
6. Harry Edgecombe, interview by P.J. Murphy, 1990, and also *Land Forest Wildlife* Winter; Vol 8, No. 4., 1965-66.
7. Jack Gosney, interview by Dennis Quintilio, Hinton, AB., 2003.
8. Phil Nichols, interview by P.J. Murphy, Salt Prairie, AB., 17 April 1998.
9. Jay Champion, conversation and e-mail correspondence with Bruce Mayer, March - May, 2005.
10. Janet South, notes provided to Dennis Quintilio, 2004.
11. Art and Marilyn Peter, interview with P.J. Murphy, Calgary, 12 February 1992.

Chapter Four
1. Wallace R. Delahey, "Report to Minister of Lands and Mines for Alberta 1950," Consulting Forester, Toronto, 1950.
2. Alberta, "Department of Lands and Forests Annual Report 1959-60."
3. Alberta, "Department of Lands and Forests Annual Report 1960-61."

Chapter Five
1. Canada, 1910, "Departmental Annual Report 1910."
2. Alberta, "Department of Lands and Forests Annual Report, 1932-33."
3. Eric S. Huestis, talk at Forestry Training School, Hinton, AB., 31 January 1972.
4. J. Niederleitner, "Fire Detection Study 1984," *Lookout Classification Memo* (Edmonton: Alberta Forest Service) 1984.
5. Tim Klein, personal conversations and emails with Bruce Mayer January, 2005, Edmonton, AB., 2005.
6. Alberta, "Department of Lands and Forests Annual Report 1942-43."
7. Gordon Fowlie, interview by P.J. Murphy, Edmonton, AB., 28 January 2001.
8. Alberta, "Department of Lands and Forests Annual Report 1958-59."
9. Sam Fomuk, interviews by P.J. Murphy, Edmonton, AB., series of phone conversations 1998-99.
10. Jeff Henricks, Article in *E.P.I.C.* (Fort Vermilion: Alberta Environmental Protection departmental newsletter) Vol. 3, Issue 2, Summer 1995.
11. Doris Gosney, "Letter to P.J. Murphy from Doris Gosney," Hinton, 26 October 2000.
12. Chuck Rattliff, interview by P.J. Murphy, Grande Prairie, AB., 26 March 1999.

Chapter Six
1. Chuck Rattliff, interview by P.J. Murphy, Grande Prairie, AB., 26 March 1999.
2. Alberta, "Department of Lands and Forests Annual Report 1966-67."
3. Hon. John Zaozirny, "Department of Energy and Natural Resources Annual Report 1983-84,".
4. Alberta, "Department of Lands and Forests Annual Report 1966-67."
5. C.D. Schultz & Company, *The Environmental Effects of Timber Harvesting Operations in the Edson and Grande Prairie Forests of Alberta* (Vancouver: C.D. Schultz & Company) 1973.
6. Alberta, "Department of Energy and Natural Resources Annual Report 1975-76."

7. Alberta, "Department of Energy and Natural Resources Annual Report 1980-81."
8. Environment Conservation Authority (ECA), *Forest Management in Alberta: Report of the Expert Review Panel* (Edmonton: ECA) 1979.
9. Alberta, "Department of Lands and Forests Annual Report 1968-69."
10. Nick Nimchuk, personal conversations and e-mail correspondence with Bruce Mayer, Edmonton, AB., January, 2005.
11. E.V. Stashko, 1971. "Fire Weather Meteorologist to the 1971 Intermediate Fire Behaviour course," Hinton, AB., 1971
12. Dale Huberdeau, interview by P.J. Murphy, Fort McMurray, AB., 22 July 1997.
13. Alberta, "Department of Lands and Forests Annual Report 1969-70."
14. Alberta, "Department of Lands and Forests Annual Report 1971-72."

Chapter Seven
1. Lloyd Van Camp, interview by P.J. Murphy, Columbus, Indiana., 20 September 1989.
2. Jack Grant, interview by P.J. Murphy, Edmonton, AB., 26 January 1999.
3. C.F. Platt, "Memorandum," *Forest Protection Archives* (Edmonton, 1956).
4. T.R. Hammer, "Memorandum," *Forest Protection Archives* (Edmonton, 1956).
5. Cliff Smith, interview by P.J. Murphy, Sherwood Park, AB., 16 March 1999.
6. Forest Service, "Alberta Lands and Forests," *Para Cargo Manual* First Edition, August 1974, and *Alberta RAP magazine* Volume 2, Number 19, 15 October 1973.

Chapter Eight
1. Gro Brundtland, Chair, *Our Common Future. World Commission on Environment and Development* (Oxford University Press) 1987.
2. CCFM 1992, "Sustainable Forests: A Canadian Commitment," *Canadian Council of Forest Ministers* (Ottawa: CCFM) 1992.
3. CCFM 1995, "Defining sustainable forest management: a Canadian approach to criteria and indicators," *Canadian Council of Forest Ministers* (Ottawa: CCFM) 1995.
4. Alberta, "Department of Forestry Lands and Wildlife Annual Report 1985-86."
5. Ibid.
6. Alberta, "Department of Forestry, Lands and Wildlife Annual Report 1986-87."
7. Ibid.
8. Alberta, "Department of Forestry, Lands and Wildlife Annual Report 1987-88."
9. Alberta, "Department of Forestry, Lands and Wildlife Annual Report 1986-87."
10. McDonald, Hon. Mr. Justice D.C. In the court of Queen's Bench of Alberta, Judicial District of Edmonton. *In the matter of the Forests Act, R.S.A. 1980, C.f-16, s. 16; And in the matter of Alberta rules of court, Part 56.1; Action No. 9003-23400. Between: Peter Reese, Alberta Wilderness Association, Peace River Environmental Society and Sierra Club of Western Canada, Applicants, And Her Majesty the Queen in right of Alberta, the Minister of Forestry, Lands and Wildlife and Daishowa Canada Co. Ltd. Respondents. Reasons for judgment of the Honourable Mr. Justice D.C. McDonald.* Edmonton, AB, 1992
11. B.P. Dancik, "Forest Management in Alberta: Report of the Expert Review Panel." *Alberta Energy/Forestry, Lands and Wildlife*, Publ. No. I/340. Edmonton, 1990.
12. Alberta, "Department of Forestry, Lands and Wildlife Annual Report 1990-91."
13. Alberta, "Department of Forestry, Lands and Wildlife Annual Report 1985-86."
14. Alberta, "Alberta Environmental Protection Annual Report 1992-93."
15. Alberta, "Alberta Round Table on Environment and Economy Annual Report 1992-93."
16. Alberta, "Department of Forestry, Lands and Wildlife Annual Report 1988-89."
17. I.G.W. Corns, and R.M. Annas. *Field guide to forest ecosystems of west-central Alberta.* (Edmonton: Canadian Forestry Service) 1986.
18. Alberta, "Department of Forestry, Lands and Wildlife Annual Report 1985-86."
19. Alberta, "Department of Forestry, Lands and Wildlife Annual Report 1986-87."
20. Alberta, "Department of Forestry, Lands and Wildlife Annual Report 1990-91."
21. Alberta, "Department of Forestry, Lands and

Wildlife Annual Report 1991-92."
22. Cliff Smith, interview by P.J. Murphy, Sherwood Park, AB., 1 April 1999.
23. Alberta, "Department of Forestry, Lands and Wildlife Annual Report 1990-91."
24. Con Dermott, interview by P.J. Murphy, Sherwood Park, AB., 14 March 1999.
25. Alberta, "Department of Forestry, Lands and Wildlife Annual Report 1986-87."
26. Ibid.
27. Information provided by those named in conversations with Bruce Mayer, 2004-05.

Chapter Nine
1. Cliff Smith, interview by P.J. Murphy, Sherwood Park, AB., 1 April 1999.
2. Alberta, "Department of Environmental Protection Annual Report 1992-93."
3. Ibid.
4. Alberta, "Department of Environmental Protection Annual Report 1995-96."
5. Alberta, "Department of Environmental Protection Annual Report 1994-95."
6. Alberta, "Department of Environmental Protection Annual Report 1995-96."
7. Alberta, "Department of Environmental Protection Annual Report 1996-97."
8. Alberta, "Department of Environmental Protection Annual Report 1997-98."
9. FireSmart, *Protecting Your Community from Wildfire; Partners in Protection,* (Quincy: Firewise) 1999.
10. Ibid.
11. Alberta, "Department of Forestry Lands and Wildlife Annual Report 1991-92."
12. Alberta, "Department of Environmental Protection Annual Report 1994-95."
13. Alberta Government, "Sustainable Resource Development," Edmonton, AB., Available from http://www3.gov.ab.ca/srd/, 2005.
14. Cliff Smith, interview by P.J. Murphy, Sherwood Park, AB., 1 April 1999.
15. Cliff Henderson, interview by P.J. Murphy, Edmonton, AB., 24 October 2003.
16. Ibid.
17. Ibid.
18. Sustainable Resource Development, News Release, "Province to Reduce Industrial Footprint Northwest of Chinchaga River," Edmonton, AB., 25 November 2003.
19. Alberta, "Department of Sustainable Resource Development Annual Report 2003-04."
20. Craig Quintilio, interview with P.J. Murphy, Edmonton AB., 25 March 2004.

Chapter 10
1. J.A. Doucet, "Timber conditions in Little Smoky River Valley, Alberta and adjacent territory." *Forestry Branch Bulletin No. 41.* (Ottawa: Department of the Interior) 1914.
2. Bruce Haig, *Southern Alberta Bicentennial, A Look at Peter Fidler's Journal, Journal of a Journey over Land from Buckingham House to the Rocky Mountains in 1792 & 3* (Lethbridge: Historic Trails West, Ltd., 1991).
3. Manitoba, *Records of the Hudson's Bay post at Fort Edmonton 1812* (Manitoba: Provincial Archives) 1812.
4. I.M. Spry, *The papers of the Palliser Expedition 1857-1860* (Toronto: The Champlain Society) 1968.
5. Ibid.
6. Henry Youle Hind, "Reports of progress: together with a preliminary and general report on the Assiniboine and Saskatchewan exploring expedition made under instructions from the Provincial Secretary, Canada," Legislative Assembly, Toronto, 1859.
7. J.C. Southesk, Earl of, *Saskatchewan and the Rocky Mountains – a diary and narrative of travel, sport, and adventure, during a journey through the Hudson's Bay Company's territories in 1859 and 1860* (Vermont: Charles E. Tuttle) 1969.
8. V. Milton, and Walter Cheadle, *The North-West Passage by Land* (Toronto: Prospero Canadian Collection) 2001.
9. Henry Stelfox, *Rambling thoughts of a wandering fellow: a natural history of wildlife, native peoples, homesteading, and conservation in western Alberta: 1906-1968* Edited by John G. Stelfox and Betty Stelfox. (Edmonton: I.D.B. Press) 1972.
10. Albert Hanson, interview by P.J. Murphy, Grande Prairie, AB., 7 June 1993.
11. J.G. MacGregor, *Land of Twelve Foot Davis: a*

history of the Peace River country (Edmonton: Applied Arts Products) 1952.
12. James McEvoy, "Geological Survey of Canada," Ottawa. 1900.
13. Hugh M. Stutfield, and Norman Collie, *Climbs & Exploration in the Canadian Rockies* (London: Longmans Green and Co.) 1903.
14. J.S. Beck, *No Ordinary Woman: The Story of Mary Schaffer Warren* (Calgary: Rocky Mountain Books) 2001.
15. Edward Moberly, interview by P.J. Murphy, Hinton, AB., 29 August 1980.
16. Crowsnest Pass Historical Society, *Crowsnest and its people* (Calgary: Friesen Printers) 1979.
17. K. Johnstone, *Timber and Trauma: 75 years with the federal forestry service, 1899-1974* (Ottawa: Forestry Canada) 1991.
18. W.C. Taylor, *Tracks across my trail. Story of Curly Philips, guide and outfitter* (Jasper: Yellowhead Historical Society) 1984.
19. Eric S. Huestis, interview by P.J. Murphy, Edmonton, AB., 26 September 1983.
20. Rod Gregg, "15 cents an hour and grub," Newspaper clipping, Alberta, P.J. Murphy files, 1933.
21. Eric S. Huestis, talk at Forestry Training School, Hinton, AB., 31 January 1972.
22. Eric S. Huestis, interview by P.J. Murphy, Edmonton, AB., 26 September 1983.
23. British Columbia, "Report on the Phillips Fire," *B.C. Forest Service Archives* (Victoria: British Columbia) 1936.
24. R.M. Patterson, *The Buffalo Head* (Toronto: Macmillan) 1972.
25. Ibid.
26. J. Dexter Champion, interview by P.J. Murphy, Hinton, AB., 27 May 1986.
27. K. Johnstone, *Timber and Trauma: 75 years with the federal forestry service, 1899-1974* (Ottawa: Forestry Canada) 1991.
28. Eric S. Huestis, interview by P.J. Murphy, Edmonton, AB., 26 September 1983.
29. Crowsnest Pass Historical Society, "Account of the 1936 Pass Creek Fire," *Archives of the CPHS* Blairmore, 1936. Joe Kovach, interview by Kim Clark, Chase, BC, 18 March 1989.
30. J.R.H. Hall, "Rocky Mountain House fire story," *Pulp and Paper Magazine of Canada* March 1942.
31. Evelyn M. McGuire, "Fire along the Smoky," *Western Producer/Western People* Saskatoon, June 1988.
32. P.J. Murphy, and C. Tymstra, "The 1950 Chinchaga River fire in the Peace River region of British Columbia/Alberta: preliminary results of simulating forward spread distances," *Third Western Region Fire Weather Committee Scientific and Technical Seminar* Edmonton, 4 February 1986.
33. Robert J. Adams, *Beyond the Stump Farm* (Spruce Grove: Megamy Publishing) 1999.
34. Gilliat, Neil W.W., *Watch Over the Forest* (Edmonton: Brightest Pebble Publishing) 1999.
35. Jock McLean and Blane Coulcher, *Seven Days in May – Meteorological Factors Associated with the Alberta Forest Fires of May 18-25* (Edmonton: Alberta Forest Service) 1968.
36. Dennis Quintilio, 2001 - personal communication
37. M.A. Alexander, B. Janz and D. Quintilio, "Analysis of extreme wildfire behaviour in east-central Alberta: a case study," *Seventh conference on fire and forest meteorology April 25-29, 1983* (Boston: American Meteorological Society) 1983.
38. Council of Assiniboia, "Fire ordinance," *Proceedings of Council* (Fort Garry: Council of Assiniboia) 1882.
39. B.J. Stocks, M.E. Alexander, R.A Lanoville, "Overview of the International Crown Fire Modelling Experiment (ICFME)," *Canadian Journal of Forest Research* 34(8): 153-154, 2004.

Chapter 11
1. Alberta, "Department of Lands and Mines Annual Report 1947-48."
2. Eric S. Huestis, interview by P.J. Murphy, Edmonton, AB., 26 September 1983.
3. Robert G. Steele, interview by P.J. Murphy, Qualicum Beach, B.C., 7 August 1997.
4. Fred McDougall, interview by P.J. Murphy, Edmonton, AB., 9 November 1998.
5. Alberta, "Alberta Departments Annual Reports between 1978 and 1985."

6. Alberta, "Forest Industry Development," Position Paper, Government of Alberta, Edmonton, 1985, revised 3 February, 1986.
7. Cliff Smith, interview by P.J Murphy, Sherwood Park, AB., December 1999.
8. Ken Higginbotham, interview by P.J Murphy, Vancouver, B.C., 25 January 2000.
9. Cliff Henderson, interview by P.J Murphy, Edmonton, AB., 24 October 2003.
10. Howard Gray, interview by P.J. Murphy, Edmonton, AB., 6 August 2003.
11. Craig Quintilio, interview by P.J. Murphy, Edmonton, AB., 25 March 2004.

Suggested Further Reading

Alberta. *Land Forest Wildlife* magazines – 1958 to 1967. SRD Library.

Alberta. *Lands Forests Parks Wildlife* magazines – 1967 to 1970. SRD Library.

Alberta Forestry Association. *Alberta Trees of Renown – An Honour Roll of Alberta Trees*; Alberta Forestry Association; First Edition May 1984 ISBN: 0-86499-150-9; Second Edition May 1986 ISBN: 0-86499-320-X

Bott, Robert, Peter Murphy and Robert Udell. *Learning from the Forest – A Fifty-Year Journey Towards Sustainable Forest Management*; Foothills Model Forest; 2003; ISBN: 1-894856-23-6

Boucher, Bob. *Forest Trails – Reminisce of Alberta by erstwhile Rangers and Game Guardians.* 1996. Written, researched and typeset by Bob Boucher, Sundre. Printed by Bowden Business Enterprises, Bowden, Alberta.

Canadian Institute of Forestry. *A History of the Rocky Mountain Section, Canadian Institute of Forestry 1948-1967*; G.R. Hopping and W.H. McCardell; Rocky Mountain Section, CIF; 1969

Canadian Institute of Forestry. *A History of the Rocky Mountain Section, Canadian Institute of Forestry 1968-1982*; Compiled by Ron Fytche; Rocky Mountain Section, CIF.; 1985.

Fox Creek Historical Society. *Iosegun Reflections – A History of Fox Creek*; 1992; Friesen Printers; ISBN 1-55056-077-8.

Gilliat, Neil W.W. *If Moose Could Only Talk*; Stories from the Canadian Rockies in the Early Days of the Alberta Forest Service. 1998. Brightest Pebble Publishing, Edmonton. ISBN: 0-9683627-1-0; September, 1998

Gilliat, Neil W.W. *Watch Over the Forest – More tales of the Alberta Forest Service and life in the forest communities*; 1999. Brightest Pebble Publishing, Edmonton. ISBN: 0-9683627-2-9; November, 1999

Glen, Jack. *Mountain Trails* – Memoirs of Jack Glen, Forest Ranger at Entrance, Alberta. 1920 – 1942. Printed in the Western Producer 1969. Compiled from a set of clippings prepared by AFS Librarian Stella McCreedy. SRD Library. (Edited and expanded version in process by Foothills Model Forest, 2005).

Grande Cache Historical Society. *Muskeg Ranger Station Forest Rangers 1942 – 1970*; 2004; Compiled by Jo Sharlow.

Guest, Robert. *Trail North: A Journey in Words and Pictures*; Robert Guest; 1995; Lone Pine Publishing. ISBN 1551050528 (bound), 1551050277 (pbk.).

Huth, Robin A. *Horses to Helicopters – Stories of the Alberta Forest Service;* 1980. Alberta Forest Service, Energy and Natural Resources.

Huth, Robin A. *Guardians of the Forests – the First Hundred Years*; Robin Huth; 2002; Wombat Press, Silverton, B.C. ISBN: 0-9732034-0-4

Johnstone, Kenneth. *Timber and Trauma – 75 Years with the Federal Forestry Service 1899-1974*; Forestry Canada, Ottawa. ISBN: 0-662-17616-2

Murphy, Peter J. *History of Forest and Prairie Fire Control Policy in Alberta*; Alberta Energy and Natural Resources, Forest Service; January 1985; ENR Report Number: T/77; ISBN: 0-86499-208-4

Murphy, Peter J. and Robert E. Stevenson. *A Fortuitous International Meeting of Two Yale Foresters in 1908: H.R. McMillan and W.N. Millar.* Forest History Today. Forest History Society; Spring, 1999

Murphy, Peter J. and Martin K. Luckert. *The Evolution of Forest Management Agreements on the Weldwood Hinton Forest.* 2002. The Foothills Model Forest History Series, Volume 3. Hinton, Alberta.

Murphy, Peter J. with R.Udell, R. Stevenson and T. Peterson. *A Hard Road To Travel -- Land, Forests and People in the Upper Athabasca Region.* (2005 in press); Foothills Model Forest and Forest History Society, Inc, Durham, N.C.

Nortcliffe, Peter J. *A Forest Ranger's Stories*; 2002. Peter Nortcliffe, Blairmore, Alberta; ISBN: 0-7795-0042-3

Potter, Mike. Fire Lookout Hikes in the Canadian Rockies; 1998; Luminous Compositions Ltd., Banff. ISBN: 0-9694438-5-4

ALBERTA FOREST SERVICE

Directors of Forestry and Division Heads

	Department of the Interior Minister (Ottawa)	Deputy Minister (Ottawa)	Director of Forestry (Ottawa)	Director for Alberta
1899-00	Clifford Sifton	James Smart	Elihu Stewart	
1900-01	Clifford Sifton	James Smart	Elihu Stewart	
1901-02	Clifford Sifton	James Smart	Elihu Stewart	
1902-03	Clifford Sifton	James Smart	Elihu Stewart	
1903-04	Clifford Sifton	James Smart	Elihu Stewart	
1904-05	Clifford Sifton	James Smart	Elihu Stewart	
1905-06	Frank Oliver	James Smart	Elihu Stewart	
1906-07	Frank Oliver	W.W. Cory	Elihu Stewart	
1907-08	Frank Oliver	W.W. Cory	Robert Campbell	(Abraham Knechtel)
1908-09	Frank Oliver	W.W. Cory	Robert Campbell	(Abraham Knechtel)
1909-10	Frank Oliver	W.W. Cory	Robert Campbell	(Abraham Knechtel)
1910-11	Frank Oliver	W.W. Cory	Robert Campbell	A. Helmer
1911-12	Robert Rogers	W.W. Cory	Robert Campbell	W.N. (Willis) Millar
1912-13	William Roche	W.W. Cory	Robert Campbell	W.N. (Willis) Millar
1913-14	William Roche	W.W. Cory	Robert Campbell	W.N. (Willis) Millar
1914-15	William Roche	W.W. Cory	Robert Campbell	Ernest Finlayson
1915-16	William Roche	W.W. Cory	Robert Campbell	Ernest Finlayson
1916-17	William Roche	W.W. Cory	Robert Campbell	Ernest Finlayson
1917-18	Arthur Meighen	W.W. Cory	Robert Campbell	Ernest Finlayson
1918-19	Arthur Meighen	W.W. Cory	Robert Campbell	Ernest Finlayson
1919-20	James Lougheed	W.W. Cory	Robert Campbell	Ernest Finlayson
1920-21	James Lougheed	W.W. Cory	Robert Campbell	Charles H. Morse
1921-22	Charles Stewart	W.W. Cory	Robert Campbell	Charles H. Morse
1922-23	Charles Stewart	W.W. Cory	Robert Campbell	Charles H. Morse
1923-24	Charles Stewart	W.W. Cory	Robert Campbell	Charles H. Morse
1924-25	Charles Stewart	W.W. Cory	Ernest Finlayson	Charles H. Morse
1925-26	Charles Stewart	W.W. Cory	Ernest Finlayson	Charles H. Morse
1926-27	Charles Stewart	W.W. Cory	Ernest Finlayson	Charles H. Morse
1927-28	Charles Stewart	W.W. Cory	Ernest Finlayson	Charles H. Morse
1928-29	Charles Stewart	W.W. Cory	Ernest Finlayson	Charles H. Morse
1929-30	Charles Stewart	W.W. Cory	Ernest Finlayson	Charles H. Morse

MINISTERS OF LANDS & MINES

HUGH W. ALLEN
1933 - 1935

CHARLES C. ROSS
1935 - 1936

ROBERT G. REID
1930 - 1933

PROVINCE OF ALBERTA

Provincial Archives of Alberta, A2454

NATHAN E. TANNER
1936 - 1952

ALBERTA
MINISTERS OF LANDS & FORESTS

IVAN CASEY
1952 - 1955

NORMAN A. WILLMORE
1955 - 1965

HENRY A. RUSTE
1965 – JULY 1968
DEC. 1968 – MAY 1969

ALFRED J. HOOKE
JULY 1968 – DEC. 1968

DR. J. DONOVAN ROSS, B.A., M.D.
1969 - 1971

DR. ALLAN A. WARRACK, P.Ag.
1971 - 1975

Provincial Archives of Alberta, A2459

Alberta – Department of Lands and Mines

	Minister	Deputy Minister	Director of Forestry	Assistant Director of Forestry	Chief Timber Inspector	Radio Superintendent
1930-31	R.G. Reid	L.C. Charlesworth /John Harvie[0]	T.F. (Ted) Blefgen		F.W. (Frank) Neilson	
1931-32	R.G. Reid	John Harvie	T.F. (Ted) Blefgen		F.W. (Frank) Neilson	
1932-33	R.G. Reid	John Harvie	T.F. (Ted) Blefgen		F.W. (Frank) Neilson	
1933-34	Hugh W. Allen	John Harvie	T.F. (Ted) Blefgen		F.W. (Frank) Neilson	
1934-35	Charles C. Ross	John Harvie	T.F. (Ted) Blefgen	J.A. (Hutch) Hutchison	F.W. (Frank) Neilson	
1935-36	Charles C. Ross	John Harvie	T.F. (Ted) Blefgen	J.A. Hutchison	F.W. (Frank) Neilson	
1936-37	Nathan E. Tanner	John Harvie	T.F. (Ted) Blefgen	J.A. Hutchison	F.W. (Frank) Neilson	
1937-38	Nathan E. Tanner	John Harvie	T.F. (Ted) Blefgen	J.A. Hutchison	F.W. (Frank) Neilson	
1938-39	Nathan E. Tanner	John Harvie	T.F. (Ted) Blefgen	J.A. Hutchison[1]	F.W. (Frank) Neilson	
1939-40	Nathan E. Tanner	John Harvie	T.F. (Ted) Blefgen	E.S. (Eric) Huestis a/[2]	F.W. (Frank) Neilson	
1940-41	Nathan E. Tanner	John Harvie	T.F. (Ted) Blefgen	E.S. (Eric) Huestis a/	F.W. (Frank) Neilson	
1941-42	Nathan E. Tanner	John Harvie	T.F. (Ted) Blefgen	E.S. (Eric) Huestis a/	F.W. (Frank) Neilson	A.E. (Tony) Earnshaw[3]
1942-43	Nathan E. Tanner	John Harvie	T.F. (Ted) Blefgen	E.S. (Eric) Huestis a/	F.W. (Frank) Neilson	Tony Earnshaw
1943-44	Nathan E. Tanner	John Harvie	T.F. (Ted) Blefgen	E.S. (Eric) Huestis a/	F.W. (Frank) Neilson	Tony Earnshaw
1944-45	Nathan E. Tanner	John Harvie	T.F. (Ted) Blefgen	E.S. (Eric) Huestis a/	F.W. (Frank) Neilson	Tony Earnshaw
1945-46	Nathan E. Tanner	John Harvie	T.F. (Ted) Blefgen	E.S. (Eric) Huestis a/	F.W. (Frank) Neilson[4]	Tony Earnshaw
1946-47	Nathan E. Tanner	John Harvie	T.F. (Ted) Blefgen	J.A. Hutchison a/[5]	J.L. (Jack) Janssen[6]	Tony Earnshaw
1947-48	Nathan E. Tanner	John Harvie	T.F. (Ted) Blefgen (ill)/ E.S. (Eric) Huestis	unknown[7]	J.L. (Jack) Janssen	Tony Earnshaw
1948-49	Nathan E. Tanner	John Harvie	E.S. (Eric) Huestis	unknown	J.L. (Jack) Janssen	Tony Earnshaw

Alberta – Department of Lands and Forests

	Minister	Deputy Minister	Director of Forestry	Assistant Director of Forestry (Forest Management)	Chief Timber Inspector (Fire)	Forest Surveys Superintendent
1949-50	Nathan E. Tanner	John Harvie	Eric Huestis	unknown	Jack Janssen	Reginald D. Loomis[8]
1950-51	Nathan E. Tanner	John Harvie	Eric Huestis	Herb Hall[9]	Jack Janssen	Reg Loomis
1951-52	Ivan Casey	H.G. Jensen	Eric Huestis	Herb Hall	Jack Janssen	Reg Loomis
1952-53	Ivan Casey	H.G. Jensen	Eric Huestis	Herb Hall	Jack Janssen	Reg Loomis
1953-54	Ivan Casey	H.G. Jensen	Eric Huestis	Herb Hall	Jack Janssen[10]	Reg Loomis
1954-55	Norman Willmore	H.G. Jensen	Eric Huestis	Herb Hall	Ted Hammer[11/12]	Reg Loomis
1955-56	Norman Willmore	H.G. Jensen	Eric Huestis	Herb Hall	Ted Hammer	Reg Loomis
1956-57	Norman Willmore	H.G. Jensen	Eric Huestis	Herb Hall	Ted Hammer	Reg Loomis
1957-58	Norman Willmore	H.G. Jensen	Eric Huestis	Herb Hall	Ted Hammer	Reg Loomis
1958-59	Norman Willmore	H.G. Jensen	Eric Huestis	Herb Hall	Ted Hammer	Reg Loomis
1959-60	Norman Willmore	H.G. Jensen	Eric Huestis	Reg Loomis (1 July 1959)	Ted Hammer	Robert G. Steele (1 July 1959)
1960-61	Norman Willmore	H.G. Jensen	Eric Huestis	Reg Loomis	Ted Hammer	Bob Steele
1961-62	Norman Willmore	H.G. Jensen	Eric Huestis	Reg Loomis	Ted Hammer	Bob Steele
1962-63	Norman Willmore	H.G. Jensen	Eric Huestis	Reg Loomis	Ted Hammer	Bob Steele
1963-64	Norman Willmore	Eric Huestis	Bob Steele[14]	Reg Loomis	Ted Hammer	Stan Hughes
1964-65	Willmore/Ruste	Eric Huestis	Bob Steele	Reg Loomis	Stan Hughes	John Hogan
1965-66	Henry Ruste	Eric Huestis	Bob Steele	Reg Loomis	Stan Hughes	John Hogan
1966-67	Henry Ruste	V.A. Wood	Bob Steele	Reg Loomis	Stan Hughes	John Hogan
1967-68	Henry Ruste	V.A. Wood	Bob Steele	Reg Loomis	Stan Hughes	John Hogan
1968-69	Ruste/Alfred Hooke	V.A. Wood	Bob Steele	Reg Loomis[15]	Stan Hughes	John Hogan

Branch Re-structuring

	Minister	Deputy	Director	Timber Management	Forest Protection	Forest Surveys[16]
1969-70	Ruste/Donovan Ross	V.A. Wood	Bob Steele	F.W. McDougall	Stan Hughes	J.J. Lowe
1970-71	Ross/Allan Warrack	V.A. Wood	Bob Steele	F.W. McDougall	Stan Hughes	J.J. Lowe
1971-72	Allan Warrack	V.A. Wood	Bob Steele	F.W. McDougall	Stan Hughes	J.J. Lowe
1972-73	Allan Warrack	V.A. Wood	Bob Steele	F.W. McDougall	Stan Hughes	J.J. Lowe
1973-74	Allan Warrack	Bob Steele	Fred McDougall	D.H. Fregren (May 1974)	Stan Hughes	J.J. Lowe

Radio Superintendent	Forestry Training School
Tony Earnshaw	
Tony Earnshaw	(Banff School)
Tony Earnshaw	Victor Heath
Tony Earnshaw	Victor Heath
Tony Earnshaw	John Hogan
Tony Earnshaw	John Hogan
Tony Earnshaw	John Hogan
Tony Earnshaw	Peter J. Murphy
Tony Earnshaw	Peter Murphy
Tony Earnshaw	Peter Murphy
Tony Earnshaw	Peter Murphy
Tony Earnshaw	Peter Murphy
Tony Earnshaw[13]	Peter Murphy
	Peter Murphy
	Peter Murphy
	Peter Murphy
	Peter Murphy
	Peter Murphy
	Peter Murphy
	Peter Murphy

Forest Land Use[17]	Construction	Training
G.M. Smart	John Hogan	Peter Murphy
Gordon Smart	J.F. Hogan[18]	Peter Murphy
Gordon Smart		Peter Murphy
Gordon Smart		John Wagar[19]/B. Simpson
Gordon Smart		Bernie F. Simpson

Directors of Forestry and Division Heads 1930 - 2005

Alberta Department of Energy and Natural Resources

		Minister	Deputy	Director of Forestry	Timber Management	Forest Protection	Forest Surveys
	1974-75	Don Getty	Bob Steele	Fred McDougall	Don Fregren	Hank Ryhanen	J.J. Lowe
	1975-76	Don Getty	Bob Steele	Fred McDougall	Don Fregren	Hank Ryhanen	J.J. Lowe
	1976-77	Don Getty	Bob Steele	Fred McDougall	Don Fregren	Hank Ryhanen	J.J. Lowe

Branch Re-structuring

	Minister	Deputy	Director of Forestry	Timber Management	Forest Protection
1977-78	Don Getty	Bob Steele	Fred McDougall	Don Fregren	Hank Ryhanen
1978-79	Merv Leitch	Bob Steele	Fred McDougall	Don Fregren	Hank Ryhanen
1979-80	Merv Leitch	Fred McDougall	Al Brennan (ADM)	Don Fregren	Hank Ryhanen
1980-81	Merv Leitch	Fred McDougall	Al Brennan	Don Fregren	Ryhanen/ C.B. Smith[23]
1981-82	Merv Leitch	Fred McDougall	Al Brennan	Don Fregren	Cliff Smith
1982-83	John Zaozirny	Fred McDougall	Al Brennan	Don Fregren	Cliff Smith
1983-84	John Zaozirny	Fred McDougall	Al Brennan	Don Fregren	Cliff Smith
1984-85	John Zaozirny	Fred McDougall	Al Brennan	Don Fregren[24] (Aug. 1984)	Cliff Smith

Department of Forestry, Lands and Wildlife

	Minister	Deputy	ADM Forestry	Timber Management	Forest Protection	Reforestation & Reclamation
1985-86	Zaozirny/Sparrow	Fred McDougall	Cliff Smith	Con Dermott	John Benson	John Drew
1986-87	Don Sparrow	Fred McDougall	Cliff Smith	Con Dermott	John Benson	John Drew
1987-88	Sparrow/ Fjordbotten	Fred McDougall	Cliff Smith	Con Dermott	John Benson	John Drew
1988-89	LeRoy Fjordbotten	Fred McDougall[25]	Cliff Smith	Con Dermott	John Benson	John Drew (June 88) Cliff Henderson
1989-90	LeRoy Fjordbotten	McDougall (June 89) Cliff Smith (Jan. 90)	Smith/Ken Higginbotham	Con Dermott	John Benson	Cliff Henderson
1990-91	LeRoy Fjordbotten	Cliff Smith	Ken Higginbotham	Con Dermott	John Benson	Cliff Henderson
1991-92	LeRoy Fjordbotten	Cliff Smith	Ken Higginbotham	Con Dermott	John Benson	Cliff Henderson

Department of Environmental Protection

	Minister	Deputy	ADM Land & Forest Service[27]	Timber Management	Forest Protection	Reforestation & Reclamation
1992-93	Fjordbotten/Evans (Dec. 1992)	Cliff Smith[29]/ Melnychuk	Ken Higginbotham	Con Dermott	John Benson	Cliff Henderson
1993-94	Brian Evans	Peter Melnychuk	Ken Higginbotham	Cliff Henderson[31] (May 93)	Con Dermott (May 93)	

Branch Re-structuring

	Minister	Deputy	ADM Land & Forest Service	Forest Management	Forest Protection	Land Administration
1994-95	Brian Evans/Ty Lund (Oct. 1994)	Peter Melnychuk	Ken Higginbotham	Cliff Henderson	Con Dermott (Jan. 95)	Rick McDonald
1995-96	Ty Lund	Peter Melnychuk	Cliff Henderson (May 1995)	Dennis Quintilio (July 95)	Kelly O'Shea (July 93)	Rick McDonald
1996-97	Ty Lund	Peter Melnychuk	Cliff Henderson	Dennis Quintilio	Kelly O'Shea	Rick McDonald
1997-98	Ty Lund	Jim Nichols (Nov. 97)	Cliff Henderson	Dennis Quintilio	Kelly O'Shea	Craig Quintilio[32]
1998-99	Gary Mar	Jim Nichols	Cliff Henderson	Dennis Quintilio	Kelly O'Shea	Craig Quintilio

Department of Alberta Environment

	Minister	Deputy	ADM Land & Forest Service	Forest Management	Forest Protection	Land Administration
1999-00	Gary Mar	Doug Radke	Cliff Henderson	Doug Sklar[34]	Kelly O'Shea	Craig Quintilio
2000-01	Halvar Jonson	Doug Radke	Cliff Henderson	Doug Sklar	Craig Quintilio	Glenn Selland

Forest Land Use	Construction	Training
Gordon Smart	R.L. Heatherington (Dec .1974)	Bernie Simpson
Gordon Smart	R.L. Heatherington	Bernie Simpson
Gordon Smart	R.L. Heatherington	Bernie Simpson

Forest Land Use	Reforestation and Reclamation[20]	Program Support	Forest Research	Training/FTS[21]
Gordon Smart[22]	Con Dermott	Chuck Geale	Joe Soos (Apr. 1978)	Bernie Simpson
Gordon Smart	Con Dermott	Chuck Geale	Joe Soos	Bernie Simpson
Gordon Smart	Con Dermott	Chuck Geale	Joe Soos	Bernie Simpson
John Benson (Jan '81)	Con Dermott	Chuck Geale	Joe Soos	Bernie Simpson
John Benson	Con Dermott	Chuck Geale	Joe Soos	Bernie Simpson
John Benson	Con Dermott	Chuck Geale	Joe Soos	Bernie Simpson
John Benson	Con Dermott	Chuck Geale	Joe Soos	Bernie Simpson
John Benson	Con Dermott	Chuck Geale	Joe Soos	Bernie Simpson

Forest Land Use	Program Support	Forest Research	Training/ FTS	Forest Industry
Don Fregren (Aug. 1985)	Chuck Geale	Joe Soos	Bernie Simpson	Al Brennan Executive Officer
Don Fregren	Chuck Geale	Stan Navratil (Nov. 1986)	Bernie Simpson	Al Brennan
Don Fregren	Chuck Geale	Stan Navratil	Bernie Simpson	Al Brennan
Don Fregren	Chuck Geale	Ken Higginbotham	Bernie Simpson	Al Brennan
Don Fregren	Chuck Geale	Higginbotham to Jan 1990[26]	Bernie Simpson	Al Brennan
Don Fregren	Chuck Geale		Simpson/ Dennis Quintilio	Al Brennan
Don Fregren	Chuck Geale		Dennis Quintilio	Al Brennan

Forest Land Use	Land Administration	Program Support	Training ETC[28]	Forest Industry
Don Fregren[30]	Rick McDonald	Murray Turnbull/R.B. (Rod) Simpson	Dennis Quintilio	Al Brennan
	Rick McDonald (May 93)	Carson McDonald	Dennis Quintilio	

Program Support			Training ETC	Forest Industry
Carson McDonald			Dennis Quintilio	
Carson McDonald/C. Quintilio (Aug. 1995)			Ross Risvold	
Craig Quintilio			Ross Risvold	
			Risvold/Podlubny (Nov. 1997)	
			Don Podlubny	

Ecological Landscape[33]			Training ETC	Forest Industry
Dennis Quintilio			Don Podlubny	Howard Gray[35]
Dennis Quintilio			Don Podlubny	Howard Gray

Department of Sustainable Resource Development[36]

	Minister	Deputy	ADM Land & Forests	ADM Forest Protection	ADM Public Lands	ADM Fish & Wildlife	Training ETC
2001-02	Mike Cardinal	Dr. R.J. Fessenden	Howard Gray	Cliff Henderson	Craig Quintilio	Morley Barrett	Don Podlubny
2002-03	Mike Cardinal	Bob Fessenden	Howard Gray	Cliff Henderson	Craig Quintilio	Ken Ambrock	Don Podlubny
2003-04	Mike Cardinal	Bob Fessenden	Howard Gray	Cliff Henderson	Craig Quintilio	Ken Ambrock	Rob Thorburn (HTC)[37]

Department of Sustainable Resource Development (restructure 2004-05)

	Minister	Deputy	ADM Strategic Forestry Initiatives[38]	ADM Forest Protection	ADM Public Lands and Forests Division	ADM Fish & Wildlife Division	ADM Strategic Corporate Services
2003-04	Mike Cardinal	Bob Fessenden	Howard Gray	Cliff Henderson	Craig Quintilio	Ken Ambrock	Stew Churlish
2004-05[39]	David Coutts	Brad Pickering	Howard Gray	Cliff Henderson	Craig Quintilio	Ken Ambrock	Stew Churlish

End Notes

0. L. C. Charlesworth was appointed Acting Deputy Minister October 1, 1930; John Harvie was appointed Director of Provincial Lands October 6, 1930. John Harvie was subsequently appointed Acting Deputy Director of Lands and Mines January 2, 1931. John Harvie was later appointed Deputy Minister of Lands and Mines November 26, 1931.

1. Hutchison enlisted in WW II

2. Huestis named Acting Director to replace Hutchison during WW II

3. Tony Earnshaw was hired in 1942 when the first radios were introduced for AFS communications

4. Frank Neilson retired – had previously been Timber Inspector at Westlock

5. J.A. Hutchison resigned to become Superintendent of Banff National Park, later Director of Parks in Ottawa. Eric Huestis was listed as Fish & Game Commissioner in the Annual Report

6. Jack Janssen was promoted from Superintendent of the Lesser Slave Forest Reserve

7. There was no Assistant Director of Forestry listed

8. Reg Loomis was a forester hired to take charge of the first forest inventory that had been initiated by Huestis

9. Herb Hall was promoted from Forest Superintendent in Rocky Mountain House

10. Jack Janssen retired, became a Timber Buyer for North Western Pulp and Power before retiring again

11. Ted Hammer was promoted from Forest Superintendent in Grande Prairie

12. Frank Platt was promoted in 1954 from Timber Inspector at Entwistle to Assistant in Forest Protection, retired 1974 – major contributor to the Forest Protection program

13. AFS Radio was transferred to Alberta Government Telephones, Earnshaw moved as Consultant until retirement

14. Bob Steele was among the foresters hired by Huestis in 1949, served in Forest Surveys then Forest Superintendent in Rocky Mountain House

15. Reg Loomis retired to his farm at Sandy Lake, west of Morinville

16. Forest Surveys became part of the Forest Inventory Section and was transferred to the Timber Management Branch under Fred McDougall

17. Forest Land Use elevated to full Branch status, staff largely from Forest Surveys, first branch head Gordon Smart

18. The Construction section was transferred to the Department of Public Works in July 1971

19. John Wagar was Senior Instructor at FTS, was appointed Head in 1973 and died of a coronary condition shortly after. He was active in planning the new school buildings, and the Gymnasium was named after him in recognition of his contributions

20. Was a section under the T/M Branch until being formed as its own branch under Con Dermott in 1977

21. Forestry Training School was renamed to the Forest Technology School in the fall of 1965

22. Forest Surveys moved to Forest Land Use Branch from Timber Management

23. C.B. (Cliff) Smith was appointed in January 1981

ADM Strategic Corporate Services

Stew Churlish

Stew Churlish

Stew Churlish

---

24. Don Fregren was Director of Program Coordination Branch from August 1984 to August 1985

25. Fred McDougall retired, later becoming Vice-President for Weyerhaeuser Alberta operations

26. The Forest Research section became a part of both the Timber Management and Reforestation and Reclamation Branches

27. During the 1992 reorganization the Alberta Forest Service was renamed to Land & Forest Service. From 1993 - 1995 changes were made to reduce the 10 Forests and 40 Districts to 6 Regions and 24 Districts and eventually to 18 Districts. (In August 1996 Fort Vermilion became part of Upper Hay / High Level District bringing the number down to 17)

28. The Forest Technology School name was changed to the Environmental Training Centre

29. Cliff Smith retired and became a Consulting Forester

30. Don Fregren retired May 1993. Murray Anderson acting until branch moved to Timber Management Branch

31. Forest Land Use and Reforestation & Reclamation Branches were rolled under the Timber Management Branch as of May 1993. The new division was called Land and Forest Services. This was the end of the name Alberta Forest Service.

32. The Program Support Branch was amalgamated within the Land Administration Branch under Craig Quintilio

33. The Ecological Landscape Division was created on May 4, 1999 to lead Integrated Resource Management. This was in response to Premier Klein and Cabinet's endorsement of the *'Alberta's Commitment to Sustainable Resource and Environmental Management'* policy document

34. Doug Sklar was appointed Director of Forest Management Division on June 1, 1999

35. Howard Gray became Executive Director of the Forest Products Branch of Economic Deveopment in March 1998. This program was then transferred to Resource Development, Department of Energy in 1999.

36. The Department of Sustainable Resource Development was created on March 15, 2001 from components of three ministries – Environment, Agriculture, Food and Rural Development, and Resource Development

37. The Environmental Training Centre name changed to the Hinton Training Centre and was transferred to the Forest Protection Division

38. In July 2003 the Strategic Forestry Initiatives Division was created under the direction of Howard Gray

39. Provincial election held November 2004 and Minister Coutts appointed as new minister of Sustainable Resource Development

ALBERTA FOREST SERVICE

Executive and Forest Superintendents
1930 to 2005

Alberta – Department of Lands and Mines Forests and Superintendents

	Director	Chief Timber Inspector (EFRD)	Cypress Hills	Crowsnest	Bow River	Clearwater	Brazeau
1930-31[1]	T.F. Blefgen	F.W. Neilson	H.A. Parker	J.P. Alexander	J.A. Hutchison/A.G. Smith	E.S. Huestis	C.E. White
1931-32[2]	T.F. Blefgen	F.W. Neilson	G. Ambrose	J.P. Alexander	A.G. Smith	E.S. Huestis	C.E. White/ F.G. Edgar

AFS Restructured. Forest Reserves and Eight Divisions[3]

	Director	Chief Timber Inspector (NAFD)	Cypress Hills Forest Reserve	Bow-Crow Forest	Clearwater Forest	Brazeau-Athabaska Forest (Coalspur)	Edmonton (Breton)
1932-33	T.F. Blefgen	F.W. Neilson[5]	G. Ambrose	A.G. Smith	J.P. Alexander	F.G. Edgar	R.S. Wyllie
1933-34	T.F. Blefgen	F.W. Neilson	G. Ambrose	A.G. Smith	J.P. Alexander	F.G. Edgar	R.S. Wyllie
1934-35	T.F. Blefgen	F.W. Neilson	G. Ambrose	A.G. Smith	J.P. Alexander	F.G. Edgar	R.S. Wyllie
1935-36	T.F. Blefgen	F.W. Neilson	G. Ambrose	J.P. Alexander	F.G. Edgar	J.R.H. Hall	R.S. Wyllie
1936-37	T.F. Blefgen	F.W. Neilson	G. Ambrose	J.P. Alexander	F.G. Edgar	A.G. Smith	J.R.H. Hall (Edmonton)
1937-38	T.F. Blefgen	F.W. Neilson	G. Ambrose	J.P. Alexander	F.G. Edgar	R.S. Wyllie	J.R.H. Hall
1938-39	T.F. Blefgen	F.W. Neilson	G. Ambrose	J.P. Alexander	F.G. Edgar	R.S. Wyllie (E. Huestis[6])	J.R.H. Hall

Department of Lands and Mines Forest Divisions combined, reduced to five

	Director	Chief Timber Inspector	Cypress Hills Forest Reserve	Bow-Crow Forest Reserve	Clearwater Forest Reserve	Brazeau-Athabaska (Coalspur)	Western Division[7]
1939-40	T.F. Blefgen	F.W. Neilson	G. Ambrose	J.P. Alexander	F.G. Edgar	C. McDiarmid (E.Huestis)	C. McDiarmid
1940-41	T.F. Blefgen	F.W. Neilson	G. Ambrose	F.G. Edgar	J.R.H. Hall	C. McDiarmid	C. McDiarmid

Department of Lands and Mines Forest Districts restructured, expanded to nine

	Director	Chief Timber Inspector (NAFD)	Cypress Hills	Bow-Crow	Clearwater	Brazeau-Athabaska	Edmonton/Breton
1941-42	T.F. Blefgen	F.W. Neilson	T.D. Best	F.G. Edgar	J.R.H. Hall	D. Buck	H.E. Noble
1942-43	T.F. Blefgen	F.W. Neilson	T.D. Best	F.G. Edgar	J.R.H. Hall	D. Buck	H.E. Noble
1943-44	T.F. Blefgen	F.W. Neilson	J.D. Champion	F.G. Edgar	J.R.H. Hall	D. Buck	T.F. Somers
1944-45	T.F. Blefgen	F.W. Neilson	J.D. Champion	J.P Alexander & F. Edgar	J.R.H. Hall	D. Buck	T.F. Somers
1945-46	T.F. Blefgen	F.W. Neilson	J.D. Champion	J.P Alexander & F. Edgar	J.R.H. Hall	D. Buck	T.F. Somers
1946-47	T.F. Blefgen	F.W. Neilson/ J.L. Janssen	R. Mackey	J.P Alexander & F. Edgar	J.R.H. Hall	D. Buck[10]	J. Burleigh
1947-48	Blefgen/Huestis	J.L. Janssen	R. Mackey	J.P Alexander & F. Edgar	J.R.H. Hall	D. Buck	J. Burleigh
1948-49	E.S. Huestis	J.L. Janssen[11]	R. Mackey	J.P Alexander & F. Edgar	J.R.H. Hall	D. Buck	R. Smuland

432 Alberta Forest Service

Athabasca	Lesser Slave Forest Reserve
T.C. Burrows	J.R.H. Hall
T.C. Burrows	J.R.H. Hall

Edson	Athabasca	Bonnyville	McMurray	Slave Lake[4]	Grande Prairie	Peace River
J.R.H. Hall	Axsel Smith	D.A. McKay	H.D. McDonald	C.H. MacDonald	D. Buck	D. Minchin
J.R.H. Hall	C. Ranche	D.A. McKay	H.D. McDonald	C.H. MacDonald	D. Buck	D. Minchin
J.R.H. Hall	C. Ranche	D.A. McKay	H.D. McDonald	C.H. MacDonald	F.S. Truby	C. Mc Diarmid
J.R.H. Hall	C. Ranche	D.A. McKay	H.D. McDonald	C.H. MacDonald	D. Buck	F.E. Smith
R.S. Wyllie	C. Ranche	D.A. McKay	H.D. McDonald	D. Buck	V. Mitchell	F.E. Smith
R.S. Wyllie	C. Ranche	D.A. McKay	H.D. McDonald	D. Buck	V. Mitchell	F.E. Smith
R.S. Wyllie	F.E. Smith	D.A. McKay	H.D. McDonald	D. Buck	V. Mitchell	C. Ranche

Edson Division	McMurray Division	Slave Lake Division	Peace River Division
C. McDiarmid	H.D. McDonald*	D. Buck	V. Mitchell
C. McDiarmid	H.D. McDonald*	J.L. Janssen	V. Mitchell*

Edson	Carrot Creek/ Entwistle	Athabasca/Westlock[8]	Athabasca/Calling Lake	McMurray (Lac La Biche HQ)	Slave Lake	Grande Prairie	Peace River
D. Buck	H. Morden	F.E. Smith	C. Carter	T.R. Hammer	J.L. Janssen	H.D. McDonald	L. West
D. Buck	H. Morden	F.E. Smith	C. Carter	T.R. Hammer	J.L. Janssen	H.D. McDonald	V. Mitchell
D. Buck	H.E. Noble	F.E. Smith	C. Carter	J.V. Logan	J.L. Janssen	T.R. Hammer	V. Mitchell
D. Buck	H.E. Noble	F.E. Smith	C. Carter	J.V. Logan	J.L. Janssen	T.R. Hammer	V. Mitchell
D. Buck	H.E. Noble	F.E. Smith[9]	C. Carter	J.V. Logan	J.L. Janssen	T.R. Hammer	V. Mitchell
H.E. Noble	J.D. Rogers	W.M. Wood/R.F. Krause	C. Carter	F.V. Keats/L.P. Gauthier	J.L. Janssen/ W.M. Wood	T.R. Hammer	V. Mitchell
H.E. Noble	C.F. Platt	**Whitecourt** R.F. Krause	C. Carter	L.P. Gauthier	W.M. Wood	T.R. Hammer	V. Mitchell
H.E. Noble	C.F. Platt	R.F. Krause	C. Carter	L.P. Gauthier	W.M. Wood	T.R. Hammer	V. Mitchell

Alberta – Department of Lands and Forests Forest Reserves and Forest Superintendents[12] & Forest Divisions and Timber Inspectors[13]

	Director	Crowsnest	Bow River	Clearwater	Rocky[14]	Edson	Whitecourt
		J.P. Alexander – Sr. Supt. of Forest Reserves 1949-52[15]					
1949-50	E.S. Huestis	F. Edgar	L. West	J.R.H. Hall	R. Smuland	D. Buck & H. Parnall	R.F. Krause
1950-51	E.S. Huestis	N. Lind	L. West	F.V. Keats	R. Smuland	D. Buck & H. Parnall	R.F. Krause
1951-52	E.S. Huestis	N. Lind	L. West	F.V. Keats	R. Smuland	D. Buck & H. Parnall	R.F. Krause
1952-53	E.S. Huestis	N. Lind	L. West	F.V. Keats	R. Smuland	D. Buck & H. Parnall	R.F. Krause
1953-54	E.S. Huestis	N. Lind	S.R. Hughes	F.V. Keats	R. Smuland	D. Buck & H. Parnall	R.F. Krause

Alberta – Department of Lands and Forests Forest Reserves and Divisions & Forest Superintendents[16]

	Director	Crow	Bow	Clearwater	Rocky (Asst. Supt)	Edson	Whitecourt
1954-55	E.S. Huestis	N. Lind	S.R. Hughes	F.V Keats	H.M. Ryhanen	D. Buck	R.F. Krause
1955-56	E.S. Huestis	N. Lind/J. Hogan	S.R. Hughes/ F.V. Keats	F.V. Keats / N. Lind	H.M. Ryhanen	D. Buck	R.F. Krause
1956-57	E.S. Huestis	J.F. Hogan	F.V Keats	R.G. Steele	H.M. Ryhanen	D. Buck	R.F. Krause
1957-58	E.S. Huestis	J.F. Hogan	F.V Keats	R.G. Steele	R. Sund	H.M. Ryhanen	R.F. Krause
1958-59	E.S. Huestis	J.F. Hogan	F.V Keats	R.G. Steele	R.Sund/ N. Gilliat	H.M. Ryhanen	R.F. Krause
1959-60	E.S. Huestis	J.F. Hogan	F.V. Keats	Steele/ G.A. Longworth	N.W.W. Gilliat	H.M. Ryhanen	R.F. Krause
1960-61	E.S. Huestis	J.F. Hogan	F.V. Keats	G.A. Longworth	N.W.W. Gilliat	H.M. Ryhanen	R.F. Krause
1961-62	E.S. Huestis	J.F. Hogan	F.V. Keats	G.A. Longworth	N.W.W. Gilliat	H.M. Ryhanen	R.F. Krause
1962-63	E.S. Huestis	J.F. Hogan	F.V. Keats	G.A. Longworth	N.W.W. Gilliat	H.M. Ryhanen	R.F. Krause

	Director	Crow	Bow	Rocky-Clearwater[17]	Edson	Whitecourt	Lac La Biche
1963-64	R.G. Steele	J.F. Hogan	F.V. Keats	G.A. Longworth	H.M. Ryhanen	R.F. Krause	W. E. Coast
1964-65	R.G. Steele	G.A. Longworth	F.V. Keats	F.E. Sutherland	H.M. Ryhanen	R.F. Krause	W. E. Coast
1965-66	R.G. Steele	G.A. Longworth	F.V. Keats	F.E. Sutherland	H.M. Ryhanen	R.F. Krause	W. E. Coast
1966-67	R.G. Steele	G.A. Longworth	F.V. Keats	F.E. Sutherland	H.M. Ryhanen	R.F. Krause	W. E. Coast
1967-68	R.G. Steele	G.A. Longworth	L.P. Gauthier	F.E. Sutherland	H.M. Ryhanen	R.F. Krause	W. E. Coast
1968-69	R.G. Steele	G.A. Longworth	L.P. Gauthier	F.E. Sutherland	J.E. Benson	D.H. Fregren (Oct. 1969)	W. E. Coast
1969-70	R.G. Steele	G.A. Longworth	L.P. Gauthier	F.E. Sutherland	J.E. Benson	D.H. Fregren	W. E. Coast
1970-71	R.G. Steele	G.A. Longworth	L.P. Gauthier	F.E. Sutherland	J.E. Benson	D.H. Fregren	W. E. Coast
1971-72	R.G. Steele	G.A. Longworth	L.P. Gauthier	F.E. Sutherland	J.E. Benson	D.H. Fregren	W. E. Coast
1972-73	R.G. Steele	Merged w/ Bow[20]	L.P. Gauthier	F.E. Sutherland	J.E. Benson	Dick Radke (Dec. 1, 1972)	W. E. Coast
1973-74	R.G. Steele	L.P. Gauthier		F.E. Sutherland	J.E. Benson	Dick Radke	W. E. Coast

Lac La Biche	Slave Lake	Grande Prairie	Peace River	Edm - Breton	Edm - Entwistle	Calling Lake – Athabasca
W.E. Coast	W.M. Wood	T.R. Hammer	L.P. Gauthier	R. Smuland	C.F. Platt	J.D. Champion
W.E. Coast	W.M. Wood	T.R. Hammer	L.P. Gauthier	R. Smuland	C.F. Platt	J.D. Champion
W.E. Coast	W.M. Wood	T.R. Hammer	L.P. Gauthier	R. Smuland	C.F. Platt	J.D. Champion
W.E. Coast	W.M. Wood	T.R. Hammer	L.P. Gauthier	R. Smuland	C.F. Platt	J.D. Champion
W.E. Coast	W.M. Wood	T.R. Hammer	L.P. Gauthier	R. Smuland	H.M. Ryhanen	J.D. Champion

Lac La Biche	Slave Lake	Grande Prairie	Peace River
W.E. Coast	W.M. Wood	R. Smuland	L.P. Gauthier
W.E. Coast	W.M. Wood	R. Smuland	L.P. Gauthier
W.E. Coast	W.M. Wood	R. Smuland	L.P. Gauthier
W.E. Coast	W.J. MacGregor	R. Smuland	L.P. Gauthier
W.E. Coast	W.J. MacGregor	R. Smuland	L.P. Gauthier
W.E. Coast	W.J. MacGregor	R. Smuland	L.P. Gauthier
W.E. Coast	W.J. MacGregor	R. Smuland	L.P. Gauthier
W.E. Coast	W.J. MacGregor	R. Smuland	L.P. Gauthier
W.E. Coast	W.J. MacGregor	R. Smuland	L.P. Gauthier

Athabasca (Ft. McMurray)[18]	Slave Lake	Grande Prairie	Peace River	Footner Lake[19]
	N.W. Gilliat	R. Smuland	L.P. Gauthier	
L.G. Babcock	N.W. Gilliat	R. Smuland	L.P. Gauthier	
L.G. Babcock	N.W. Gilliat	R. Smuland	L.P. Gauthier	
L.G. Babcock	N.W. Gilliat	R. Smuland	L.P. Gauthier	H.R. Winn
L.G. Babcock	N.W. Gilliat	R. Smuland	Dick Radke	H.R. Winn
L.G. Babcock	H.R. Winn	R. Smuland	Dick Radke	H.R. Winn
L.G. Babcock	H.R. Winn	R. Smuland	Dick Radke	C.H. Geale
L.G. Babcock	H.R. Winn	R. Smuland	Dick Radke	C.H. Geale
L.G. Babcock	H.R. Winn	R. Smuland	Dick Radke	C.H. Geale
L.G. Babcock	H.R. Winn	D.H. Fregren (Jan. 1, 1972)	G.A. Longworth	C.H. Geale
L.G. Babcock	H.R. Winn	D.H. Fregren	G.A. Longworth	C.A. Dermott

		Director/Asst. Deputy Minister[21]	Bow-Crow	Rocky-Clearwater	Edson	Whitecourt	Lac La Biche
Department of Energy and Natural Resources Forests and Forest Superintendents	1974-75	F.W. McDougall	C.A. Dermott (Sept. 1974)	F.E. Sutherland	J.E. Benson	Dick Radke	W.E. Coast
	1975-76	F.W. McDougall	C.A. Dermott	F.E. Sutherland	J.E. Benson	Dick Radke	W.E. Coast
	1976-77	F.W. McDougall	C.A. Dermott	F.E. Sutherland	J.E. Benson	Dick Radke	W.E. Coast
	1977-78	F.W. McDougall	J.E. Benson	F.E. Sutherland	N.R. Rodseth (May 1, 1977)	Dick Radke	L.G. Huberdeau (Oct. 1977)
	1978-79	**ADM** J.A. Brennan	J.E. Benson	F.E. Sutherland	N.R. Rodseth	Dick Radke	L.G. Huberdeau
	1979-80	J.A. Brennan	J.E. Benson	F.E. Sutherland	N.R. Rodseth	Dick Radke	L.G. Huberdeau
	1980-81	J.A. Brennan	A.J. Peter	F.E. Sutherland	N.R. Rodseth	C.J. Henderson (May 15, 1980)	L.G. Huberdeau
	1981-82	J.A. Brennan	A.J. Peter	F.E. Sutherland	N.R. Rodseth	C.J. Henderson	L.G. Huberdeau
	1982-83	J.A. Brennan	A.J. Peter	F.E. Sutherland	N.R. Rodseth	C.J. Henderson	L.G. Huberdeau
	1983-84	J.A. Brennan	A.J. Peter	F.E. Sutherland	N.R. Rodseth	C.J. Henderson	L.G. Huberdeau
	1984-85	J.A. Brennan	A.J. Peter	F.E. Sutherland	N.R. Rodseth	C.J. Henderson	L.G. Huberdeau

		ADM Forestry[22]	Bow-Crow	Rocky-Clearwater	Edson	Whitecourt	Lac La Biche
Department of Forestry, Lands and Wildlife Forests and Forest Superintendents	1985-86	C.B. Smith	A.J. Peter	F.E. Sutherland	N.R. Rodseth/ W. Fairless	C.J. Henderson	L.G. Huberdeau /B. Ward
	1986-87	C.B. Smith	A.J. Peter	Lorne Goff	W. Fairless	C.J. Henderson	B. Ward
	1987-88	C.B. Smith	A.J. Peter	Lorne Goff	W. Fairless	C.J. Henderson	B. Ward
	1988-89	C.B. Smith	A.J. Peter	Lorne Goff	W. Fairless	C.J. Henderson	B. Ward
	1989-90	C.B. Smith	A.J. Peter	Lorne Goff	W. Fairless	G. Bisgrove (Jan. 1989)	B. Ward
	1990-91	K.O. Higginbotham	A.J. Peter	Lorne Goff	W. Fairless	G. Bisgrove	B. Ward
	1991-92	K.O. Higginbotham	A.J. Peter	Lorne Goff	W. Fairless	G. Bisgrove	B. Ward
	1992-93	K.O. Higginbotham	A.J. Peter	Lorne Goff	W. Fairless	G. Bisgrove	B. Ward

		Assistant Deputy Minister Land & Forest Service	Bow-Crow	Rocky-Clw.	Edson	Whitecourt	Lac La Biche
Department of Environmental Protection[23] Forests and Forest Superintendents	1993-94	K.O. Higginbotham	A.J. Peter	Lorne Goff	Bill Fairless	Gordon Bisgrove	Brydon Ward
	1994-95	K.O. Higginbotham	Kelly O'Shea (April 94)	Lorne Goff	Bill Fairless	Gordon Bisgrove (April 95)	Brydon Ward

Department Re-structuring[24]

	Assistant Deputy Minister Land & Forest Service	Southern East Slopes Region (Rocky Mountain House)	Northern East Slopes Region (Whitecourt)	Northwest Boreal Region (Peace River)
1995-96	C.J. Henderson (July 95)	Lorne Goff/Patrick Guidera (Jan. 96)	Jerry Sunderland	Howard Gray (July 95)
1996-97	C.J. Henderson	Patrick Guidera	Jerry Sunderland	Howard Gray
1997-98	C.J. Henderson	Patrick Guidera	Jerry Sunderland	Howard Gray/Rory Thompson[25]
1998-99	C.J. Henderson	Patrick Guidera Parkland/Bow/Prairie Region (Created Oct. 1, 1998)	Jerry Sunderland	Dan Wilkinson

		Assistant Deputy Minister Land & Forest Service	Southern East Slopes Region (Rocky Mountain House)	Northern East Slopes Region (Whitecourt)	Northwest Boreal Region (Peace River)
Department of Alberta Environment[26]	1999-00	C.J. Henderson	Patrick Guidera	Michael Poscente (May 1999)	Dan Wilkinson/Ken McCrae
	2000-01	C.J. Henderson	Patrick Guidera	Michael Poscente	Ken McCrae

Athabasca (Ft. McMurray)	Slave Lake	Grande Prairie	Peace River	Footner Lake (High Level)	Pine Ridge Forest Nursery
L.G. Babcock	H.R. Winn	C.B. Smith	G.A. Longworth	C.A. Dermott (Feb. 1, 1974)	
L.G. Babcock	H.R. Winn	C.B. Smith	G.A. Longworth	N.R. Rodseth (Dec. 1974)	
A.J. Peter (May 1, 1977)	H.R. Winn	C.B. Smith	G.A. Longworth	N.R. Rodseth	
A.J. Peter	C.S. McDonald (May 1977)	C.B. Smith	L.G. Babcock (May 1, 1977)	C.J. Henderson	H.R. Winn
A.J. Peter	C.S. McDonald	C.B. Smith	L.G. Babcock/ A.J. Peter	C.J. Henderson	H.R. Winn
L.D. Goff	C.S. McDonald	C.B. Smith	A.J. Peter	C.J. Henderson	H.R. Winn
L.D. Goff/J. Skrenek (Jan 1, 1981)	C.S. McDonald	C.B. Smith	L.D. Goff	C.W. Leary	H.R. Winn
J. Skrenek	C.S. McDonald	D.M. Timanson (March 1981)	L.D. Goff	C.W. Leary	H.R. Winn
J. Skrenek	C.S. McDonald	D.M. Timanson	L.D. Goff	C.W. Leary	H.R. Winn
J. Skrenek (Sept. 1984)	C.S. McDonald	D.M. Timanson	L.D. Goff	C.W. Leary	H.R. Winn
W. Fairless	C.S. McDonald	D.M. Timanson	L.D. Goff	C.W. Leary	H.R. Winn

Athabasca	Slave Lake	Grande Prairie	Peace River	Footner Lake	Pine Ridge Forest Nursery
W. Fairless	H.W. Gray	D.M. Timanson	C.W. Leary (1986)	J. Johnston (1986)	C.S. McDonald (Dec 1985)
G. Armitage	H.W. Gray	D.M. Timanson	C.W. Leary	J. Johnston	C.S. McDonald
G. Armitage	H.W. Gray	D.M. Timanson	C.W. Leary	J. Johnston	C.S. McDonald
G. Armitage	H.W. Gray	D.M. Timanson	C.W. Leary	J. Johnston	C.S. McDonald
G. Armitage	H.W. Gray	D.M. Timanson	C.W. Leary	J. Johnston	C.S. McDonald
G. Armitage	H.W. Gray	D.M. Timanson	C.W. Leary	J. Johnston	C.S. McDonald
G. Armitage	H.W. Gray	D.M. Timanson	C.W. Leary	J. Johnston	C.S. McDonald
G. Armitage	H.W. Gray	D.M. Timanson	C.W. Leary	J. Johnston	C.S. McDonald

Athabasca	Slave Lake	Grande Prairie	Peace River	Footner Lake	Pine Ridge Forest Nursery
Gordon Armitage	Howard Gray	Mort Timanson	Carl Leary	Jorden Johnston (Aug. 1993)	C.S. McDonald
Gordon Armitage	Howard Gray	Mort Timanson	Carl Leary (June 95)	Carl Leary	Neil Barker (May 1, 1995)

Northeast Boreal Region (Lac La Biche)
B. Ward/N. Barker (June 96)
Neil Barker
Neil Barker
Neil Barker

Northeast Boreal Region (Lac La Biche)
Neil Barker
Neil Barker

		ADM Forest Protection Division ADM Land & Forest Division	Prairie-Bow-Parkland Region (Rocky Mountain House)	Northern East Slopes Region (Whitecourt)	Northwest Boreal Region (Peace River)
Department of Sustainable Resource Development[27]	2001-02	C.J. Henderson; H.W. Gray	Patrick Guidera	Michael Poscente	Ken McCrae
		ADM Forest Protection Division ADM Land & Forest Division ADM Public Lands Division	Southwest Region (Rocky Mountain House)	Southeast Region (Calgary)	Northwest Region (Peace River)
	2002-03	C.J. Henderson; H.W. Gray; J.C. Quintilio	Patrick Guidera	Domenic Ruggieri	Ken McCrae
Department of Sustainable Resource Development	Department Re-structuring-July[28]	ADM Forest Protection Division ADM Strategic Forestry Initiatives ADM Public Lands and Forests Division	Southwest Region (Rocky Mountain House)	Southeast Region (Calgary)	Northwest Region (Peace River)
		2003-04 Cliff Henderson; Howard Gray; Craig Quintilio	Patrick Guidera	Domenic Ruggieri	Ken McCrae
	Department Re-structuring-November[29]	2004-05 Cliff Henderson; Howard Gray; Craig Quintilio	David Christensen (Acting)	Domenic Ruggieri	Ken McCrae

End Notes

[1] Alberta assumed responsibility for natural resources, including forests, on 1 October 1930. For the rest of the fiscal year 1930-31, the former Dominion Forestry Branch organization was continued. There were then four Forest Reserves and the Edmonton Fire Ranging District. The Forest Reserves included the Lesser Slave Forest Reserve and the Rocky Mountains Forest Reserve which comprised five Forests: Crowsnest, Bow River, Clearwater, Brazeau and Athabasca. Two smaller Forest Reserves, Cooking Lake and Cypress Hills were run as ranger districts. The rest of the forested area lay within the Edmonton Fire Ranging District (EFRD).

[2] In the spring of 1931 the Alberta Forest Service (AFS) took over responsibility for the *Prairie Fires Act* from the Department of Agriculture. The entire province was then declared a Fire District, and new fire regulations were passed. This was to try to effect more unified control over the fire program.

[3] In 1932 the administration and organization were restructured. Until this time, the AFS was responsible for all protection and timber on the Forest Reserves and protection on the EFRD. Timber berths were issued under authority of the *Public Land Act* and Timber Inspectors worked under the Lands Division to oversee timber sales and logging operations. Most Timber Inspectors were transferred to AFS in 1932 and AFS became responsible for timber as well. Under the new organization, the EFRD was renamed the Northern Alberta Forest District (NAFD, a title that would remain until 1956). The NAFD was divided into eight Divisions and a Timber Inspector was placed in charge of each. The Divisions were staffed by a sparse network of seasonal rangers. Administration of the Forest Reserves remained essentially the same.

[4] Forest Reserve status for Lesser Slave was dropped in 1932, the former forest reserve was combined with other NAFD lands to form the Slave Lake Division.

[5] Frank Neilson was Chief Timber Inspector, a position in which he continued to be responsible for the Northern Alberta Forest District. This position later evolved to head of forest protection. He worked through a network of 8 Timber Inspectors whose positions later evolved to Forest Superintendents. In addition to the ones listed, D.A McKay held the position of Timber Inspector at Bonnyville/Wasketanau from 1932 to 1954.

[6] E.S. (Eric) Huestis noted during his interview that in 1938 and 1939 he was appointed Supervisor of the Brazeau-Athabaska Forest and moved to Edson. At that time headquarters of the combined forests was at Coalspur so he arranged to have it changed to Edson, which enabled more effective train connection with both the Coal Branch and main line to Hinton-Entrance. He likely worked through the two listed "Inspectors" during these years. He moved to Edmonton to be acting assistant director under Ted Blefgen when J.A. Hutchison enlisted.

[7] The Western Division included Alder Flats, Winfield, Muldoon, Drayton Valley, Carrot Creek, Whitecourt and Fort Assiniboine. Peace River Division included the Grande Prairie area. The McMurray Division included Athabasca and Lac La Biche. McMurray was a forestry administrative centre during the pre-1930 Dominion days, so it was a logical site to choose in 1930 – headquarters changed to Lac La Biche in 1941.

[8] Westlock was a forestry administrative centre for the DFB, and AFS until 1947 when headquarters was moved to Whitecourt.

[9] F.E. Smith transferred to Bowden where he served as Timber Inspector until retirement in 1954.

[10] There were 2 senior positions in Edson – D. Buck as Forest Superintendent of the Brazeau-Athabaska Forest, and H.E. Noble, Timber Inspector for the Edson Division, the area that lay outside the Forest Reserve.

[11] J.L. (Jack) Janssen remained as Chief Timber Inspector in Edmonton, functioning as head of forest protection. His name is listed among the 'branch' heads on the separate list.

[12] The Eastern Rockies Forest Conservation Board (ERFCB) was set up in 1948 under federal and provincial acts to increase protection on the important watersheds of the South and North Saskatchewan Rivers. Policies and administration of the three southern Forests – Crowsnest, Bow River and Clearwater – were directed by a federal-provincial board. The previously combined Bow-Crow

Northeast Boreal Region (Lac La Biche)

Neil Barker

Northeast Region (Lac La Biche)

Neil Barker

Northeast Region (Lac La Biche)

Neil Barker

Neil Barker

Forest was again split into its original parts. The three Forests were each led by a Forest Superintendent.

13. Forest areas in the NAFD were headed by Timber Inspectors located at the locations identified.

14. The forested lands to the east of the Clearwater Forest lay outside of the Forest Reserve so were not covered by the ERFCB. The Forest Superintendent at Rocky Mountain House also acted as head of what was called the Rocky Forest, for which an Assistant Timber Inspector, later Assistant Forest Superintendent was assigned.

15. When the three southern Forests were managed by the ERFCB, the Senior Forest Superintendent in Calgary supervised the three Forest Superintendents – in this case, J.P. (Jack) Alexander who was previously Superintendent of the combined Bow-Crow Forest held the position until he retired in 1952.

16. The new organization was explained by Huestis in a memo to all field personnel dated 2 November 1953. He described that there would be 6 Divisions, each to be led by a Forest Superintendent. It was understood that the three southern Forests continued under the policy arrangement with the ERFCB. In addition, the Rocky Forest would be run by the Forest Superintendent at Rocky Mountain House.

17. When the Eastern Rockies (ERFCB) agreement expired, administration of the Clearwater Forest was combined with the Rocky Forest into what became known as the Rocky-Clearwater Forest. Clearwater-Rocky was considered, but AFS wanted each forest name to start with a different letter, and Crowsnest already had the 'C'.

18. Athabasca Forest was created in 1964 by splitting off the northern portion of the Lac La Biche Forest. Consideration was given to naming the area the 'Fort McMurray' forest since historically the Athabasca forest was located in the Edson area. However, AFS wanted to save the letter 'F' for the proposed Footner Lake Forest.

19. Footner Lake Forest was created in 1966 by splitting off the northern portion of the Peace River Forest.

20. Crowsnest Forest was combined with Bow River in 1973, called Bow-Crow Forest.

21. When Bob Steele retired in 1978, Fred McDougall was appointed Deputy Minister of Natural Resources to replace him. The position of Director of Forestry was re-named Assistant Deputy Minister for Forestry and J.A. (Al) Brennan was recruited to head the Alberta Forest Service.

22. In the new Department of Forestry Lands and Wildlife, Al Brennan was appointed to head the new Forest Industry Development Division (FIDD) to encourage investment in the sector. C.B. (Cliff) Smith was promoted to Assistant Deputy Minister for Forestry. Fred McDougall was named Deputy Minister of the new department.

23. Department restructuring – Department of Environmental Protection created.

24. Field organization for Environmental Protection for the province as a whole restructured to six Regions, of which four were predominantly in the forested areas. In December 1994 the division was renamed Land and Forest Service from Land and Forest Services (for a short period in December, the division was named Lands and Forests Services).

25. Howard Gray went to Economic Development as the Executive Director of the Forest Products Branch. Ken McCrae was Acting Director until the appointment of Rory Thompson.

26. Department and Regions re-named

27. Department of Sustainable Resource Development was created March 15, 2001. Divisions of Forest Protection, Lands and Forests, Fish and Wildlife, and Strategic Corporate Services were created.

28. Division of Strategic Forestry Initiatives was created.

29. Provincial election held November 2004 and Minister Coutts appointed as new minister of Sustainable Resource Development.

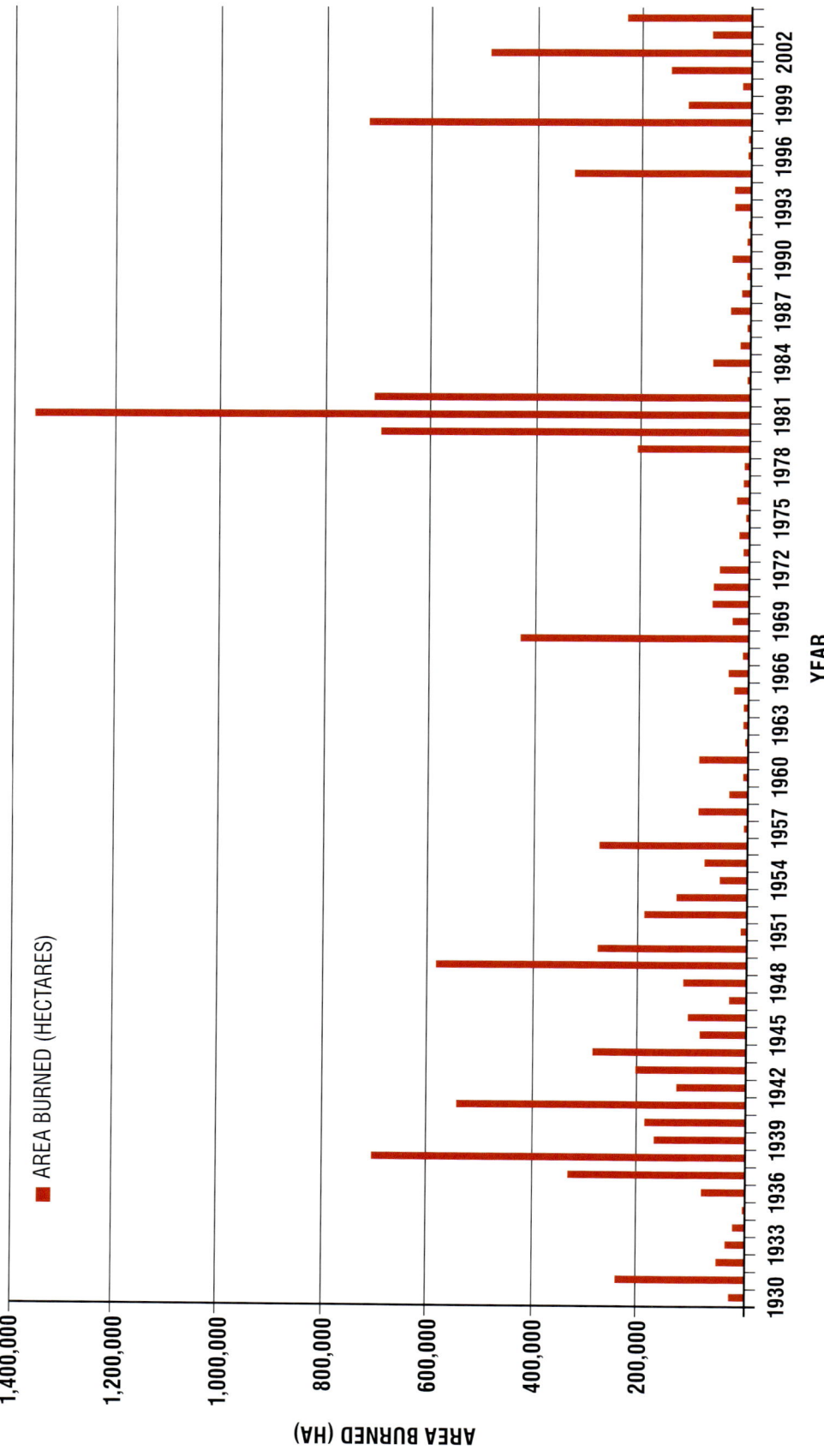

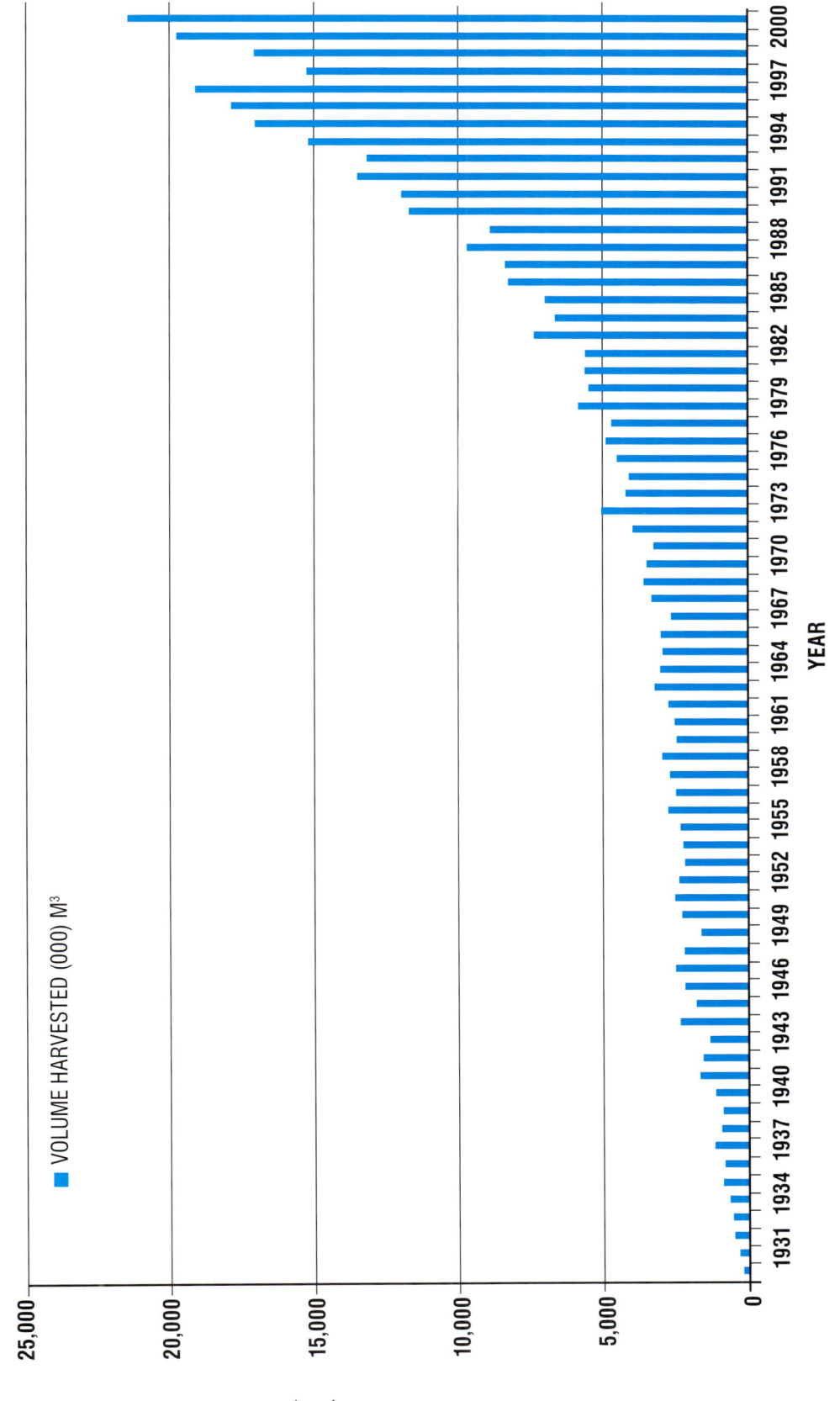

Alberta Crown Area Harvested
1937 to 2003

Note: Significant deciduous harvest began in the early 1990s

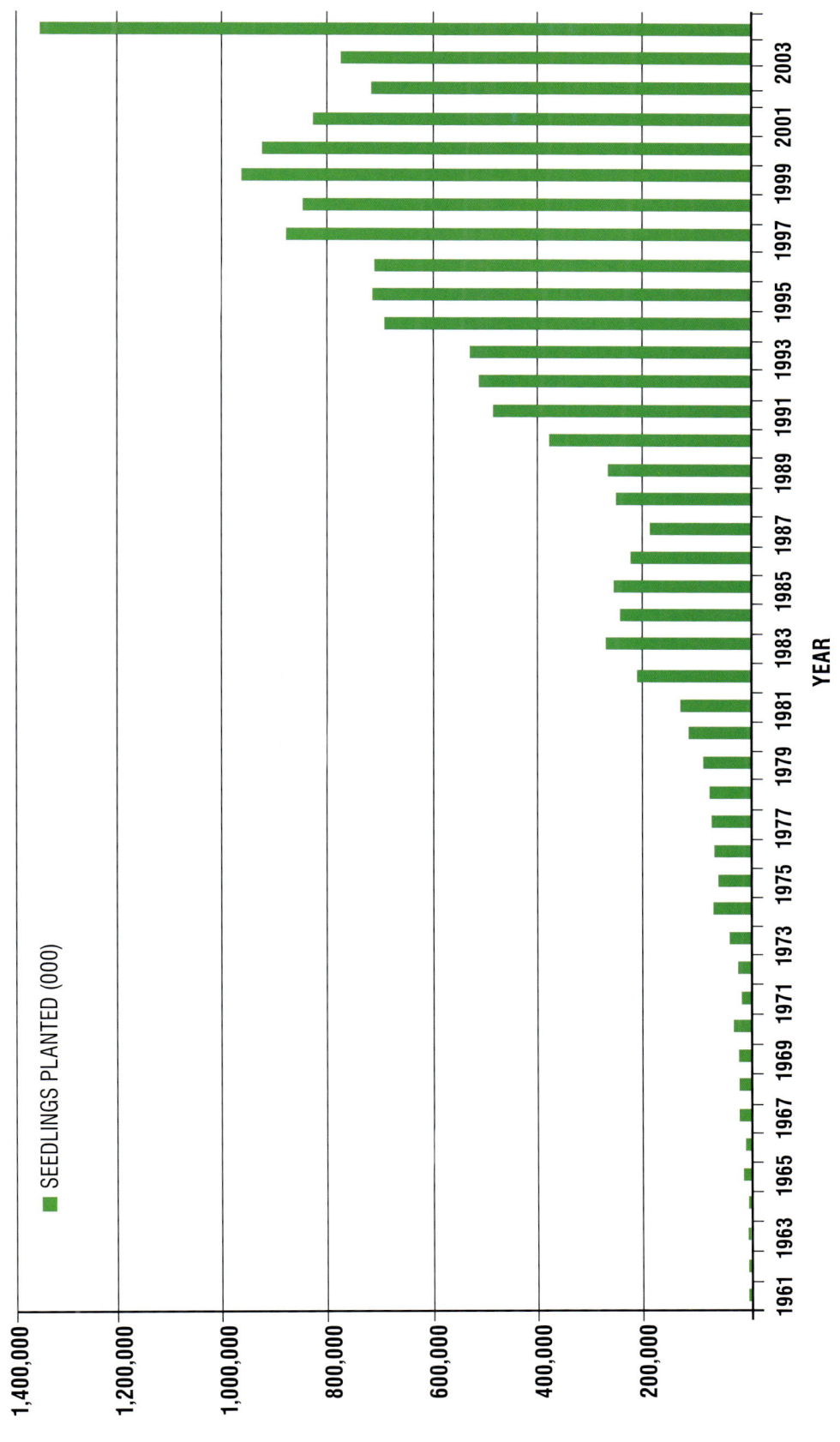

Ranger Stations and Cabins

John Edward (Jack) Bell began his career as Assistant Forest Ranger in the Elbow River District in August, 1925, retiring as the building construction and maintenance foreman for the Rocky Mountains Forest Reserve in 1963. In the 1950s the Eastern Rockies Forest Conservation Board began a major capital program for roads and facilities. Jack's skills were subsequently utilized beginning with the construction of the Blairmore office and garage in the Crowsnest Forest, following with supervision of the construction of new ranger stations on all districts in the Rocky Mountains Forest Reserve. The local ranger assisted Jack in construction duties, learning carpentry fundamentals through a short course before construction began. Jack Bell married Dolly King, sister of the famous King Brothers ranching team of Lundbreck. Jack is quoted as saying "if you've got a day's work to do, a good time to start is early in the morning".
Alberta Government, Land Forest Wildlife Magazine

Ranger Stations and Cabins

The construction of a cabin was one of the first jobs of a Ranger when hired by the Dominion Forest Service, later the Alberta Forest Service. Cabins were approximately 14 feet wide by 16 feet long made of logs, with sawn planks for roofs and floors and were located one day's ride apart. The earliest ranger stations were constructed in the early 1910s – Jumping Pound Ranger Station, Athabasca Forest Headquarters, Red Deer and Elbow Ranger Stations. The following series of photos illustrates the history of both cabins and ranger stations and is supplemented by a list of 'known' cabin and ranger station names. The list is a start – the editors recognize considerable improvements can be made.

All photos Alberta Government, AFHPC unless otherwise noted

Ranger Stations / Ranger Districts / Cabins

Bow River Forest / Bow Forest / Bow Crow Forest / Crowsnest Forest
East Porcupine
West Porcupine
Willow Creek
Highwood
Turner Valley
Skyline
Sentinel
Big Horn
Spray Lakes - Elk River Cabin
Canmore - Whitegoat Cabin
Elbow - Bighorn Cabin
Coleman
Kananaskis *(Boundary, Ribbon Creek)*
Gap *(Livingstone)*
Jumping Pound
Aura RS *(later called Ghost)*
Blairmore
Calgary
Castle River District (Castlemount originally)
Pincher Creek *(might have been the Beaver Mines site)*

Morley RS
Sheep River *(west of Turner Valley)* *(Big Horn District?)*
Lynx Creek or Carbondale
Mill Creek *(east of the Castle)*
Elk Lodge

Brazeau/Athabaska Forest / Edson Forest
Grave Creek RS *(photo 6452)*
Edson
Hinton
Robb
Grande Cache
Gregg River *(stop over cabin?)*
Moberly
Lovett
Entrance *(was DFB HQ for Athabasca Forest Reserve)*
Mountain Park
Coalspur
Rock Lake
Drayton Valley
James River
Cabin Creek RS *(between Hinton and Muskeg)*
Muskeg *(open til early 1970s)*
Cadomin *(Leyland)*
Brule
Medicine Lodge
Wildhay
Little Berland

Cabins - Eagles Nest, Big Grave Flats, Smoky River, Sheep Creek, Winter Creek, Polecat Creek, A la Peche, Adams Creek, Summitt

Rocky/Clearwater Forest / Rocky Forest
Rocky Mountain House
Alder Flats   Whitegoat - stopover cabin
StrachanRam Falls, Bighorn cabin
Nordegg
Prairie Creek
Red Deer
Sundre
Clearwater
Lodgepole
Ricinus
Shunda
Saunders Creek
Rimby
Meadows *(east of Ram Tower on North Ram River)*
Upper North Saskatchewan *(now under Lake Abraham)*
Breton
Winfield
Swan Lake Forestry Cabin
Wilson

Whitecourt Forest
Berland
Fox Creek
Pine
Whitecourt
Cold Creek
Blue Ridge
Fort Assiniboine
Swan Hills
MacKay *(Cold Creek?)*
Breton

Evansberg
Drayton Valley
Westlock

Peace River Forest
Peace River
Hines Creek
Worsley *(amalgamated with Hines Creek in 1983 - Clearwater District. Remained open until 1996)*
Keg River *(amalgamated with Manning and Dixonville approx 1979/80)*
Three Creeks *(closed in 1972 or 73 with staff moved to Peace River)*
McLennan *(amalgamated with Three Creeks to become East Peace - stayed open as satellite until 1991)*
Dixonville
Manning

Grande Prairie Forest
Grande Prairie
Grovedale
Valleyview   Little Smoky Cabin
Spirit River   Cabin Demmit *(Goose Lake)*
Debolt
South Wapiti
Wanham
New Fish Creek
Woking
Watino
Hythe

Footner Forest
Footner
Fort Vermilion
North Vermilion
Little Red River
High Level
Rainbow Lake
Steen River
Habay *(Chateh/Assumption)* 1957-1966
Meander River *(Upper Hay)*

Lac La Biche Forest
Lac La Biche
Beaver Lake
Calling Lake
Wandering River
La Corey
Conklin
Chard
Athabasca (old and new)

Slave Lake Forest
Wabasca
Slave Lake
High Prairie
Red Earth
Mirror Landing
Smith
Salt Prairie
Kinuso

Grouard
Sunset House
Faust
Swan River

Athabasca Forest
Fort McMurray
Waterways *(old and new)*
Fort McKay
Fort Chipewyan
Anzac
Embarras
Chard

Ranger Districts as of 1971 (Alberta Forests magazine)

Footner Lake Forest, HQ at Footner Lake
F1 - Fort Vermilion
F3 - High Level
F5 - Meander River
F6 - Rainbow Lake

Athabasca Forest, HQ at Fort McMurray
A1 - Anzac
A2 - Fort McMurray
A3 - Fort McKay
A4 - Fort Chipewyan

Peace River Forest, HQ at Peace River
P1 - McLennan
P2 - Dixonville
P3 - Hines Creek
P4 - Worsley
P5 - Manning
P6 - Peace River
P7 - Keg River

Grande Prairie, HQ at Grande Prairie
G1 - South Wapiti
G2 - Grovedale
G3 - Debolt
G4 - Valleyview
G5 - Spirit River

Edson Forest, HQ at Edson
E1 - Robb
E2 - Hinton
E3 - Grande Cache
E4 - Edson

Whitecourt Forest, HQ at Whitecourt
W1 - Fox Cree
W2 - MacKay (Cold Creek)
W3 - Blue Ridge
W4 - Fort Assiniboine
W5 - Swan Hills

Slave Lake Forest, HQ at Slave Lake
S1 - Smith
S2 - Slave Lake
S3 - Kinuso
S4 - High Prairie
S5 - Salt Prairie
S6 - Wabasca
S7 - Slave Lake - Red Earth

Lac La Biche Forest, HQ at Lac La Biche
L1 - La Corey
L2 - Lac La Biche
L3 - Calling Lake
L4 - Wandering River

Clearwater - Rocky Forest, HQ at Rocky Mountain House
R1 - Ricinus
R2 - Rocky Mountain House
R3 - Nordegg
R4 - Prairie Creek
R5 - Strachan
R6 - Rocky Mountain House
R7 - Alder Flats
R8 - Shudna
R9 - Lodgepole

Bow River - Crowsnest Forest, HQ at Calgary
B1 - Turner Valley
B2 - Bragg Creek
B3 - Canmore
B4 - Cochrane
B5 - Sundre
C1 - Blairmore
C2 - Blairmore
C3 - Blairmore

Ranger Districts as of 2004
Northwest Region - HQ at Peace River
High Level - sub offices in Rainbow Lake and Fort Vermilion
Peace River - sub office in Hines Creek, Manning
Grande Praire - sub office in Valleyview
Slave Lake - sub offices in High Prairie and Wabasca
Northeast Region - HQ at Lac La Biche
Fort McMurray - sub office in Fort Chipeywan
Lac La Biche - sub offices in Athabasca, Wandering River and Bonnyville
Southwest Region - HQ in Rocky Mountain House
Rocky Mountain House - sub offices in Nordegg and Drayton Valley
Calgary - sub office in Blairmore
Edson/Hinton - sub office in Hinton, Grande Cache, Cold Creek
Whitecourt - sub office in Fox Creek, Swan Hills

Lynx Creek cabin, Crowsnest Forest, Rocky Mountains Forest Reserve, 1911

Jumping Pound Ranger Station, Bow River Forest, Rocky Mountains Forest Reserve, 1912

Ranger Station in the Highwood (Sentinel) District, Bow River Forest, Rocky Mountains Forest Reserve, 1910s

Moberly Creek Ranger Station, Athabasca Forest, Rocky Mountains Forest Reserve, in the early 1910s or 1920s

Ranger Stations and Cabins 447

The first Hinton Ranger Station, Athabasca Forest, Rocky Mountains Forest Reserve, 1913

Dominion Rangers conducting a Forest Survey overnight at the Southesk Cabin, Brazeau Forest, Rocky Mountains Forest Reserve, 1913

Ranger J.P. (Jack) McDonald and family at the Gap Ranger Station, Bow River Forest, Rocky Mountains Forest Reserve, 1913 (west of Granum)

Grave Creek RS, Brazeau Forest, Rocky Mountains Forest Reserve, 1913

Ranger and wife outside the Nordegg Ranger Station, Clearwater Forest, Rocky Mountains Forest Reserve, 1914

Remnants of the Dead Horse Cabin, February, 2002. This was a Dominion Forest Service cabin located on the Klondike Trail, Township 65, Range 7, West 5th Meridian
Brad Mckenzie

Entrance Ranger Station, Athabasca Forest, Rocky Mountains Forest Reserve, 1917. The Entrance Ranger Station was the headquarters for the Athabasca Forest and tended to be a central hub in the northern rockies

Kananaskis Ranger Station (Boundary cabin), Bow River Forest, Rocky Mountains Forest Reserve, July, 1917

Ranger Stations and Cabins 449

North Fork Cabin, Bow River Forest, Rocky Mountains Forest Reserve, 1920

Cypress Hills Headquarters office and department vehicle, Cypress Hills Forest Reserve, 1920

Fire Rangers Headquarters at Fort McMurray. The buildings were located along the Snye, a side channel of the Athabasca River. Raphrew Canning was the Chief Fire Ranger from 1921 to 1933. In 1936, a flood demolished these buildings. A new location on Main Street between Manning and Fraser was chosen for the new headquarters. Fire Rangers Headquarters, Forestry Branch, Department of the Interior, 1921

Gap Ranger Station and outbuildings, Crowsnest Forest, Rocky Mountains Forest Reserve, 1920s

Cooking Lake Ranger Station, Cooking Lake Forest Reserve, 1926

Elbow Ranger Station house, Bow River Forest, Rocky Mountains Forest Reserve, November, 1926

East Porcupine Ranger Station, Crowsnest Forest, Rocky Mountains Forest Reserve, 1920s. The sign above the gateway instructs visitors to register

West Porcupine Ranger Station, Crowsnest Forest, Rocky Mountains Forest Reserve, late 1920s

Ranger Stations and Cabins

Elbow Ranger Station, Bow River Forest, Rocky Mountains Forest Reserve, 1933
Jay Champion

Boundary Ranger cabin, Kananaskis District, Bow River Forest, 1938

Harold Creek Cabin, Bow River Forest, Rocky Mountains Forest Reserve, 1939
Jay Champion

Sheep Creek Forestry Cabin, Brazeau-Athabasca Forest, 1941. Ranger Gordon Watt standing by the door

Castlemount Ranger Station, west of Pincher Creek, Crowsnest Forest, Rocky Mountains Forest Reserve, 1943
Jay Champion

McMurray Ranger Station, Northern Alberta Forest District, 1940s. Parts of the buildings that survived the flood in the 1920s were used in construction of this new ranger station
Neil Gilliat

Lynx Creek Ranger Station with registration check-in box, Crowsnest Forest, 1940s

Fort McMurray Ranger Station, Northern Alberta Forest District, 1948. Mrs. Doris Roy (Ranger Jack Roy's wife) and two children Blane and Beverly with Ranger Gunner Brauti in front of the Ranger Station. On August 18, 1949 Ranger Brauti died when two railway speeders crashed in heavy fog
Mrs. Jack Roy

Gap Ranger Station, Crowsnest Forest, Rocky Mountains Forest Reserve, 1940s

Little Berland Ranger Station, Brazeau-Athabasca Forest, 1949. Rangers Neil Gilliat (L) and Rex Winn standing in front of Ranger Bill Smith's new jeep
Neil Gilliat

The Breton cache was like a number of caches throughout the Northern Alberta Forest District – they contained a half dozen shovels, axes and pulaski's for use by firefighters
Bill McPhail

Whitegoat stopover cabin, Clearwater Forest, Rocky Mountains Forest Reserve, 1940s

Old Rock Lake Ranger Station, Brazeau-Athabasca Forest, 1940s

Moberly Creek Ranger Station, Edson Forest, 1951
Neil Gilliat

Sketch of log cabin
Artist Lorna Bennett

Ranger Bill McPhail outside the Breton Ranger Station and cache, Northern Alberta Forest District, 1950s
Bill McPhail

Leyland Ranger Station in the Cadomin area, Brazeau-Athabasca Forest, 1950s
Neil Gilliat collection

Ranger Stations and Cabins 455

New Fish Creek Ranger Station, Grande Prairie Division, 1950s
Bill McPhail

Peter Murphy at the door of the Adams Creek Cabin, Brazeau-Athabasca Forest, 1950s

Meadows Ranger Station house, Clearwater Forest, Rocky Mountains Forest Reserve, 1950

Eagles Nest Cabin, Athabasca Forest, Rocky Mountains Forest Reserve, 1951. There were numerous line cabins along the Mountain Trail. Starting at Rock Lake the Eagles Nest was 12 miles to the north. These cabins were a basic shelter. Most had two spring cots which would have been difficult to pack on a horse. The two mattresses were to be rolled up and hung by a rope from the rafters when not in use, otherwise mice would dismantle them within a couple of weeks. There was a wash basin on a pole stand and a similar makeshift table at which we would sit, on pack boxes. Outside was a hitching rail and large spikes were driven into logs to hang up saddlery. It was important to hang up anything of leather, as the porcupines loved the leather salty from the horse sweat. A tin stove and in some a small airtight heater hooked up to tin stove pipes. Most had a kerosene lamp, but where absent, people relied on candles. The golden rule was to always leave wood and kindling when you left. It was another 14 miles through Eagles Nest Pass to Rock Creek and on to "Summit" or Mile 58 cabin. Some spectacular scenery on the way and a must was a picture with Cathedral Mountain in the background
Neil Gilliat collection – photo and story

Meadows Patrol Cabin, Clearwater Forest, Rocky Mountains Forest Reserve, 1950s

Fort Vermilion Ranger Station, District 30, Northern Alberta Forest District, June, 1956
Provincial Archives of Alberta, PA2755-1

Swan Ranger Station and residence, Slave Lake Division, 1950s
Lou Foley

Whitegoat Patrol Cabin, Clearwater Forest, Rocky Mountains Forest Reserve, 1950s

Ranger Bill Willows at the Smoky River cabin, Grande Prairie Division, 1950s
Bill McPhail

Ranger Stations and Cabins

Entrance Ranger Station, Brazeau-Athabasca Forest, 1950s

Ranger Ernie Stroebel at the Little Berland Ranger Station, Brazeau-Athabasca Forest, late 1950s

Hay River Ranger Station, Brazeau-Athabasca Forest, 1950s

Little Berland cabin, Brazeau-Athabasca Forest, 1950s

Whitecourt Division Ranger Station, Northern Alberta Forest District, June, 1956
Provincial Archives of Alberta, PA2754-1

Upper Saskatchewan Ranger Station, Clearwater Forest, Rocky Mountains Forest Reserve, 1950s

Lovett Ranger Station, Brazeau-Athabasca Forest, 1954

Lodgepole Ranger Station, Rocky Forest, 1957

Peace River Division Ranger Station, Northern Alberta Forest District, June, 1956
Provincial Archives of Alberta, PA2753-1

Cadomin Ranger Station, Brazeau-Athabasca Forest, May 1958

Ranger Stations and Cabins

Visitors, recreationists and industrial workers were required to check-in before entering the Forest Reserve along the eastern slopes. This allowed a ranger to keep in touch with users and have an understanding of the activity taking place in the area he was tasked to protect. Bow River Forest, 1955

North Vermilion Ranger Station, District 62, Peace River Division, late 1950s

Little Red Ranger Station, District 44, Peace River Division, late 1950s

Old Grave Flats Cabin, Brazeau-Athabasca Forest, 1950s

Rangers Angus Crawford and Bert Prowse at the Red Cap or Mile 10 Cabin, Brazeau-Athabasca Forest, January 1958. This was the first trip by vehicle to the cabin

Mountain Park Ranger Station Headquarters, Brazeau-Athabasca Forest, 1958

Calling Lake Ranger Station, ranger house and office, garage and cache, buildings facing west, Lac La Biche Forest, 1959
Ernie Stroebel

Mirror Landing Ranger Station, District 31, Northern Alberta Forest District, November, 1959. Mirror Landing was just north of the Athabasca River from Smith, the location of the future ranger station
Cliff Smith

Clearwater Ranger Station, Clearwater Forest, September, 1960

Cabin Creek Ranger Station, Brazeau-Athabasca Forest, 1960s

Steen River Ranger Station, north of High Level, Peace River Division, early 1960s
Bruce MacGregor

Medicine Lodge Ranger Station, Edson Forest, 1960s

Leyland Ranger Station, Brazeau-Athabasca Forest, 1960s

Rock Lake Ranger Station, Edson Forest, 1960s

Aerial view of the Red Deer Ranger Station, Bow River Forest, September, 1960

Habay Ranger Station, west of High Level, Peace River Division, 1961

Ranger Stations and Cabins 463

Porcupine Ranger Station garage, Crowsnest Forest, mid-1960s

Ranger house at the Prairie Creek Ranger Station, Rocky Forest, mid-1960s

Clearwater Ranger Station and houses, Clearwater Forest, 1960s

Prairie Creek Ranger Station, Rocky Forest, mid-1960s

Meander River Ranger Station (also called Upper Hay), north of High Level, Peace River Division, early 1960s
Bruce MacGregor

Whitecourt Division Headquarters office, Whitecourt Forest, mid-1960s

Strachan Ranger Station, Rocky Clearwater Forest, 1960s
Bill McPhail

Aerial view of the Worsley Ranger Station, Peace River Forest, 1962
Ken McCrae

Skyline Ranger Station, Crowsnest Forest, July 1, 1963. Photo taken to record the overnight snowfall
Harold Ganske

Ranger and wife at the Saskatoon Ranger Station office/residence, Grande Prairie Division, 1963

Habay Ranger Station, west of High Level, Peace River Division, flood of May, 1963
Larry Huberdeau

Ranger Art Peter standing on porch of Elk River Cabin, Whitecourt Division, fall, 1963

Swan Hills Ranger Station, Whitecourt Division, January, 1964

Slave Lake Headquarters office, Slave Lake Division, 1964
Neil Gilliat

High Level Ranger Station, Peace River Division, December, 1964

Aerial view of the Footner Lake Ranger Station, Peace River Forest, 1965

Highwood Ranger Station office, house, garage and cache, Bow River Forest, 1965

Edson Headquarters warehouse building, Edson Forest, 1965

Brule Ranger Station, Edson Forest, August, 1965

Blairmore Headquarters office, Crowsnest Forest, 1965

Red Deer Ranger Station, Rocky Clearwater Forest, August, 1965

Ranger Stations and Cabins

Willow Creek Ranger Station garage, Crowsnest Forest, mid-1960s

Entrance Ranger Station garage and warehouse buildings, Edson Forest, mid-1960s

Rocky Mountain Forest Headquarters garage facility, Rocky Clearwater Forest, mid-1960s

Skyline Ranger Station, Crowsnest Forest, mid-1960s

Alder Flats Ranger Station house, Rocky Clearwater Forest, mid-1960s

Castle Ranger Station houses under construction Crowsnest Forest, 1960s

Willow Creek Ranger Station cache and bunkhouse, Crowsnest Forest, mid-1960s

Sundre Ranger Station with the Sundre Crawl Tower in the background, Rocky Forest, August, 1965

Hinton Ranger Station, Edson Forest, mid-1960s

Old residence at the Livingstone (Gap) Ranger Station, Crowsnest Forest, 1960s

Assistant Rangers house, Livingstone Ranger Station, Crowsnest Forest, mid-1960s

Superintendent's house in Slave Lake, Slave Lake Forest, 1967
Neil Gilliat

Fox Creek Ranger Station, Whitecourt Forest, fall, 1968
Don Welsh

Aerial view of the Fort Assiniboine Ranger Station, Whitecourt Forest, 1968
Bruce MacGregor

Aerial view of the Worsley Ranger Station, Peace River Forest, 1967
Glen Gache office collection

Aerial view of the Steen River Ranger Station, Footner Lake Forest, 1969
Ross Graham

Meander River Ranger Station, Footner Lake Forest, fall, 1969
Ross Graham

Fort McKay Ranger Station, Lac Athabasca Forest, spring, 1969
Corinne Huberdeau

Moberly Cabin, Edson Forest, 1968. The Moberly Creek cabin was dismantled and relocated to the Alberta Forest Service Museum in Hinton in 1971

Red Earth Ranger Station (on the left) and residence trailer for Rangers, Slave Lake Forest, March 20, 1971
Jamie McQuarrie

Calgary Headquarters, Bow River Forest, 1972

Lodgepole Ranger Station office with black bear at door, Whitecourt Forest, July, 1972
Ross Graham

Ranger Stations and Cabins

Hines Creek Ranger Station,
Peace River Forest, 1970s
Glen Gache office collection

Canmore Ranger Station was located on the north side of Highway 1,
Bow River Forest, 1970s

Moving the old Red Earth
Ranger Station, Slave Lake
Forest, June, 1977. A new
facility was constructed and
opened in spring, 1977
Ken Orich

Aerial view of the Keg River Ranger Station, Peace River Forest, summer, 1978
Ross Graham

Aerial view of the Slave Lake Forest Headquarters site and buildings, Slave Lake Forest, 1980
Dave Brown

Aerial view of the old Nordegg Ranger Station, Rocky Clearwater Forest, summer, 1980
Ken Orich

The new Nordegg Ranger Station office, Coliseum Mountain in the background, Rocky Clearwater Forest, September, 1980
Ken Orich

Red Deer Ranger Station, Bow River Forest, 1980s. Bertie Beaver fire danger sign in foreground
Butch Shenfield office collection

Aerial view of the Prairie Creek Ranger Station, Rocky Clearwater Forest, 1980s
Butch Shenfield office collection

Ranger Stations and Cabins 473

Aerial view of the Strachan Ranger Station, Rocky Clearwater Forest, 1980s
Butch Shenfield office collection

Kananaskis Ranger Station, Bow Crow Forest, 1981. This ranger station is also referred to as the Boundary Ranger Station and Ribbon Creek Ranger Station

Aerial view of the Robb Ranger Station, Edson Forest, 1981
Ross Graham

Peace River Ranger Station bunkhouse, Peace River Forest, 1980s
Terry Van Nest

Alberta Forest Service

Peace River Forest Headquarters, Peace River Forest, 1980s
Terry Van Nest

Sketch of Moberly Cabin by Jeanette Osborne, Word Processing Operator at the Forest Technology School, 1980s

Rainbow Lake Ranger Station, Footner Lake Forest, early 1980s
Corinne Huberdeau

Rainbow Lake Ranger Station office burned down December 7, 1983 - temperature was minus 35C. Fire started from an electrical short, Footner Lake Forest
Jamie McQuarrie

Ranger Stations and Cabins

Fort McMurray Headquarters office, Athabasca Forest, 1984
Ken McCrae

Aerial view of the Ghost Ranger Station, Bow Crow Forest, 1985

Sketch of Shand-Harvey's Cabin at Entrance, 1989. James Shand-Harvey was a guide and outfitter, former Dominion Forest Ranger
Artist Robert Guest

Aerial view of the Fox Creek Ranger Station, Whitecourt Forest, fall, 1990

Aerial view of the Ghost Ranger Station, Bow Crow Forest, summer, 1991

Aerial view of the Wabasca Ranger Station, Slave Lake Forest, 1991

Cold Creek Ranger Station office, just outside of Nojack, Edson Forest, 1990s
Ross Graham

Sketch of old cabin
Artist Lorna Bennett

Ranger Stations and Cabins 477

Aerial view of the old Meadows Ranger Station site, Rocky Clearwater Forest, 1990s
Butch Shenfield office collection

Aerial view of Calling Lake Ranger Station, Lac La Biche Forest, 1994

Ranger standing at boundary marker, north-east corner of Section 13, Township 21, Range 7, Bow River Forest Reserve, Rocky Mountains Forest Reserve, 1940s

Newly constructed trail marker in the Ram District, Clearwater Forest, Rocky Mountains Forest Reserve, 1940s. This marker was quite possibly made by Ranger Wally Richardson while he was the ranger in charge of the Ram District
Jack Richardon

Remains of an old forestry trail marker along the Ram River, Clearwater Area, 2002
Dave Ferster

Whitecourt Forest Division boundary sign, late 1950s

Slave Lake Forest Division boundary sign, fall, 1959
Cliff Smith

Grande Prairie Forest Division boundary sign, October 23, 1959
Cliff Smith

Peace River Forest Division boundary sign, fall, 1959
Cliff Smith

AFS Museum

The Alberta Forest Service Museum in Hinton was established in 1971 to preserve a history of forestry in the Province of Alberta. The museum consists of three log structures. The first is the equipment shed which houses heavy equipment used by the Forest Service for transportation, construction and maintenance. The second is the Moberly Creek Ranger Station, used between 1922-1959 as the ranger headquarters for the Moberly Creek Ranger District, approximately 40 km northwest of Hinton. The cabin was relocated to this site in 1971. The third is this main museum building housing displays reflecting the work performed and artifacts used by the early rangers. Various saddles, telephones, radios, uniforms, instruments and maintenance equipment are on display, along with an excellent display of fire tools, including fire pumps dating back to 1910. The museum opened to the general public in 1974.

Alberta Forest Service Museum, Hinton, 1980s

Museum being constructed in the early 1970s

Alberta Forest Service Museum, Hinton, Alberta

Miles Moberly (L) and Fred Plante were the two individuals who, under the direction of Harry Edgecombe, built the museum; a 26' by 40' log building. Felix Plante, Fred's father, was involved in the construction of the Moberly Ranger Station cabin in 1922. Early 1970s, Hinton

Harry Edgecombe supervised the construction of the Alberta Forest Service Museum. Harry Edgecombe began as a ranger with the Alberta Forest Service in 1946 serving in the Clearwater Forest and the Fort McMurray, High Level and Grande Prairie districts before moving to Hinton where he led the fire management training program. Harry retired in 1979

Every ranger station had a cache where tools were stored and maintained – the tool cache was essential. A wide assortment of commonly used items has been assembled for display. AFS Museum, Hinton, 1980s

Until 1947 two-way radios were not commonly available and the telephone was the backbone of forestry communications. By 1947, the Forest Service maintained over 950 miles of telephone line. Various telephones, radios and maintenance equipment are on display. AFS Museum, Hinton, 1980s

Tack room in the museum – display of saddles and packs, AFS Museum, Hinton, 1980s

Display at the AFS Museum, Hinton, 1980s

Forge and tools used for branding and shoeing horses, AFS Museum, Hinton, 1980s

Radio's and equipment used for telephone line maintenance – climbing spurs, linesman belt; AFS Museum, Hinton, 1980s

The 'stitching horse' was used to hold a ranger's saddle while he made repairs to it

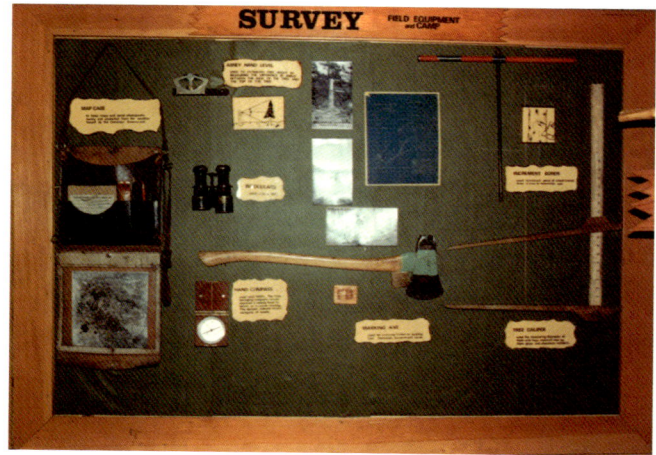

Display of forest survey equipment, AFS Museum, Hinton, 1980s

Display of firefighting pumps and hand tools, AFS Museum, Hinton, 1980s

Seven Rules for the Prevention of Forest Fires, AFS Museum, Hinton, 1980s

This is the old forest ranger station on the Moberly District. Moberly Ranger Station was located on the Lower Trail – the pack trail from Entrance to Muskeg and Grande Cache. Moberly was about 22 miles along the trail – two days ride from Entrance. This trail linked the forestry cabins and rangers at Winter Creek, Wildhay River, Moberly, Cabin Creek and Muskeg. The original ranger station formed part of the Athabasca Forest Reserve managed by the Dominion Forest Service before the transfer of resources in 1930.

This cabin was built in 1922-23 by Felix Plante and Louie Holmes. It was last used in 1959 when John Currat, who was District Ranger, retired from the Alberta Forest Service. John Currat later supervised the dismantling of the cabin in 1970 and its reconstruction of the Forest Technology School in Hinton in 1971, with the assistance of Junior Forest Ranger crews. The cabin has been restored as nearly as possible to its original condition

'Rocky Mountain John' Currat was a seasonal ranger with the Dominion and Alberta Forest Services from the 1920s until appointed to a permanent position in 1936, retiring in 1959. Currat came to the Hinton area in 1946 working the Rock Lake, Hay River and Moberly areas. John Currat was a resident of the Moberly Ranger Station cabin from 1950 until his retirement in 1959

Norman B. Nelsen, an employee of the Forest Surveys Branch from 1956 to 1974, constructed a model depicting a forest survey tent camp for a Lands and Forests conference in 1959. The model is a feature exhibit at the Alberta Forest Service museum. Shown here during the dedication of the model is (L to R): Frank Lewis, Chief Ranger, Hinton District; Bill Fairless, Superintendent, Edson Forest; and Bernie Simpson, Director, Forest Technology School. Mr. Nelsen's widow Margaret was on hand to dedicate the model her husband built. May, 1989

History of Posters

Series of fire prevention and resource management posters from 1930 to present.

Alberta Forest Service

History of Posters

  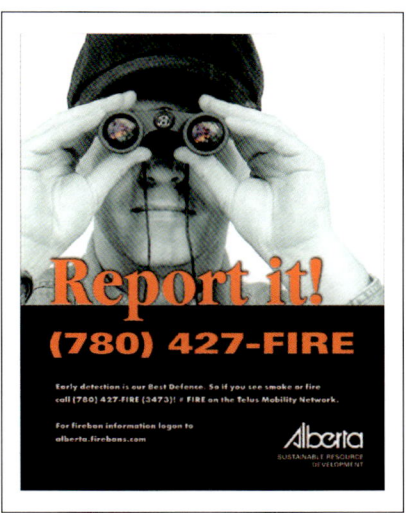

  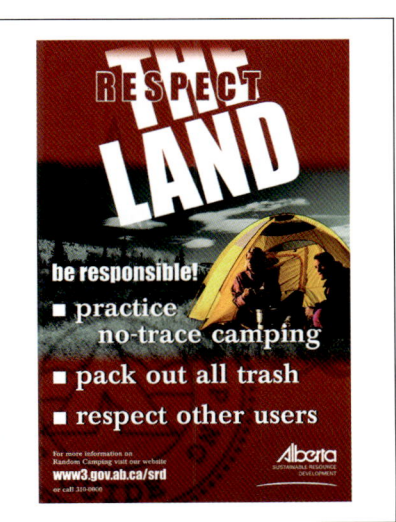

History of Posters